ON A MISSION

Sally Ride views the Earth from the pilot's seat on Space Shuttle *Challenger*, June 1983. *NASA*

ON A MISSION

|||||||||||||||||||||||||||||||||||||||||||||||||||||||||||||||||||||||||||||||||||||||||||||||||||||||||||

The Smithsonian History of US Women Astronauts

Valerie Neal

Smithsonian Books
WASHINGTON, DC

Dedicated to these trailblazing, inspired, empowered women

Published by Smithsonian Books
PO Box 37012, MRC 513
Washington, DC 20013
smithsonianbooks.com

Director: Carolyn Gleason
Senior Editor: Jaime Schwender
Production Editor: Julie Huggins
Digital Imaging Technician: Bill Whitcher

Edited by Gregory McNamee
Designed by David Griffin

This book may be purchased for educational, business, or sales promotional use. For information please write the Special Markets Department at the address or website above.

Library of Congress Catalog Control Number 2025024059.

Hardcover ISBN: 978-1-58834-776-3
eBook ISBN: 978-1-58834-777-0

Printed in the United States, not at government expense
29 28 27 26 25   1 2 3 4 5

CONTENTS

In 1978 NASA announced the first six women astronaut candidates, almost twenty years after America's first astronauts, all men, were introduced and ten years after men went to the Moon. Not yet thirty years old, I was finishing my doctoral degree in interdisciplinary American studies and anticipating an academic career. I was one of only three young women in my high school graduation class to pursue a PhD, while a number of our male peers flocked to medical and law schools at a time when few women were admitted. I identified with those six women astronauts as fellow midcentury girls—all born within a few years of 1950, coming of age in the social and cultural milieu of the 1960s, and preparing for careers with graduate education in the 1970s. We were a virtual cohort on the cusp of adventures and possibilities.

I especially took note of Margaret Rhea Seddon, also from a small town in a southern state. She had the temerity not only to be popular and smart in high school but also to leave Tennessee to attend the University of California, Berkeley, a foreign land to a sheltered southern belle, and then go to medical school and become a surgeon. All six candidates chosen that year had similar histories of early achievement, atypical educational interests, and experiences as the "first," "only," or "one of a few." I was close enough to them in age that we might have been in school together, childhood playmates, teen best friends, or friendly competitors. Like them, I aspired to a career, left home to study elsewhere, and found myself something of an anomaly with lofty, untraditional ambitions. Like them, I wanted to test myself in the broader world by pushing boundaries to see what I could accomplish. I have followed their careers, and those of their many successors, with interest and admiration while thriving in my own career related to space exploration.

In our rosy youths, we were largely unaware that certain social and legal barriers might impede our futures. Until well into the 1970s, as females we could not attend most Ivy League schools, and certainly not military academies. We could not obtain a loan to buy our first car or get a credit card in our own name; we needed a father or husband to assume responsibility for our payments. We were not allowed to serve on a jury. We were typically paid less than men for jobs, but we paid more for health insurance. We had no recourse against sexual harassment. We were quizzed and sometimes mocked for wanting to do something adventurous with our lives. Although our ambitions were untraditional, we soon learned that social traditions complicated our journeys. We had to be persistent when others discouraged or doubted our ambitions.

Since that first group of women, a total of sixty-one American women have earned astronaut pins and served in every position in the NASA astronaut corps, on land and in space. They have flown on more than a hundred space shuttle missions and more than thirty long-duration stays on the Russian Space Station *Mir* and the International Space Station, and they aren't finished. They have flown on every spacecraft in operation since the 1980s—shuttle, Soyuz, SpaceX Crew Dragon—and one woman is already assigned as a crewmember of the new Orion craft on a planned Artemis mission to the Moon. They have spacewalked, operated robotics, carried out scientific research, operated and repaired spacecraft systems, and commanded missions. Their achievements have equaled, and in some instances surpassed, those of their male peers in the United States and abroad. They have dispelled any doubts about the ability of women to have successful careers as astronauts and have inspired one another, as well as countless others, to aim high.

In preparing this book, I consulted a variety of published and archived historical sources and extant oral history interviews for all sixty-one women. In addition, I had electronic interviews and conversations with thirty of the thirty-two living former US women astronauts who responded generously to my outreach with original material and insight. Unfortunately, I was not allowed to contact any of the twenty-three current women astronauts directly, although NASA informed them of my project so they could contact me voluntarily. None did, and I assume they were too busy with their own work. What this means is that perspectives from the youngest of America's women astronauts are missing here; they might have been an interesting counterpoint to the experiences and reflections of the earlier women. To tell the stories of the younger women selected in the 2000s and the seven deceased women astronauts, I relied on their official NASA biographical fact sheets and information already in the public domain, including published interviews and podcasts.

Only seven former women astronauts—Rhea Seddon, Kathryn Sullivan, Shannon Lucid, Nicole Stott, Eileen Collins, Jan Davis, and Cady Coleman—have thus far published their memoirs, and some have published short lives for a young audience. Typically, other books for juvenile and young adult readers focus only on inspirational "first women" astronauts—Sally Ride, Mae Jemison (African American), Ellen Ochoa (Hispanic), Kalpana Chawla (Indian American), and Eileen Collins (shuttle pilot and commander)—whose stories can be found in multiple versions, and the few adult books on women in space are selective. Most of the women have not received such literary attention and thus are not as well known. I am hopeful that more women astronauts will publish memoirs and that other authors will produce biographies as masterful as Lynn Sherr's *Sally Ride: America's First Woman in Space* or as intimate as Jean-Pierre Harrison's memoir of his wife, *The Edge of Time: The Authoritative Biography of Kalpana Chawla*. Each woman has a distinctive story based on her own journey to becoming an astronaut, the experiences she had on Earth and in space, and how she has lived her life after being an active-duty astronaut. Each has wisdom to share, not to mention humor, hope, insight, and a few regrets. We would be fortunate to have more of their stories told in their own words.

America's women astronauts may not agree with everything I have written in this book, but I discussed its themes and perspectives with many of them, and they offered facts, clarifications, anecdotes, suggestions, and constructive criticism, for which I am grateful. Crafting this examination of the context for their journeys to NASA and then into space and on to other ventures, I aimed to be fair and accurate, to cast more than a superficial glow on these highly accomplished women, and to present some of the substance that makes them exceptional. I also address some of their difficulties along the way because, despite their accolades, they too are normal beings whose lives include such moments. For all their achievements, they are in some ways ordinary women who excel in extraordinary circumstances, and they have much in common with other women. My goal is a respectful consideration of who they are, what they do, and their place in space history and our nation's history.

Several years ago, in a conversation with space shuttle pilot and commander astronaut Pamela Melroy, I remarked on my admiration for the women who have become America's astronauts. I mused that with my bent toward history and literature, I would never have met the science and math requirements to become an engineer or qualify as an astronaut. Her generous reply took me by surprise: "Ah, but we haven't done what you can do." It was a gratifying reminder that we all have our abilities and roles. I hope that this book rewards her gracious acknowledgment of those who study and present spaceflight without having left Earth. We are participating in exploration, too, but with different tools and talents.

Twenty-four women gathered at sunset at a lakeside restaurant, atmospheric and sedate enough for special occasions. From the shady terrace of Villa Capri in Seabrook, Texas, the landscaped lawn spread to the quiet waterfront. Nothing in their demeanor or attire betrayed the women's identities as they mingled outdoors. They shared warm greetings and hugs as a sisterhood of friends and colleagues coming together to remember one of their own.

They were astronauts.

It was September 17, 2012, and the next morning they would join in a NASA ritual—planting a live oak tree and laying red roses in the memorial grove on the nearby Johnson Space Center (JSC) campus. News of Sally Ride's death at her home in California in July had caught this group by surprise. No one knew she had been fighting cancer. NASA and the nation had already mourned the passing of America's first woman in space. The evening would be the group's time of remembrance. Tomorrow her tree would start to grow beside those of other legends whose time on Earth and in space had passed.

The NASA Astronaut Office had tapped Anna Fisher to make arrangements for the ceremony. She was the only active astronaut left who had worked with Sally as classmates back in 1978 when the first six women were newcomers to the astronaut corps. She and Cady Coleman, a younger astronaut inspired by Sally, arranged for this all-woman get-together on the eve of the ceremony as a gesture of hospitality to Sally's sister, Karen "Bear" Ride; her spouse, Susan Craig; and her adult son and daughter, who had come from California for the tree planting. Rhea Seddon had flown in from her home in Tennessee, and newly retired Shannon Lucid was there, both of them also Sally's classmates. Carolyn Huntoon attended, the only woman to serve on the Astronaut Selection Board for the first recruitments of shuttle astronauts, who advised and counseled the women astronauts about settling into their new roles in NASA's traditionally masculine, quasi-military culture. A Russian cosmonaut training at JSC, Yelena Serova, also came to pay her respects.[1]

The group moved into a private dining room and took seats around a single, large table. Three generations of women astronauts sat at the table: those who started flying in the 1980s, those who came in the 1990s, and the newest ones arriving in the 2000s. As dinner progressed, they took turns sharing anecdotes about Sally Ride and her significance to them. At least one woman from all of the astronaut classes of 1978 to 2009—except 1987—was present and spoke. Some of their remarks about Sally were affectionately humorous; all were affirming. The younger women were incredulous about what it was like for the first cohort, and the older ones were touched to hear how

Remembering Sally Ride, September 17, 2012: (*front, from left*) Carolyn Huntoon, Ellen Baker, Mary Cleave, Rhea Seddon, Anna Fisher, Shannon Lucid, Ellen Ochoa, Sandy Magnus; (*rear, from left*) Jeanette Epps, Mary Ellen Weber, Marsha Ivins, Tracy Caldwell Dyson, Bonnie Dunbar, Tammy Jernigan, Cady Coleman, Janet Kavandi, Serena Auñón-Chancellor, Kate Rubins, Stephanie Wilson, Dottie Metcalf-Lindenburger, Megan McArthur, Karen Nyberg, Lisa Nowak. *Photo by Whit Scott, courtesy Ellen Ochoa*

important they and Sally were as inspiring role models who had shown the younger ones that aiming for spaceflight was an attainable goal. There was a keen sense that the first ones had cleared the way for the next generations of women to become astronauts.

Indeed, Sally's peers were especially struck by the passage of time. Sally was already a legend when some of the then-current astronauts were children. The youngest among them was born in 1978, after which women astronauts were a reality. Three of the astronauts present were the fiftieth, fifty-first, and fifty-third women in space, and others not there would be flying soon. The women gathered there reflected on the scope of their presence in the astronaut corps and spaceflight. They had a history and legacy. In the press of relentlessly intense work, it hadn't jointly dawned on them that so many American women had now flown and accomplished so much in space. They saw in one another the lineage from the first to the latest women astronauts and their union in something grand and significant. They had proved that women belong in space.

As of the end of 2024, forty-six years since the first ones arrived in the class of 1978, US women astronauts numbered sixty-one in all, with twenty-three then active in NASA management or ready for flight assignments. The youngest among them were born about thirty years after the first six women who became astronauts. Most of the eldest are now retired but actively engaged in post-NASA pursuits—traveling for public appearances and consulting, teaching, writing books, working as executives in the aerospace industry, serving on boards—and some are enjoying life with grandchildren. They remain role models. The cycle of life so familiar to women everywhere plays out even in space-flight, as each generation yields to the next.

People born and living about the same time are thought of as members of a generation. A social generation is typically considered to be a span of fifteen to twenty years, in contrast to a familial generation of twenty-five to thirty years. Neither term has an exact scientific definition, but they are useful popular concepts for organizing time and interpreting change.[2] Social generation is a broad term that embraces not just ages of a cohort but also the social environment in which they grew from childhood into adults, such as Baby Boomers, Generation X, or Millennials. People within a social generation are thought to share certain attitudes, values, and tastes arising from their shared history as youth. We can thus posit three generations of astronauts, each influenced by a changing social milieu. They are somewhat arbitrary and imprecise, but still illustrative.

The first generation of America's women astronauts was selected from 1978 through 1990 specifically to integrate the astronaut corps and to fly on the space shuttle. All were born before 1960, grew up in the 1950s, and came of age during the 1960s and 1970s. Women's educational and professional opportunities were limited in their youth, yet they had ambitious goals for their adult life and set out to attain them, starting with college and postgraduate education. They were on the cusp of societal changes toward greater equality of the sexes but young enough to have met some resistance in places where women were uncommon. They had to prove themselves to earn acceptance. Let's call them the Trailblazers.

The next generation was selected from 1992 through 2000 as NASA was both operating the space shuttle, planning the International Space Station, and beginning to assemble and use it. Most of this group flew on the shuttle's regular missions and those to the space station. Most of these new astronauts were born in the 1960s, finished high school and college in the 1980s, and started launching their careers in military or civilian work after women began to be visible as astronauts. The world of their youth was transformed by changing attitudes and opportunities brought about by the civil rights and women's rights movements. Doors of opportunity were opening, they were entitled to enter, and thus they were more readily welcomed wherever they headed. Let's call them the Inspired.

The 2004 class was a bridge to the third generation selected from 2009 through 2021 and possibly beyond. Like the 2000 class, the 2004 class would fly on the last shuttle missions, complete building the space station, and rotate on and off the station as residents. The final space shuttle missions flew in 2011, after which that transportation system was retired and the shuttle program ended. Born in the late 1970s through 1980s, the third-generation astronauts were chosen to become the post-shuttle astronaut corps. They would have stints on the International Space Station and become the crews of the new NASA and commercial spacecraft to follow the shuttle. In the world of their youth, women could aspire to be anything—fighter pilots and commanders of space missions, heads of large companies and federal agencies, candidates for the highest political offices, and successful professionals in any field. They were able to prepare to be astronauts more extensively than their predecessors and entered the astronaut corps with formidable resumes. Let's call them the Empowered.

In 2020, as plans further crystallized, NASA designated a group of eighteen astronauts as "the Artemis generation" to train for missions to the Moon. This group of nine women and nine men was drawn primarily from the 2013 and 2017 classes, with a few from 2009, 2004, and even 1996. By 2024, the Artemis team expanded to include all forty-seven active-duty astronauts, including those selected in 2021, with a total of twenty women. They ranged in age from early thirties to late fifties, and most had already flown one to three missions in low-Earth orbit. Several were veteran spacewalkers. Assuming new recruitments in the coming years, this chronology-spanning Artemis group may become a larger fourth astronaut generation, possibly to be followed by a future Mars generation.[3]

This book delves into the experiences of all sixty-one American women who had chosen a career as astronauts as of 2024. Their experiences as astronauts occurred within a panorama of social changes that expanded opportunities for American women since the 1950s and also within the US space program as it evolved since its founding more than sixty years ago. It was no accident or coincidence that women had to wait twenty years after men to make their way into space; both implicit and explicit barriers and biases kept them out. Their achievements become more remarkable when understood in the context of America's social history over the past century. The no longer favored term "manned spaceflight" is a testament to the way things were and to the revolution that followed. It has since become "human spaceflight."

Sally Ride received so much publicity that her fame eclipses that of equally worthy women who made their careers in space. This book aims to recognize all of America's women astronauts from 1978 through 2024, from the first group until the date this manuscript was submitted for publication. Their stories constitute forty-six years of achievement on Earth, in the air, and in space. During their time as astronauts, American society changed to their advantage, and so did NASA, military aviation, and the aerospace industry. The job of astronaut also changed as NASA moved from crewing the space shuttle to operating the space station, encouraging the development of commercial spaceflight, and preparing crews for missions to the Moon and eventually Mars.[4]

Please note a few caveats before heading into the book proper.

The first caveat: the term "women astronauts" or "female astronauts" is somewhat problematic. Women who are astronauts are not a formal subset of the astronaut corps, although they have always been in the minority. They account for about 19 percent of all US astronauts serving since 1978, when women were first accepted.[5] In 2024, twenty

"Sally Ride and generations of girls and women," cartoon by Jeff Parker. *Florida Today* and the *Fort Myers News-Press*

women made up 42 percent of the corps of forty-seven astronauts on active flight status. If the twelve astronauts in management positions, three of whom are women, are counted, women now constitute more than one-third of the total number of working astronauts.

Men and women astronauts train and perform technical work exactly the same, fly the same kinds of missions, are assigned the same roles in flight and on the ground, and do the same jobs as co-residents on the space station. There is no difference in expectations. All that matters is whether one can do the job. Competence is everything, and with that, it matters not whether one is male or female. For those reasons, the women today and earlier prefer to be known simply as astronauts, with no gender tag. The astronauts consulted about using "women astronauts" as a group descriptor have been tolerant of the term for the purpose of this book, but they never identify themselves, nor do they like being referred to, as a woman astronaut or female astronaut.

A note of caution about these statements: "The only thing that matters is whether one can do the job. Competence is everything, and with that, it matters not whether one is male or female." This is the consensus of women astronauts who loved their career, thought of it as a privilege and honor, admired their colleagues, and felt that NASA was committed to their success. However, in conversations, some may start a sentence with "But there was this time when . . ." or "But one guy . . ." or "I couldn't believe what I was hearing." These comments often lead to admissions that being a woman sometimes did matter. Such anecdotes, some of which appear later in the book, suggest perceived or latent biases that did not magically disappear the moment women entered the astronaut corps. Bias may not have been blatant or widespread, but it still signaled to some of the women that they had to prove themselves worthy of equal treatment. The Astronaut Office was not paradise, and there were some tendencies that women had to resist and set right.

Women are the focus of this book because their history is distinct from that of most male astronauts. They had more nuanced journeys and obstacles, and often fewer opportunities and choices than men, and that experience must be acknowledged and appreciated. The same can be said of African American, Hispanic, Asian American, or Native American astronauts, or any other identity that has been marginalized. American society is not fully equitable, and the paths to their dreams and success have been more challenging for some individuals (including some men) than for others. It took more years than necessary for women to become qualified for and welcomed into spaceflight, and it is important that the context for their belated arrival be understood. That they have persisted and excelled for most of the space agency's sixty-plus years is reason to celebrate.

The second caveat: this book focuses on women astronauts selected into the NASA astronaut corps to be employees of the agency and fly on missions or work on projects that NASA manages. They are civil servants or members of the armed forces of the United States, serving and representing the nation. This book does not focus on the increasing number of private space flyers in the commercial aerospace sector who work for SpaceX, Blue Origin, Virgin Galactic, Axiom, Boeing, and other companies, or on women not employed by those companies who have flown as passengers. It also does not include female astronauts from other countries who have flown in space, even on NASA missions, or American women who flew on the space shuttle or International Space Station without being members of the astronaut corps. Their stories are part of a different historical narrative.

The third caveat: most of the women considered in this book have been "first" or "only" at some point in their lives or their NASA career, sometimes more often than once. Some readers may feel that acknowledging "firsts" or "onlys" diminishes the real achievement, which is the inherent talent and ability that make it possible to gain such distinctions. What of the women who are second, fifth, or tenth? Is their achievement not equally laudable? The label "first" or "only" isn't as important as the self-discipline and competence. Being first is only one achievement; it is the complete woman who matters. Another perspective is that a first is a mark in time that defines what had not been achieved previously and thus is a useful historical indicator. Although all are driven to excel, most of these women never set out to be the first or the only one to achieve anything but their own goals. Fueled by ambition and perseverance, they often found themselves the first or only woman in classes and academic majors, in the military service academies, in flight school and test pilot school, as pilots of a particular aircraft or in combat, serving in certain military units or positions in the civilian workforce, and even achieving firsts as astronauts. In this book, firsts are noted to signify historical progress and personal achievement, with no intent whatsoever to diminish any woman's accomplishments.

The fourth and last caveat: this book is not a directory of biographies. Each woman's official biographical fact sheet is readily available on the NASA website,[6] and most of the women included here have online reference pages and their own websites where the curious can find more details of their careers. Instead, the book is organized thematically, and profiles of the women appear in their respective sections.

Chapter 1 offers a survey of the status of women in aviation, science, mathematics, and engineering in the late 1950s through 1970s, the first two decades of the space age. As they were beginning to seek careers in these professions and aspiring to become astronauts, the harsh truth was that they were not qualified and could not become qualified under the strictures at the time. Highly motivated women took the first steps toward their spaceflight ambitions but were just enough "ahead of the times" to be less than successful in achieving their untraditional goals. Legislative, policy, and social advances and the activists urging them began to force open the doors that had been barred against women. This prehistory is essential context for understanding what ambitious women faced and what changes were necessary for them to be considered of astronaut caliber.

Within that restrictive environment, at least three brief interludes occurred that illustrated women's potential capabilities for spaceflight. One has become well-known since the 1990s as the so-called Mercury 13, more accurately participants in the Lovelace Women in Space research program in 1961–1962.[7] These were women aviators who individually passed the same physical and mental tests given to male astronaut candidates. The other two interludes are less widely known but arguably more influential in NASA's eventual acceptance of women as astronauts. In 1970 a crew of five women (four scientists and an engineer) completed a two-week "Man-in-the-Sea" mission in an underwater research station.[8] NASA cosponsored this expedition as an analog space station mission comparable to Skylab and led the behavioral research program. The women aquanauts' behavior and productivity were so impressive that NASA actually invited the team leader to consider becoming an astronaut well before 1978.[9] The third episode occurred in 1975 at NASA's Marshall Space Flight Center. Three scientists and an engineer, all women working there, were selected to simulate a laboratory research mission during the planning for Spacelab, a research facility sometimes installed in the space shuttle's payload bay. After they completed the week-long laboratory "mission" in isolation, they took part in astronaut familiarization training in the neutral buoyancy simulator (underwater immersion tank) and KC-135 zero-gravity aircraft flights. Two of these women went on to submit applications for the 1978 class of astronauts.[10] Neither the all-women aquanauts' expedition nor the all-women laboratory "mission" has been well recorded or adequately recognized in the annals of NASA history. It is not an exaggeration to call them unknown or forgotten.

After this history, three chapters follow. Each chapter addresses one of the generations—Trailblazers, Inspired, Empowered—from the pivotal entry of women to the latest arrivals into the NASA astronaut corps. Each generation is situated in relation to social and cultural hallmarks of the period and to NASA's evolving spaceflight program. Each selection group or class of astronauts is discussed in general, and every woman is profiled to recognize their individual and collective achievements during the next four and a half decades. During this long period, women's place in space proved to be varied, valued, and fully effective in every role, including command and leadership.

Chapter 5 is a deep dive into the actual job—the various roles on the ground, in the air, and in space. There is no distinction between the work that male and female astronauts do or the roles they assume. All perform the rigorous work necessary to make spaceflight as safe and productive as possible. An introduction to the variety of jobs that astronauts must do, and do exceedingly well, may give the reader a heightened appreciation for the caliber of individuals who can master such knowledge and skills and play such different roles concurrently.

Chapter 6 examines several aspects of spaceflight that uniquely or especially affect women and can hamper their participation in some roles. Most notably, the reported sizing issues with the extravehicular activity (EVA) spacesuits, which do not fit the smallest women, preclude them from being assigned to spacewalks. For those who choose marriage and children, the intensity of their demanding work can raise work/life balance challenges, and even more so for astronaut couples. Single women who are astronauts also find that the job unexpectedly affects their domestic and social life. Each woman has to figure out how to negotiate such challenges while also maintaining her physical and behavioral health against the stresses and risks that come with her job. Astronauts are not unique in these challenges. Many other working women, wives, and mothers likewise experience the balancing acts their work and family roles entail.

NASA announced in 2019 its intent to land the first woman and next man, later revised to first woman and first person of color, on the Moon in the mid-2020s. This lunar program is called Artemis, for the twin sister of Apollo in Greek mythology, and as of 2024, twenty women are eligible for flight. NASA has assigned one, Christina Koch, to the first four-person Artemis mission, a lunar fly-around without a landing, that may happen in 2026. If the mission succeeds, an American woman could walk on the Moon in 2028, near the fiftieth anniversary of her first predecessors' debut as astronauts. Presumably the NASA astronaut corps will continue to be replenished by the caliber of highly motivated, highly competent women whose achievements marked the space shuttle and space station eras. And if the longer-range plan is realized, a woman may someday step out on Mars, the first in a new generation to live and work on another planet. Perhaps the next book on America's women astronauts will start on the Moon or Mars.

CHAPTER 1

Making Space for Women: The Early Space Age

Seven men in business suits strode to the platform and took their seats behind a draped table. A US flag stood at each end; a round red, white, and blue NASA logo dominated the floor-to-ceiling drapery behind them. Scale models of a rocket and a spacecraft leaned against the center front of the table. Rows of NASA officials and members of the media faced them, eager to learn about these men and their new job. They were America's first astronauts, introduced to the public in a lively press conference held in a Washington, DC, ballroom on April 9, 1959.[1] They were the top-ranked men to emerge from the most grueling physical, medical, mental, and psychological evaluations ever conducted for government employees. All were white, test pilots, and between the ages of thirty-two and thirty-seven years. They were soon featured in a *Life* magazine cover story, already celebrities before they had fully settled into their work. NASA had not yet celebrated its first anniversary as an agency, and it was busy creating a new cultural hero for the space age—the astronaut.

A fascinated public was quick to idolize them, although no one was quite sure of what these astronauts were meant to do. Space travelers populated science fiction, but a real-world job description and qualifying criteria did not exist until 1958. They had to be invented. Selection of these first astronauts—the Mercury 7—resulted from a process designed to identify the most capable men to face the rigors, hazards, and uncertainties of spaceflight.[2] They also were highly capable engineers who, as test pilots, not only flew the aircraft but also developed test plans and analyzed test data. As astronauts they would spend most of their time on the ground working on technical problems and monitoring contractors' work on their spacecraft.[3] The NASA public relations office, the admiring media, and the astronauts themselves were influential in defining what an astronaut was, but these concepts also were influenced by the social and cultural milieu of the late 1950s and 1960s.

The iconic figure of the American astronaut that emerged was a heroic male. Two decades would pass before women would join them. This book explores the barriers that kept women out for so long in the context of the evolving culture of NASA, the status of women pilots in civilian and military aviation, the status of women in science and engineering, the civil rights of women, and what might be called inadequate understanding of the female body. Changing the status quo came about slowly through concerted efforts to make space for women where they had not been welcome. During the struggle to overcome those barriers, a few scientific trials of women as potential astronauts held promise for their eventual acceptance into the profession.

Of Course Astronauts Were Men

Men had all the advantages at the dawn of the space age from the late 1950s into the 1970s. The first was that they were suitably educated. Men held the degrees in engineering, physics, chemistry, and mathematics—the primary disciplines for the design and operation of rockets and spacecraft—and in astronomy and geology, two keys to celestial and planetary exploration. Most physicians and physiologists, who were needed to study the health and protection of human space travelers, were male. With the benefit of the GI Bill of Rights, the older among them had filled colleges, universities, and technical schools after their military service in World War II and Korea, and they were ready to take positions in the emergent aerospace industry and the new National Aeronautics and Space Administration (NASA) established by Congress in 1958. Like NASA, the National Defense Education Act, also passed in 1958, was a response to the Soviet Union's launch of the Sputnik satellite a year earlier. The aim of this law was to jumpstart scientific education in the United States by increasing the number of professors and students in technical fields to bolster aerospace and other vital Cold War industries. Women, however, were not yet welcomed into the masculine realm of scientific and technical education.[4]

When NASA opened for business in late 1958, it gained almost ten thousand personnel transferred from the prior National Advisory Committee on Aeronautics (NACA), the Naval Research Laboratory (NRL), and the Jet Propulsion Laboratory (JPL), as well as new hires. Men accounted for 97 percent of the NASA workforce by 1960, a third of them engineers and scientists. The 3 percent of women were mainly secretaries and clerks, with an occasional scientist or group of mathematicians known as "computers" or "programmers."[5]

NASA immediately wrestled with the question of what qualifications would make the best astronaut candidates.[6] After considerable discussion of the desirable characteristics of the first people to face the uncertainties and risks of riding rockets and maneuvering space capsules, NASA settled on a short list of requirements and with President Dwight Eisenhower decided that military test pilots were the ideal candidates. It was not an arbitrary decision. Test pilots were a readily identifiable cadre, they were engineers, they had the most experience flying and decision making under extreme stress, they had security clearances and the rigorous discipline of military service, and their performance and health records were readily accessible. These men had the advantage of experience, having already proven their abilities in combat and on the edge of space. They were accustomed to high-risk scenarios, new flight technologies, and pushing the boundaries of flight. Under the pressure of time in the accelerating space race, it was deemed more efficient to locate a qualified group of seasoned military test pilots than train civilians to be the first Americans in space. All military test pilots at the time were men; women did not fly at the military test pilot ranges until the 1980s, and no women could serve as combat pilots until the 1990s.[7]

Another subtle advantage that some men, especially military pilots, may have had was a passion for flight arising from their childhood interests in military aircraft and the combat exploits of pilots in World War II and the Korean War. As boys, they built or collected model airplanes and were well read in accounts of heroic wartime battles in the air.[8] Flight was "in their blood" and their mindset. It was not a giant step for them to become military aviators when they came of age and then to prepare for flight in space as a logical next step. Women who became passionate about flight generally had a different entrée, not through interests in military aviation but what has been called the "romance" of flight. They idolized early female aviators and wanted to experience the freedom of being in the air, enjoying the panoramic views, performing in air shows and competing in air races, and feeling a sense of technical accomplishment in flying. These early differences in childhood interests may have been formative in first enticing boys onto a path that ultimately led to space. Girls caught up later as they forged their own paths in general aviation before military aviation was an option.

Likewise, boys tended to be more attracted than girls to early popular science fiction, written by men and populated with male characters such as Buck Rogers, Flash Gordon, and Tom Gordon in comic book and television series, and the more serious and visionary novels of Isaac Asimov, Robert Heinlein, and others.[9] A series of popular Walt Disney television programs in the 1950s presented space travel as possible and imminent. At the time, though, the culture of space science fiction was dominantly masculine, and it shaped a vision of the future that initially seemed more possible for men than women. Had toymakers, writers, and producers more consciously appealed to girls and women, females might have been as culturally conditioned at a young age as males to imagine themselves in, and prepare themselves for, aviation and spaceflight.

Other requirements for astronauts included age and height (under forty years and less than 5 ft. 11 in. tall); a bachelor's degree in engineering or physical sciences; graduation from a military test pilot school; 1,500 flight hours; and expertise in high-performance jet- and rocket-powered aircraft. NASA screened 508 military test pilot personnel records, found only 110 men who met the minimum standards, tested and interviewed fifty-six of them, and selected only seven Project Mercury astronauts in 1959 to make the first solo flights in space. It was such a clear assumption that the candidates would be men that it is unlikely anyone noticed that the education and aviation requirements implicitly excluded women.

NASA selected fifty-nine more astronauts during the 1960s to fly the two-man Gemini missions and three-man Apollo missions, and to occupy Skylab, the nation's first space station. NASA also acquired seven more astronauts, all of whom were white men, from the Air Force when its Manned Orbiting Lab program was canceled. NASA modified the requirements for astronauts slightly during those years, seeking civilian as well as military pilots, dropping (but still preferring) the requirement for test pilot status, and reducing the number of flight hours to 1,000. These changes might have opened a door to women pilots, but NASA did not directly seek them.

Responding to pressure from the scientific community, in 1965 the agency made a major exception and recruited scientist-astronauts who were not required to be pilots. The call for applicants did not specify men, but the assumption was evident in the requirements to submit transcripts "from all institutions of higher education which he has attended" and forms "completed by his physician."[10] The group of six male astronauts selected in 1965 from about nine hundred applications included two engineers, two physicians, an astrophysicist, and a geologist, who would later receive flight training. NASA recruited eleven more scientists and engineers in the class of 1967, also all male. The invitation to apply was directed to the nation's young scientists and engineers of exceptional ability, with assurance that anyone selected would be able to "maintain his scientific competence" and "continue his growth as a productive scientist in his field of interest."[11] These two expansions of the astronaut corps could have been an opportunity to recruit women scientists, but the scientific community at the time was predominantly male, and all those involved in discussions about admitting scientist-astronauts were men. Allegedly, some women applied but did not make the cut. It is unclear whether any documentation of such female applicants exists to verify this claim; none has been cited.[12]

Seventy-three men had entered the NASA astronaut corps by 1970, before recruitments ceased as the Apollo missions to the Moon came to an end. NASA made no explicit effort to recruit women scientists for the 1965 and 1967 classes, a missed opportunity that might have moved forward the inclusion of women into spaceflight by a decade or more. Nor did NASA make an effort to seek qualified Black pilots or scientists. In the race to reach the Moon, diversity of the astronaut corps was not a priority. More than enough qualified men were available, and manned spaceflight remained a white man's world.

By education, aviation experience, and presumption, men had all the advantages to become astronauts during the pioneering first years of spaceflight. It was inequitable but not necessarily intentional. These advantages were embedded in long-held traditions and beliefs about the proper roles of men and women in American society. Men pursued education and remunerative work to provide for their families; women took care of the home and children. Men did hazardous work, served in the military, and handled complex or dangerous technology; women nurtured and protected families. These presumptions colored much broader conceptions of masculinity and femininity, affecting how people should look, dress, act, and engage in sports, politics, the media, business, and a host of other gendered domains.

During a 1962 congressional hearing into whether NASA was discriminating against women in selecting astronauts, America's venerated first man in orbit, John Glenn, testified, "The fact that women are not in this field is a fact of our social order." No one laughed or argued with him. He claimed that a woman who could meet the high standards would be welcome, but qualifications should not be reduced. It was not an unusual gender stereotype to assume that women's participation would mean lowering standards.[13]

It would take concerted efforts by individuals, organizations, and organized movements to make space for women and people of color to have the same opportunities as white men. Demands for changes in law, policy, education, and the military services yielded results in the 1960s and 1970s that challenged prejudices, removed obstacles, and opened paths that would eventually lead to a more diverse astronaut corps that better reflected the US populace. Changes were not instant; they came through a long social struggle to overcome sex and race discrimination during years marked by anger, hope, resistance, disappointment, and incremental progress. Society gradually absorbed changes in attitudes and practices, but the work of equalizing opportunity remained unfinished well into the next century.

Why Not Astronauttes?

The idea of women as astronauts was not unheard of as America entered the space age, but it was not taken seriously. The language itself was dismissive: journalists and others referred to "lady astronauts," "girl astronauts," "astronettes."

A few people, however, began to argue that women eventually would travel into space and therefore should be taken seriously and prepared for it. In 1962, the *New York Times* published a long article by a physician asking, "Why Not 'Astronauttes' Also?" Although he betrayed chauvinism disguised as humor more than once ("If man is to colonize the planets, if celestial housekeeping is ever to be instituted, the 'second sex' must have booking on future space flights."), he compared the sexes in various physical and mental abilities and concluded that there was no reason, except bias and habit, to exclude women from spaceflight. In essence, an astronaut must have a healthy mind and body and certain acquired skills. Given opportunities to acquire those skills, both women and men could meet that standard.[14] By 1970, a NASA executive in manned spaceflight admitted this in a public forum: average people in good health, including women, would be able to fly into space.[15]

Prominent writer, politician, and ambassador Clare Booth Luce wrote an outraged piece for *Life* magazine immediately after the Soviet Union launched Valentina Tereshkova as the first woman in space in 1963. Luce saw the failure of the United States to recognize her significance as a costly Cold War blunder. Too many American male space experts dismissed Tereshkova's flight as a stunt or claimed that she had been hysterical or hadn't performed well. Luce instead praised it as a symbol that Russian women "actively share (not passively bask, like American women) in the glory of conquering space." Already astronauts and cosmonauts had become symbols of the way of life of their nations, and she found it shameful that the United States did not yet recognize and elevate "the inherent equality of men and women."[16]

Girls and young women began to write to NASA during the 1960s, asking how they could prepare to become astronauts. This caused some consternation as the agency's public information officers asked how to respond. Early answers to such letters were blunt: NASA had no plans for women astronauts "because of the degree of scientific and flight training, and the physical characteristics, which are required."[17] Yet there was a dawning awareness that women probably would be included eventually. An inquiry sent a decade later received a much more encouraging response. Director of Flight Crew Operations and longtime chief of the Astronaut Office "Deke" Slayton wrote to university-level aerospace engineering student and pilot Marsha Ivins in 1970, "The exact time when we would seriously consider women is indefinite, but I am sure it is inevitable." Meanwhile, she should do the best she could in her studies and apply to NASA when the next astronaut selection program was announced. She applied, was selected to be an astronaut in 1984, and flew on five space shuttle missions.[18]

By the time men started going to the Moon, the idea of women having a place in space was becoming more common but still not taken seriously. A 1968 *Washington Post* article led with the question, "Is NASA for co-eds?" and stated, "Some scientists wonder if the time has come to give the weaker sex a whirl at it." Interviewing some men attending a technical conference, the reporter heard these reasons that it wasn't time yet for women in space: the mixture of the sexes in a spaceship could create an ethical problem, separate onboard facilities would be needed [for privacy], spacecraft were not designed for mixed sexes, men's sexual needs on long flights needed to be considered, and "there's just too much difference between men and women [who] really don't speak the same language." Some suggested that all-woman crews might be a better solution than mixed-sex crews.[19]

By 1971 some of these arguments were waning. Popular science writer Isaac Asimov published a scathing article in *Ladies' Home Journal* castigating NASA's men to give up "their stubborn male pride." Doubting that it was impossible to find a qualified woman in a nation of 200 million people, he claimed, "It is just that women are not wanted in the U.S. space program. Period. Yet they should be." He cited their smaller and lighter size, resistance to stress and some common physical disorders, ready adaptation to weightlessness, and longer lifespan as reasons to move beyond NASA's focus on masculinity. He also argued that single-sex long-term missions on the Moon or to Mars were unrealistic, so women should be included as scientists and engineers, not as companions for the relief of men. He concluded that it wasn't worth the money, effort, or dreams to continue a "Men Only" space program. "Call if off, call it *all* off, or open it to the human race, the *whole* human race."[20]

Some women, of course, flew airplanes, attended universities, attained graduate degrees, entered fields traditionally occupied by men, and followed their interests and talents to achieve their goals regardless of societal expectations, but they were exceptions. It would take more women seeking opportunities and pushing against boundaries for the latent potential of women to be demonstrated and even more assertive tactics, both social and legal, for that potential to be recognized and nourished. Individual and organized efforts in the 1960s through 1970s initiated social changes that would start opening the way for women to enter into coveted domains and, by the end of the century, bring women to the fore as bona fide astronauts.

Women Pilots

Women were interested in flying airplanes from the outset. Katharine Wright, Wilbur and Orville's younger sister, was involved in their enterprise to invent, fly, and sell aircraft. Within a decade of the men's historic 1903 flight, in the United States and Europe, women had flown as passengers and the first women had earned pilot's licenses, made solo flights over land and water, parachuted from an airplane, competed in an air race and an air show, become a test pilot, established the first flying school, and opened a business to rent and sell airplanes. The achievements kept mounting as more women took to the sky, some flying in air shows and barnstorming to show off their daredevil piloting skills and others pushing the boundaries of distance and duration in flight to win races and prizes. It was clear that women could fly aircraft, although not everyone was convinced that they should.[21]

Flying became popular among women in the first half of the twentieth century as they sought adventure and freedom in flight. It was satisfying to do something as bold as men did, to compete for trophies as men did, to set speed and altitude records as men did, and to enjoy the renown that came with being an aviatrix. Flying was a modern pastime and a powerful way to demonstrate their competence and independence. Aviation began to change attitudes about what women could do. Aircraft companies began to hire women for advertisements and as stewardesses to demonstrate how safe air travel was, using women to boost passenger traffic, airline growth, and revenue, but the corporate role of pilot remained reserved for men. Much of their history and many of the names of early women aviatrixes were not remembered beyond their time in the limelight, but that was remedied and celebrated by the turn of the twenty-first century as women's historians sought to make them visible. The few names that come readily to mind in general—Bessie Coleman, Amelia Earhart, Jackie Cochran, Jerrie Cobb, Jerrie Mock—stand for thousands more who are less well known but signify the increasing presence and abilities of women in aviation.

World War II created a dramatic opportunity for white American women to leave aviation as a performance and enter aviation as a real job. Such opportunities were rare for women of color because of racial bias. As men flooded into military service and the war effort demanded vast production of military materiel, women stepped into the gaps throughout the economy. Thousands of women held jobs in aviation, engineering, and heavy and light industry for the first time. Some of the aircraft companies, accepting the necessity of hiring women, offered training programs, such as the Curtiss-Wright Company's Cadettes program, to quickly place "girls" in entry-level aeronautical engineering positions so they could assemble planes. Women trainers taught male pilots how to fly by instruments. Most famously, the Women Airforce Service Pilots (WASP), about one thousand pilots organized by famous aviator Jacqueline "Jackie" Cochran, tested and delivered military aircraft during the war years, towed targets for antiaircraft artillery training, and flew other missions as needed. They flew fighters, bombers, and other military operational aircraft. Women were essential in keeping the nation supplied with aircraft and materiel to win the war, and their service was vital to the wartime economy. Many women had their first taste of earning an income, shouldering responsibility outside the home, and contributing as citizens to a national need—their first experience of serving in a man's world—and they liked it.

Despite their heroic wartime efforts, women were released from employment when the men returned. WASP women were not granted status as veterans until 1977, nor were they offered the same educational and economic benefits as men who served. The windows of opportunity that had opened suddenly closed, and in the late 1950s and early 1960s women began to agitate for them to open again. Women were making headway in some aviation roles, noticeably as flight attendants and private pilots in general and commercial aviation, but the real prize—a seat in the cockpit of large aircraft—was not yet attainable.

One way to open the cockpit doors was to increase the profile of women aviators through competition. Setting records and winning trophies earned respect and proved that women could handle aircraft with precision at high speeds and altitudes. The premier woman aviator in the United States for almost forty years was Jackie Cochran, who broke and set many absolute records in competition with men. Although she wasn't as interested in setting women's records (but she indeed set many), she was eager to be the first woman to achieve a given milestone. She was one of the first women to fly jets, the first woman to complete a transatlantic jet flight, the first woman to fly a bomber across the Atlantic to Britain, and in 1953, flying an F-86 Sabre, the first woman to break the sound barrier. In 1964 she set a new record as a woman pilot by exceeding Mach 2, flying an F-104G Starfighter at 1,429 miles per hour. Cochran was the first woman to gain special access to military jets to set some of her speed and altitude records. She thought that if she could fly the most advanced aircraft, so could other women.

During the 1950s and 1960s, Cochran demonstrated that a woman could fly almost anything, hold more records than anyone else, and win the most prestigious trophies. Six times she received the annual Harmon Trophy as the world's best aviatrix. Cochran held commercial and transport pilot licenses and had a business in addition to being a competitor. She later rose to the rank of colonel in the Air Force Reserve. Cochran approached aviation as her career, and as a leader in aviation organizations she encouraged women pilots. Her influence was unparalleled in bringing women into aviation and helping to change attitudes about women pilots. Her exceptional piloting skill in many different aircraft made credible the ability of a woman to fly as one of the best.[22]

Two other women pilots achieved high acclaim and influence by the 1960s. Geraldyn "Jerrie" Cobb had amassed seven thousand hours in flight and three world records before she turned thirty. Like Cochran, she also gained access to military aircraft on occasion. Cobb desperately wanted the United States to put the first woman in space and would have been thrilled to be that woman. Jerrie Mock, a wife and mother rather than a full-time aviator, who was sometimes called the "Flying Housewife," flew solo around the world in 1964. She was the first woman to do so, becoming the first woman to fly across both the Atlantic and Pacific Oceans and making those flights in her second-hand single-engine Cessna. Mock also set various speed and distance records. Both women received many honors and awards, including a Harmon Trophy for Cobb, who leveraged her fame to promote women's right to be astronauts, whereas Mock modestly tried to avoid publicity and underappreciated the significance of her achievement. Together they inspired many young women to follow their dreams into the sky.[23]

Advocacy was another way to open cockpits to women, and a number of organizations with that mission existed by the 1960s. The Ninety-Nines, an organization of women pilots formed in 1929 to see women aviators treated seriously as equals or peers to men who flew and to preserve the history of women pilots, had grown over thirty years to have more than one hundred chapters across the country. Whirly Girls, founded in 1955 as an organization for women helicopter pilots, likewise sought camaraderie and communication among women in aviation. Both shared the goal of advancing women in aviation and used educational programs and scholarships toward that end. As pressure built in the 1970s and 1980s to open more aviation roles to women, whether in the cockpit or on the ground, these organizations were visible and vocal.

Pilots in the Lovelace Woman in Space Research Program, 1960–1962

In 1960, two aerospace medicine specialists who had managed much of NASA's evaluation program for men under consideration to become astronauts had an idea. They wondered how women would perform under the same rigorous physical and mental tests, initially a battery of medical exams that took an entire week to administer and measured every conceivable parameter of health, strength, and stamina. They hatched the idea of conducting a private research program to test women exactly as the men had been tested to discover whether women might be suited to withstand the conditions of spaceflight. Comparing data from the most fit male and female pilots would challenge the notion that women were the weaker sex and perhaps give NASA a practical reason to recruit women candidates.[24]

The research program was conducted in 1960–1962 by W. Randolph Lovelace II, who was the medical director of his family's private Lovelace Medical Center in Albuquerque, New Mexico, and director of its research arm, the Lovelace Foundation. A twenty-eight-year-old pilot who was one of the best in the world, Jerrie Cobb, was the first woman in this program to undergo the same physical and psychological fitness tests that the astronauts had passed. Two women pilots already had some exposure to the testing regimen in 1959, one in cooperation with NASA for a cover story in *Look* magazine provocatively titled "Should a Girl Be First in Space?" and the other at the invitation of Air Force medical researchers at Wright-Patterson Air Force Base, but neither had the full evaluation that Lovelace planned.[25]

Cobb reported to the Lovelace Clinic in early 1960 for the first phase of stress testing, a regimen that examined every system in her body for any physiological flaws or weakness. She then moved on to a second stage of psychological stress testing and a battery of intelligence and personality tests. Cobb performed exceptionally well, and the examiners reported that "she has very much to recommend her for selection as an astronaut candidate."[26]

Lovelace and Cobb kept quiet about this accomplishment until he announced the results of her testing in August 1960. Cobb became an overnight sensation, sought after for interviews. The *New York Times, Washington Post, Life, Look*, and *Time* magazines, among others, hailed Cobb as America's first prospective woman astronaut or the first successful female astronaut candidate. Some went so far as to call her America's first woman astronaut and predicted that she would fly as soon as 1962. Even as they praised her mettle, though, they diminished her by focusing on her figure, size, and "girlish" look, and using terms like "space lady," "girl astronaut" and other trivializing labels.[27]

Meanwhile, Cobb and Jackie Cochran scoured aviation records to identify other women pilots who might be eligible and willing to volunteer to undergo the same tests. Of twenty-five women invited, nineteen participated and twelve more passed the first round of testing.[28] Three moved into the second round of stress testing, but scheduling others was delayed. The thirteen women who successfully completed most of the astronaut fitness tests, including Jerrie Cobb, were Myrtle Cagle, Jan Dietrich, Marion Dietrich, Mary Wallace "Wally" Funk, Sarah Gorelick, Jane Hart, Jean Hixson, Rhea Hurrle, Irene Leverton, Bernice "B" Steadman, Gene Nora Stumbough, and Geraldine "Jerri" Sloan.

The third and highly anticipated final phase of testing—wearing pressure suits, simulating high-altitude flight in a pressure chamber, and testing the effects of gravity ("pulling g's") by flying in high-speed Navy jets—was to occur at the Naval School of Aviation Medicine in Pensacola, Florida. Only Cobb completed this testing in a one-time arrangement that Lovelace brokered. The Navy declined his request to test the other twelve upon realizing that NASA had not endorsed the research program or these expensive tests. At that point, the Lovelace Woman in Space program came to a sudden, disappointing end.

Jerrie Cobb and Jane Hart, the pilot wife of Michigan's Senator Philip Hart, for months campaigned around Washington for continued testing of women pilots and consideration of women as prospective astronauts. They met with government officials, gave talks around town, and had an audience with Vice President Lyndon Johnson, head of the National Aeronautics and Space Council, to seek his support.[29] They even managed to secure a congressional hearing in July 1962 to examine whether NASA was discriminating against women in astronaut selection. Cobb and Hart presented their arguments for women's ability as pilots and their proven physical and mental fitness. They also argued that women were eager to do their civic service to the nation as astronauts and that the United States, as a point of pride and leadership, should send the first woman into space.

NASA countered that the space program did not yet have a requirement for women and no woman could yet meet the essential qualification of test pilot experience. The agency did not currently have a need or resources to train women for the job. Contrary to Hart and Cobb, Jackie Cochran agreed with NASA, asserting that it was not the right time to launch a women's astronaut program during the urgent space race.

All outstanding pilots, the Lovelace women made successful careers in aviation at a time when women were rare in the business. In 2021, one of the thirteen women, Wally Funk, made it to space.[30] Many years after she had applied to NASA four times without success, Jeff Bezos, the founder of Blue Origin space company, invited the eighty-two-year-old to join him on a brief suborbital flight, finally achieving her dream.

All the Lovelace women were ahead of the times and eager to aim for spaceflight. The Lovelace program was the first modern effort to take women's bodies seriously, to examine the extent of their physical and mental toughness, and to challenge beliefs about women as the weaker sex by sustained, rigorous research. This challenge originated in the medical research community but through publicity spread into public discussion about the abilities and roles of women. It also raised the issue of sex discrimination before laws were passed to ban it. Through Cobb's and Hart's public campaign for women astronauts, the possibility received favorable media attention and congressional consideration. That in itself was a remarkable accomplishment.

Even so, when NASA seriously began to consider recruiting women into the astronaut corps during the mid-1970s, the results of the Lovelace research program were not taken into account. Indeed, the Lovelace Woman in Space program faded from memory until the 1990s and 2000s, when researchers identified those pilots as pioneering women and set about to revive their untold story and significance, culminating in a 2018

Lovelace Women in Space participants at the Smithsonian's National Air and Space Museum in 1994: (*from left to right*) Sarah Gorelik Ratley, Gene Nora Stumbough Jessen, Myrtle Cagle, Jerrie Cobb, Irene Leverton, Jane Hart, Jerri Sloan Truhill, Rhea Hurrle Woltman, Bernice Steadman, Wally Funk. *Photo by Michael Althaus*

Netflix documentary, *The Mercury 13*, that told the story mostly in the participants' own words. Some writers were eager to claim that these women paved the way or opened the door for women astronauts. In fact, they had not done so in any direct way; they met a dead end. Only Shannon Lucid, an avid pilot among the earliest women to apply to NASA for the job of astronaut had ever heard of them. Eileen Collins, who came to regard them highly enough to invite them to her space shuttle launches, did not know of them until she was already a NASA pilot astronaut.[31]

As their story has been told and retold in books, documentaries, and elsewhere, some repeated misconceptions have distorted certain facts. The Lovelace research program was private, voluntary, unofficial, and independent; NASA was not involved in it, and thus NASA did not cancel it. The program was not hidden or secret but received considerable publicity at the time. The women who were evaluated never trained to be astronauts, nor were they applicants or candidates for selection as astronauts. They were highly qualified pilots, but they did not have the right kind of piloting experience that NASA required for astronauts. They were never called the Mercury 13 until a television producer coined the term in the early 1990s. Such embellishments have to some extent layered their story with myth. The facts alone are a compelling story of "the first systematic testing of women's capabilities for spaceflight" and an important chapter in the progress of women toward spaceflight.[32]

Women Military Pilots

The most important and effective way to enable women pilots to qualify as astronauts was to admit them into military aviation.[33] That was originally the core requirement for becoming an astronaut, but women were simply not allowed. The forces of the civil rights movement, women's movement, and the military's own needs in the 1960s prompted a slow, incremental approach lasting more than thirty years until women gained equal standing in military aviation. A long-held tradition of gallantry—based on beliefs that women should be protected and that military aviation was too dangerous for women—had to be eradicated, along with beliefs that women's temperaments were unstable, their bodies not strong enough, and their presence disruptive to discipline and morale. That effort resulted in three decades of "gender wars" within the military, played out through study commissions, resistance, and other strategies. Change did not come easily.

Women in the military services during the 1960s were primarily nurses and medical specialists or "typewriter soldiers."[34] As a visible minority, these early military women were more concerned with acceptance than equality. Their professional code was to be highly qualified and present a "neat, feminine appearance," properly dressed and groomed, fit and trim. The more challenging technical positions were closed as "unsuitable for ladies." As long as the draft satisfied military manpower requirements, there was no incentive to recruit more women or expand the roles open to them.

The real thrust for change had to come from within the military services. During the Vietnam War era, to meet their workforce needs, the services began thinking about expanded training and new roles for women in aircraft maintenance, air traffic control, and instruction. Most women pilots were in general aviation, flying small planes and being flight instructors, and there were so few women in aeronautical/aerospace engineering—only 1 percent of undergraduate women—as to be invisible. The services removed limits on the number of women employed in the armed forces but still barred them from being pilots.

A 1967 article in *Air Force Times* made a then-progressive argument for women pilots in noncombat flying, in response to internal pilot shortages.[35] In 1969 the US Air Force opened its Reserve Officer Training Corps (ROTC) program to collegiate women, as the US Navy did in 1972. ROTC could open a path to aviation. The first woman to complete the ROTC program was commissioned into the Air Force in 1971, coincidentally the same year that the first woman achieved the Air Force rank of brigadier (one-star) general. ROTC was a route for many young men and women to enter into military service as commissioned officers, especially in aviation, and a number of those later selected to be astronauts followed this path.

A decade of change began in 1973 as the draft ended, military services shifted to all-volunteer forces, and ratification of the Equal Rights Amendment to the Constitution seemed on the verge of success.[36] Recognizing women as a sizable recruitment pool, the services sought ways to appeal to career-minded women. They dropped most restrictions on the types of jobs women could hold and opened promotion paths, pilot training, and eventually test pilot training to women. That year, the secretary of the Navy announced that a group of eight women had been selected to enter flight training as a "pilot project" to determine whether using women pilots for noncombat duty in helicopters and transport aircraft was feasible. Six earned their wings of gold, and the Navy never looked back. By 1977 the Navy and Naval Reserve had five thousand women in aviation as pilots and in various other roles.[37]

The first women to enter flight training in the Army and Coast Guard earned their wings, too, primarily in helicopters. The Air Force lagged in bringing women into flight training because it considered all pilots to be combat pilots, and women were legally restricted from combat duty. However, it did begin to admit women into other aviation roles, and the first woman, Jane Leslie Holley, graduated as an Air Force flight test engineer in 1975. By the end of 1975, the Air Force announced that women would begin pilot training to determine how they could be used outside of combat, and in 1976 two women's classes entered its undergraduate pilot training program in T-38 Talon supersonic jets. As in the other services, the women student pilots did as well as the men and toppled concerns that they could not handle military aviation. Flying military jets was finally the opening women needed to qualify for consideration as astronauts.

The service academies—the entry point for future military leaders and alma mater of some of the first astronauts—did not open to women until 1975, when President Gerald Ford signed Public Law 94-106 requiring admission of women. The women of the class of 1980, admitted in 1976, were the first to graduate from the Air Force, Army, and Naval academies despite resistance by those wanting to preserve these bastions of masculinity.[38] Attrition rates for women and men were comparable, demonstrating that the high standards could be met by both sexes and women would not cause the rigorous curriculum to collapse. The first woman academy graduate to enter the astronaut corps, Susan J. Helms, arrived in 1990. She graduated from the Air Force Academy in 1980 and then graduated as the Outstanding Flight Test Engineer at the Air Force Test Pilot School in 1988. This was as early as it was possible for women to gain the full military piloting experience originally required of astronauts.[39]

Not until the 1980s did the military services begin to admit women into their test pilot schools, several years after—spoiler alert—NASA had decided to start recruiting women astronauts. Although the first woman, helicopter pilot Colleen Nevius, graduated from the Naval Test Pilot School in Patuxent River, Maryland, in 1983, the military test pilot pipeline was sparsely populated by women for almost ten years. Neither men nor women could jump directly into test pilot schools; they had to meet certain qualifications in assignments and time served and compete for a limited number of slots. That path could take ten or more years to navigate. Test pilot school included two tracks: flight test engineer and flight test pilot.

Other first women to emerge through the military test pilot pipeline graduated from the US Air Force Test Pilot School at Edwards Air Force Base in California. Flight test engineer Jane Leslie Holley graduated in 1975 and submitted her application in the 1977 astronaut recruitment but was not selected. Susan J. Helms followed in 1988, graduating as a flight test engineer. In 1989 the first woman graduated from Edwards as a test pilot, Jacqueline S. "Jackie" Parker, a former NASA software analyst and member of the flight controller support team in Houston who then chose a career in the Air Force. Right behind her, graduating in 1990 as the second female Air Force test pilot, was Eileen M. Collins, who applied to NASA while at Edwards and was immediately selected as a pilot astronaut candidate. Collins and Helms were selected together into the 1990 astronaut class.[40] NASA seized opportunities to add women test pilots who met the toughest requirements for the astronaut corps, but there have not been many available to apply to NASA, and of those few women test pilots, some elected to continue their military careers instead.

In the military academies and military pilot training programs, women felt pressure to perform their best to keep the door open for other young women to enter. What had been impossible for women pilots in the 1950s and 1960s was an opportunity, even a right, cherished by military women in the 1970s and 1980s, not to be squandered. Despite hazing and lingering sexism in military culture, women found that competence was the best way to earn respect. If they could meet the same standards and do the same job as well as the best, they were likely to be accepted. Still, they faced unrelenting pressure to prove themselves to earn the respect that came more readily for men. As more women came into military aviation, researchers attempted to determine if men and women pilots differed in their personality characteristics. Their findings that pilots generally were quite similar in mentality increased confidence that women would be successful military pilots.

Progress of women in military aviation continued through the 1980s and 1990s as they became more visible and less a novelty. Yet women pilots were a miniscule fraction of the force, by 1990 amounting to only 1–2 percent. Early concerns about detrimental effects of women's presence on discipline and group cohesion, fraternization, and the fear of diluting standards gradually evaporated except in one domain: combat. Women flying in combat was the last stand of resistance. The very idea challenged too many taboos and threatened the last bastion of masculine warrior culture.[41]

As early as 1967, the Defense Advisory Committee on Women in the Services had recommended abandoning barriers to women in combat, but there was widespread military and civilian resistance. However, as women participated in military actions in Grenada and Panama in the 1980s, in some instances close enough to combat that their safety was not assured, opinion began to shift.

The Persian Gulf War of 1990–1991 became the turning point both for women's wartime presence and for restrictive policies. Women pilots served on the perimeter and sometimes in range of combat operations while flying support missions to deliver troops and cargo, refuel airborne fighter aircraft, and evacuate casualties. In modern warfare, technology blurred the line between combat and noncombat zones, and commanders began to realize that it was impossible to protect women from hostile forces as they carried out their missions.

Legendary naval aviator Rosemary Mariner, the first woman military pilot to fly a tactical jet in 1974 and the first woman to serve as commanding officer of an operational air squadron during the Gulf War in 1990, was a key figure in pushing for an end to laws and policies that prohibited women from combat duty. She and other military leaders who favored change argued for a mind shift away from gender and toward using the most qualified and capable people available for a mission as a matter of national defense policy. Commanders who wanted to put the best people in positions, combat or otherwise, found it hard to counter that argument, although they still could fall back on social and psychological concerns about integrating the sexes in combat.

The Persian Gulf War experience brought to a focus mounting pressure to repeal the combat exclusion rule for women military pilots.[42] In 1991, Representative Patricia Schroeder attached to the 1992 Defense Authorization bill a provision to repeal part of the 1948 Women's Armed Services Integration act that excluded women from aircraft engaged in combat. Her provision passed in the House but set off a storm of protest in the Senate, where debate more broadly considered women in combat roles at sea, on the ground, or in the air. The Joint Chiefs of Staff unanimously opposed repealing the combat exclusion of women, and the Senate replaced the repeal provision in the bill with one to commission a study. However, Congress passed the defense bill with another version of the repeal and a study commission. This action did not immediately open combat aircraft to women, because the secretary of defense ruled out any assignment of women to combat aircraft until the study commission made its recommendations to Congress, thus delaying implementation. One woman Air Force aviator commented, "The march to equality was not steady. It was more like a game where the women started ten points behind, and every time they clawed their way forward, the opposing team ran back over them."[43]

After almost two more years of contentious debate and feasibility study, Congress passed and President Bill Clinton signed a law in 1993 repealing the combat exclusion rule for women aviators and permitting the services to decide how best to use women in other combat roles. The secretary of defense partially lifted the exclusion for aviation, and Air Force Second Lieutenant Jeannie Flynn became America's first female combat pilot. Some restrictions remained for years, despite recommendations to remove them to enable women to advance in their careers. In 2003 women finally flew fighters and bombers in the war against Iraq. The policy was further relaxed in 2012 and ended in 2013 upon unanimous recommendation of the Joint Chiefs of Staff, a final step in what the New York Times characterized as "the steady but glacial evolution of the role of American women in war."[44] By 2016 all the services had plans in place to admit women into all combat roles.

From that point forward, military women could gain the same combat experience as military men, putting them on a level field for the first time in promotion potential and in applying to be astronauts. Although combat experience was not required of astronauts, it was an advantage that most male pilots had. The standard to do the job had finally shifted from gender to ability. These gradual changes in military aviation policies eventually enabled the selection of highly qualified military women, including academy graduates, test pilots, flight test engineers, and combat veterans, into the astronaut corps from 1990 on. Women had at last met the toughest aviation criteria and were equivalent to the male military pilots who had dominated the astronaut corps for its first three decades.[45] Women were now qualified to be not only astronauts but also pilot astronauts, eligible to pilot the space shuttle and command missions. By the year 2000, Eileen Collins became the first woman to do both.

Demystifying the Female Body and Psyche

Ignorance about the female body and psyche posed another barrier to women's participation in commercial and military aviation and spaceflight. Since ancient times, women's fertility and menstruation have been viewed with fear or awe as a mystery.[46] As recently as the 1960s in American society, misconceptions existed about what happens to women in the normal course of menstruation and how it might affect their capabilities. Whenever women were poised to enter male-dominant occupations, there was usually an initial assumption that women were biologically and temperamentally unsuitable, because they were not men and they allegedly had a periodic disability.

That belief infiltrated male-dominant realms and recreations, such as politics, social clubs, and sports. An example: the Amateur Athletic Union banned women from running a distance greater than 1.5 miles because it was too strenuous. Prolonged running might damage the female reproductive system, leaving women unable to have children or the uterus might detach and fall out. Lack of evidence was no deterrent to such beliefs. It was not until 1972 that women could officially run the Boston and New York marathons and not until 1984 that women's long-distance running events, including the marathon, were added to the Olympics.[47]

The Lovelace Woman in Space Program was one of the first scientific attempts to study women's bodies to measure their physical strength and endurance. However, it was not free from mistaken assumptions about women's fitness. Only one brief scientific paper was published in a 1964 issue of the *American Journal of Obstetrics and Gynecology* from what must have been reams of data collected.[48] The authors, both male research physicians on the Lovelace staff, concluded that women made unreliable astronaut candidates because "their menstrual cycles fundamentally compromised their suitability for space." This was pure speculation, since nothing in the women's screening program that paralleled the tests given to men addressed menstruation or its effects on performance. The authors posited that women's reproductive capabilities might compromise their attentiveness, coordination, decision making, mental stability, and other crucial functions. The authors called for further physiological research on women through an entire menstrual cycle and "a full psychiatric examination or psychoanalysis for cryptic mental aberrations" before women become eligible to venture into space.

Although it may be laughable today, the idea that women's physiology was a problem that complicated things for men was not then uncommon. Even more ignorant was the idea that women somehow became different people at different times of the month. The idea of hysteria, or excessive emotion, as a female condition related to the uterus extends back thousands of years. Hysteria is no longer recognized as a mental or physical disorder, but a belief that women are different during their periods remains, with premenstrual syndrome (PMS) named the culprit. Misunderstanding menstruation fostered a rationale—"menstrual politics"—for excluding or at least discrediting women.[49]

Social taboos on frank discussion of menstruation as a normal physical function led to mistaken notions, as evidenced in a public survey published as The Tampax Report in 1981.[50] The findings at that time were that "one quarter of Americans think women cannot function normally at work while menstruating"; "one third believe that menstruation affects a woman's thinking ability"; and almost 90 percent of women compared to only two-thirds of the men surveyed thought that menstruation has no effect on job performance. An antidote to such thinking helped women, if not men, dispel misinformed beliefs. A landmark book for the women's liberation movement during this era, *Our Bodies, Ourselves*, first published in 1973 and updated in 1984, dealt frankly with gynecological matters and countered inherited cultural notions of femininity as weakness with culturally masculine notions of strength.[51]

When the idea of women as astronauts arose there was, predictably, a culturally influenced initial response that women were not suitable because females were not as strong and capable as men. A corollary was that their monthly menstrual cycle would complicate spaceflight. At a time when no one knew exactly how the human body responded to weightlessness and the high gravitational force of launch and reentry, there were more uncertainties about the female body. Some of those concerns were practical: Would menstrual blood flow out of the body in weightlessness, or would it stay pooled inside or flow into the abdominal cavity? Would weightlessness ease or exacerbate cramps, headaches, and bloating associated with periods? How many pads and tampons would have to be stowed on board a spacecraft, how much space would they take up, and how would feminine products be disposed of without a health risk to the crew? Besides such hygiene and sanitation concerns, there were the hypotheticals: What if a woman became temperamental or incapacitated during her period? How would that affect mission success?

None of these questions could be answered until women actually flew in space, and it would take some years to acquire enough data to begin to understand which, if any, were relevant. The article by the Lovelace doctors raised these concerns and suggested a possible solution—to delay menstruation during spaceflight by use of oral hormones. However, menstruation in orbit has not been researched, so it is still unknown whether weightlessness causes any real change. Anecdotally, it does not. Menstruation in space has not been a problem.

Menstruation is inextricable from women's sexuality, and that caused further concerns about admitting women into men's realms. Military leaders worried about fraternization—romances and sexual relationships—that could impair group cohesion and combat readiness, yet they bemoaned separate sleeping quarters and latrines as a burdensome expense. Astronauts' wives had the same concerns about husbands sharing cockpits, spacecraft, and travel with women. Pundits and psychologists warned of the sexual tension that would arise with both sexes in close quarters. How would that complicate crew relationships and doing the work of a mission successfully? Would men be distracted by and attracted to women crewmates? Would the onus be on women to keep things professional? Military leaders had a further concern about women aviators; military pilots were not allowed to be on medication. What about women on birth control pills? Would an exception for them be necessary when there was an aversion to making gendered exceptions? These are just some of the instances of women's femaleness seen as an inconvenience in the world of men.

The perception of women as "Other" too often has led to their treatment as other-than-men. Whether justifications were based on biology or temperament, the idea of women as incapable and unwanted where men were in charge ultimately arose from ignorance of their basic femaleness. Menstruation is a natural part of life for women, and although their experiences may differ, most healthy women find it does not affect their ability to perform their jobs or other responsibilities.

Cultivating Women Scientists and Engineers

More women began to pursue higher education in the 1950s, but in a limited range of fields, many preparing to become teachers, nurses, librarians, home economists, or social workers—the "softer" professions deemed appropriate for women. During this prosperous decade, when it was atypical for middle-class white women to work outside the home, if they did so, their primary options without an academic degree were secretarial and office work, retail and customer service work, hairdressing, and other service jobs.

As women aspiring to professional careers headed to colleges and universities, those students with technical abilities aimed for the fields matching their interests that had elevated men to professional status: medicine, science, and engineering.[52] During the 1950s increasing numbers of collegiate women sought to enroll in undergraduate programs that would prepare them for graduate studies in these fields, but their percentage of bachelor's degrees awarded was small, less than 1 percent. The sciences were marginally friendlier turf for women than engineering and medicine but still tough to enter, especially physics. Although rare compared to men, enough women scientists had already gained a foothold, even if overshadowed by the men with whom they worked, that there was somewhat less resistance to newcomers in astronomy, mathematics, and chemistry—foundational sciences for aerospace work.

Engineering and medicine were another story altogether. Women were denied admission to classes and programs based on assumptions that they couldn't handle the rigorous work, wouldn't be committed to completing the degree, would marry and have children and thus waste their education, would take a position that should go to a man, would "undermine" the quality of technical education, and on and on. Women faced both social and intellectual discrimination. If women did manage to enroll in engineering or pre-med courses, they all too often were ignored, harassed, mocked, discouraged, and treated as intruders. Many left under such hostility.[53]

Even worse, at the graduate level, women engineers in training were disadvantaged by lack of restrooms, student housing, laboratory space, financial support, mentors, women professors, honors and awards, and other resources available to men.[54] As if these indignities were not enough, women suffered being called "gal engineers," "girl engineers," or "engineeresses" and mocked as unattractive for having "manly" interests. For years, most engineering and medical schools were unwilling to make space for women. The tide finally began to turn when they needed the revenue boost that women's tuition could provide.

As historian Margaret Rossiter noted, sexism wasn't yet a word, and patterns of discrimination were not yet widely recognized.[55] It was easy to exclude, exploit, or marginalize women in technical professions until these inequities were identified, protested, and reformed. Raising consciousness to effect change was the hard work of the 1960s, aided by new and extant organizations.[56] The American Association of University Women (AAUW) was an established, longtime advocate for women's opportunities in higher education. The Society of Women Engineers (SWE) became an important early force in response to these academic conditions. Founded in 1950 as a networking community for professional and student engineers, the society rapidly established chapters on campuses to support collegiate women and advocate for changes in their treatment. SWE began to recognize common patterns of discrimination and inequality in engineering education and eventually became active in the political struggle for equal rights. Likewise, the National Organization of Women (NOW), founded in 1966, brought attention and critiques to discriminatory attitudes and practices more broadly in society, reinforcing the consciousness of sexism in education and the workplace. The Association for Women in Science (AWS) formed in 1971, engaged in networking and advocacy, fought discrimination, and promoted equal pay and recognition for women scientists.

The National Defense Education Act (NDEA) of 1958 established as a priority the education of scientists and engineers to ensure US technical strength in the Cold War era. This national security mindset prompted an uptick in government reports and bulletins encouraging women to seek careers in the physical sciences and engineering. As a largely untapped resource, women in theory could fill shortages in the technical workforce and become essential to the nation's defense. Women received a share of

NDEA graduate fellowships in some fields, but their share in physics and engineering was paltry. Universities and employers were slow to welcome or recruit women, so their status in the technical labor force did not readily change. Women remained almost invisible in the science and engineering community well into the 1960s, with the number of degrees awarded to women a miniscule fraction of the number awarded to men.

Arising in the context of the civil rights and women's movements, momentum built toward the passage of laws in the 1970s that gave women a more equitable chance to earn advanced degrees and find employment in the sciences and engineering. What amounted to a legal revolution in women's education occurred in 1972 when Congress passed and President Richard Nixon signed two landmark laws. The Equal Employment Opportunity Act of 1972 amended and extended the reach of the Civil Rights Act of 1964, which prohibited discrimination. Implementation had been slack, and the EEOA provided for necessary enforcement, making it possible to file complaints and win remedies. Title IX of the Education Amendments Act of 1972 banned sex discrimination in any program in any educational institution receiving federal funding. Now remembered largely for its effect on women's athletic programs, Title IX had a profound effect on women's education by preventing discrimination in admissions and curriculum. It opened the doors to the academic disciplines and schools that had excluded women, empowering female students to take the courses and pursue the degrees they wanted and increasingly engage with women faculty members and mentors as barriers to their employment fell.

Women's presence in graduate and postdoctoral technical education grew from the 1970s onward.[57] Legislation provided the foundation and incentive for women to enter the formerly masculine realms of science and engineering. There women found colleagues and mutual support. They networked, organized, applied pressure, and gradually brought about local reforms that began to dispel the hostile environment that earlier women had to abide. Change was not immediate or complete, but activism had forced the academic world to make space for women scientists and engineers.

Technical Women in the Early Space Program

As women were beginning to agitate to be included in spaceflight as astronauts, NASA was advertising that women were welcome elsewhere in the space program. Articles occasionally appeared in newspapers touting the number of professional, technical women in the space workforce. Women trickled into NASA technical positions in the early years, and the agency's census shows how rare their presence was.[58] In 1960, by the time the agency was two years old, NASA reported 5,030 professional grade employees, of whom 415 were highly skilled women, and three of whom were actually aeronautical research engineers. The point was to head off criticism as a male-dominant organization and to prove that NASA recognized women's qualifications. NASA could claim that it was welcoming qualified women into the space program, while at the same time defending the exclusionary requirements for its most visible employees, the astronauts.[59]

Women were indeed working in some technical positions at NASA and related institutions and companies from the beginning, but their presence was not evident to the public or, in some cases, to their coworkers.[60] In the 1950s and 1960s they were usually few and buried deep in the organizations, sometimes working in basements or annexes away from the men who occupied management and policy positions at the upper levels or engineers who were the backbone of the early space program. A 1966 book, *Women of Space*, identified a sample of thirty-five technical women to represent others working for NASA, aerospace companies, and the Air Force in various fields and locations.[61]

Despite the antagonistic environment of engineering programs at the time, women with degrees in mathematics or chemistry were finding technical employment in aerospace enterprises. Before NASA was established, women were working as "computers," doing complex calculations for rocket engines and missiles at the California Institute of Technology and the Jet Propulsion Laboratory (JPL), which soon became affiliated with NASA. These "rocket girls" worked with engineers in the design, testing, and launching of rockets. They did the math by hand with pencils or slide rules and graph paper before calculators and computers became available, calculating propellant mixtures, thrust-to-weight ratios, nozzle and nose cone shapes, staging, trajectories, and other critical measurements, and they earned the engineers' respect. Some went on to earn their own engineering degrees. As the space program evolved in the 1960s, these women expanded their work into calculating trajectories for satellites and deep space probes. Their performance, had it been widely known, would have challenged stereotypes about women's technical abilities and perhaps eased the path into engineering schools.

Other women mathematicians, also known then as "computers," were doing similar work sequestered at the Langley Research Center in Hampton, Virginia, home of the National Advisory Committee for Aeronautics (NACA) and then NASA's Langley Research Center. Introduced as "hidden figures" by the book and movie of the same name, this cadre of African American women and their white counterparts were an important complement to the male engineers they supported. Despite the indignities they suffered, they won respect for their competence and were considered more reliable calculators than machines. For some, their work evolved into computer programming and further education, again putting the lie to stereotypes about women's technical abilities.

The women who held such positions typically were very bright, earned high grades in school and college, and often were the only girls in advanced math and science courses. They loved crunching numbers and solving equations. They were creative thinkers who worked outside the box until they found solutions. Checking and rechecking their work for accuracy, they earned a reputation for reliability.

The history of the rocket girls at JPL includes detailed commentary on the social aspects of being women in a male-dominant organization and on the challenges of holding a demanding technical job while also being a wife and mother. Male applicants with college degrees would likely be hired as engineers, but women with degrees were hired as computers. Women computers did not receive credit in academic publications. For years there was no career path for women to advance; there was only one supervisory position for computers. The workdays often lasted twelve hours during rocket engine and launch tests, with data being monitored by women in dresses, stockings, and heels. Their motto was "Look like a girl, act like a lady, think like a man, work like a dog."[62] At first the women had to quit working when pregnant, so they hid their condition as long as they could. There was no maternity leave, and their positions were not held during a leave of absence. Childcare was difficult to find, especially with irregular work hours. Women loved their work, but many felt the stress of doing it and also managing their family and household. Some of their marriages foundered under the responsibilities of work and home or an unsupportive husband.

Despite these inequities and tensions, the rocket girls at JPL found their work satisfying and relished being part of something technical, inventive, and important. The computers and engineers worked together well, and the women became close-knit and mutually supportive. They felt that JPL treated them with respect and valued their work. Gradual changes, such as flex time, the introduction of calculators and electronic computers, and advanced training improved their situation. By 1970 new hires had to have engineering degrees, but the computers were redesignated as engineers even if they held degrees in mathematics or chemistry. Some of them went on to earn engineering degrees or become computer programmers and managers.

More evidence of such pioneering women within the early space community is coming to light through nationwide efforts, such as the Smithsonian American Women's History Initiative, to identify overlooked women in science and technology. NASA itself has brought to the fore many women who made impressive contributions to the space program's success in its first decades. The names of technical women who had significant roles in the space program during the 1960s are becoming well known, recognized in books, films, scholarly articles, popular blogs, and even Lego figures, and their number keeps increasing. Examples include

- astronomer Nancy Grace Roman, who joined NASA in 1959 to set up its astronomy program and plan for observatories in space, including the Hubble Space Telescope;
- Marjorie Townsend, who also started work in 1959 and became the first woman to manage a NASA satellite project and launch;

- JoAnn Morgan, hired in 1963 as the first woman engineer at the Kennedy Space Center, worked for fifteen years in buildings without a ladies' restroom, and during her forty-year career became the Kennedy Space Center's first woman executive;
- Frances "Poppy" Northcutt, who began working as a contractor engineer at the Johnson Space Center in 1965 and was among the first women allowed on duty in the Mission Control Center;
- mathematician and aerospace technologist Ivy Hooks, hired in 1965 to work on the Apollo program and then one of only two women engineers assigned to the original space shuttle design team in Houston;
- African American mathematician-computers Dorothy Vaughan, Katherine Johnson, Christine Darden, and NASA's first black woman engineer, Mary Jackson, all doing aerospace work before NASA was established and introduced in the book and film *Hidden Figures*;
- Margaret Hamilton, the MIT-based computer scientist and software engineer who led the team developing flight software for the Apollo missions;
- and Rita Rapp, physiologist, who developed appetizing, healthy, packaged foods for space crews.

Although NASA appeared to be an all-male engineering organization, individual technical women and pockets of women in technical units were making their marks on the early space program long before women were hired as astronauts.

Women Aquanauts and Scientists Simulating Space Missions, 1970–1974

During the first half of the 1970s, women made surprising but not widely recognized headway toward being considered as astronaut material. In at least three instances, science- and engineering-trained women participated in NASA exercises to simulate and test concepts for spaceflight. These episodes occurred before the agency committed to recruiting women for the astronaut corps and demonstrate that NASA was already moving toward making space for women in space.

On a tropical July day in 1970, several weeks after the Apollo 13 crew returned safely from their failed mission to land on the Moon, five young women in bright red wetsuits and scuba gear stepped off a dive platform and began their descent to an outpost on the sea floor fifty feet below. They were heading to Tektite, a modular laboratory and residence in a sheltered cove in Great Lameshur Bay on the southern coast of St John in the US Virgin Islands, where they would spend two weeks on an underwater scientific research mission. Led by the now-famous oceanographer Sylvia Earle, this team included two recent PhDs and three graduate students, all selected by the National Science Foundation. There had been previous "man-in-the sea" engineering

All-woman Tektite II aquanaut crew ready for splashdown to start their mission, 1970. *Photo by Cecil W. Stoughton, National Park Service*

projects, but this was the first-ever extended women-scientists-in-the-sea mission. Arguably, it was also the first NASA mission to include women and its only all-woman mission crew to date.[63]

Their expedition was part of a broader research program led by the Department of the Interior and cosponsored by the US Navy, NASA, and the General Electric Company Space Division, with various other partners, including universities and the Smithsonian Institution.[64] Each had its own research objectives, but they joined forces to create and operate this research station for the study of marine life, human behavior in isolation, and biomedical health in a modified atmosphere. Aquanaut crews would breathe high-nitrogen air and live in three times normal pressure in saturation diving, unable to return without spending almost a day in decompression. Tektite received broad news coverage at the time as the nation's most ambitious "manned" scientific expedition in the sea and the largest study of social behavior in an isolated habitat.

NASA's key role, carried out by medical and engineering personnel, was to study human behavior 24-7.[65] During the first mission in 1969, Tektite I, four male marine scientists set a record by staying at depth for two months without experiencing any notable physical or behavioral problems. The women's expedition was the sixth in a series of ten missions, called Tektite II, carried out in 1970. More than fifty scientists and engineers lived and worked under the sea for fourteen to twenty days at a time, rotating in five-member crews, all of them men except this one all-woman crew.

While the Apollo lunar missions were under way, NASA was thinking ahead to longer-duration flights on a possible space station, with the orbital workshop that became Skylab already looming as the next major program. Among the unknowns were how effectively crews might perform during extended isolation in a hazardous environment where they were dependent on life support systems, and what habitability and human factor issues might be important for their well-being in confinement. Also of interest were any psychological or physiological effects they might experience in such conditions. NASA was keenly interested in how these factors might affect the efficiency and success of spaceflight. Simulations in a high-fidelity model analogous to spaceflight would yield useful behavioral and biomedical information and also permit evaluation of habitat design. NASA became involved in Tektite II to use underwater missions as an analog for space station missions.[66] Although NASA did not lead this broad program or select the crews, the agency provided enough funds and support personnel—and more important, the human behavior research program—that these marine expeditions can credibly be considered NASA missions.

Sylvia Earle recalled learning of the program from a notice posted on a bulletin board at Harvard, where she was a postgraduate research fellow. The call for proposals to conduct marine science research in an underwater habitat in the Caribbean, using new diving techniques that would allow much longer dives than scuba gear did, was too alluring to pass up. "It was clear," she said, "that no one expected women to apply. But some of us did!"[67] Their research proposals passed muster, so the Tektite II program manager agreed to put together a team of four women scientists if a woman engineer-diver could be found. "Half the fish are female, so I guess we can put up with a few women," Earle remembers him admitting. It would have been unseemly or scandalous at the time to have coed crews; women were not then permitted to bunk on research vessels. Earle, the most experienced diver, led the crew of women marine scientists: Renate Schlentz True of Tulane University, who had worked with Jacques Cousteau; Ann Hurley Hartline and Alina Szmant of the Scripps Institution of Oceanography; and Margaret Ann "Peggy" Lucas, an ocean engineering student at the University of Delaware. Their two-week mission made them the first female aquanaut crew to live and work underwater, but Earle noted, "We went as scientists primarily, and as women secondarily."[68]

The all-woman crew drew avid attention before and after their expedition, not always with due respect for their professionalism. Media and VIPs flocked to the site for their splashdown to start the mission and their splashup at the end. Earle recalls that the men were always called aquanauts, but the women were called "aquababes," "aquachicks," "aquabelles," "aquanauties," "aquanettes," and even "mermaids."[69] "Nobody can believe we are aquanauts when they meet us," Earle noted.[70] In print, Miss or Mrs. rather than Dr. usually preceded names, and Earle's status as a mother of three small children was duly noted, as were the names of husbands.[71] Reporters called them attractive and published their ages, average height, and average weight: twenty-two to thirty-four years of age, 5 feet 4 inches tall, and 110 pounds. Some thought they looked fragile, not muscular as one might stereotype an aquanaut. When the likeable project manager was asked if he expected problems with women in the program, he replied jocularly (one hopes), "You can always expect behavioral problems when there are women involved."[72] Many in the media fixated on and repeated this phrase, "The one concession to femininity is a hair dryer."[73] Earle said it was useful for "water-in-the-ear issues" but never used for hair styling "to impress the fish."[74] Although they didn't primp with the hair dryer, which was also available for the men's use, they did make one concession to modesty: a shower curtain.

The Tektite research station proved to be a comfortable home and workplace, in many ways resembling Skylab, America's first space station, occupied in 1973–1974. Tektite comprised spacious twin cylinders, each with an upper and lower room. The crew quarters contained General Electric appliances for cooking and refrigerating food, a sink and dinette set, carpet, and four stacked bunks. Above it was the command center or bridge, the engineer's primary domain, which included a lab work area with instruments, small library, tape deck and music cassettes, television, cot, the closed-circuit television system by which NASA monitored the crews' activities, and audio communications equipment linked to the command center van on shore. A short connecting tunnel led to the top floor of the other module, the equipment room, with its food freezer, toilet, sink, water heater, hot freshwater shower, and clothes dryer. Below that, the wet room offered open-hatch access to the sea, dive equipment storage, and another lab bench with instruments. Crews appreciated the bubble windows around the modules and a cupola on top that allowed observation of local marine life while the scientists were cozy inside.[75]

Tektite was not a completely closed system; the crews' air, water, and electricity were piped down from a pier, and their food and trash were shuttled up and down twice a week by a support diver. NASA stocked the habitat with their meals, frozen dinners prepared by Stouffers, to be evaluated for future space missions. All teams found the accommodations, food, and options for leisure time to be pleasant and practical, with only a few suggestions for improvement.

The women's team spent most of their workdays in diving gear exploring and experimenting with the flora, fauna, and geology near Tektite—an analog to astronauts' spacewalking activities on the Moon. Inside the modules they did the normal kinds of research and domestic tasks they would do at their home bases, from examining specimens and caring for equipment to making meals and sleeping—analogous to crew activities on a space station. Cameras and audio recorders captured their every motion and utterance. NASA collected and analyzed data on their social interactions, efficiency, problem-solving, physical condition, and anything else relevant to long-duration spaceflight. The women were themselves an experiment closely observed by psychologists in the nearby mission control center.[76]

Unsurprisingly, the pre-mission project descriptions and post-mission summary reports used the language of the era: Tektite was an effort to study "highly trained men" and the "biomedical responses of men," an "opportunity for men" to conduct meaningful scientific research on "man's first long-term scientific mission into the sea," and to observe "man" under stress in confined and isolated environments. There is no mention of women except in reference to their single mission and to note that there were no discernible differences in performance.

The women proved their competence and equaled the male teams in research productivity. Earle has said that they were not consciously competing with the men but were trying to take maximum advantage of this unique opportunity to do their research.[77] They impressed all observers with their ready accommodation to isolation and confinement, camaraderie, and productivity. They logged more hours in the water than the other groups. They experienced no physical or psychological problems; one even had a normal menstrual period during the expedition. Earle logged a record eighty-six hours—more than anyone—working outside the habitat, accomplishing in two weeks the amount of research it would have taken her in two months of surface dives.[78]

When the women emerged from twenty hours in the decompression chamber at the end of their two-week underwater stay, they were greeted with long-stem red roses, a welcome back pineapple, a large piña colada with five straws, and a spate of glowing publicity.[79] Their public appearances included a luncheon at the White House hosted by First Lady Pat Nixon, a parade and award presentation in Chicago, and other accolades as reported in national news articles. Team leader Earle accepted a request to address Congress on the Tektite project and was named *Time*'s Woman of the Year for her leading role in this pioneering underwater venture.[80] The entire crew was relieved when the unexpected, almost overwhelming attention abated, and they could return to their usual work.

In 2020, during the COVID pandemic, the four women still alive and many other luminaries appeared in an online conference, Tektite2020: Women of Sea and Space, to mark the fiftieth anniversary of their historic undersea mission. In her remarks, Sylvia Earle noted that "sea-space connections were so clear back in the '70s."[81] The parallels between hydrospace and aerospace expeditions include living in a container surrounded by a hazardous environment and depending on life support systems and one another. Whether in sea or space, explorers are far away and alone. They are reliant on technology and responsible for staying mentally and physically fit, adapting to isolation and confinement, and doing good work. The Tektite missions demonstrated how astronauts and aquanauts are fellow explorers of outer and inner space. President Richard Nixon recognized this when he congratulated the crew of the record-breaking Tektite I mission, proclaiming that "the aquanauts join the astronauts as space pioneers."[82] Astronauts participating in this conference also paid tribute to Tektite's importance in anticipating long-duration space missions.

The all-woman Tektite II mission could have opened NASA's eyes to the potential of women as astronauts if word of it had spread beyond the human behavior and habitability research teams that participated. Instead there is little mention of Tektite II in NASA publications or archives, and little indication that NASA leadership was paying attention. In 1969 and 1970 the Tektite program merited just one line in the annual *Aeronautics and Space Report to the President*, where progress on Skylab was reported in detail. The primary NASA book on Skylab mentions Tektite in only one sentence and one photo caption. The Department of the Interior and the US Navy released significant reports about their involvement, but not NASA. However, buried in a conference paper presented by the NASA and Interior Department managers of the Tektite II program was this prediction about upcoming manned space flight: crews will increase in number and variety, missions will be longer, and "women will be in space."[83] Separately, in an interview for *Smithsonian* magazine, the NASA program director for Tektite predicted that "the day is not far off when women will be included as passenger-scientists in mixed spacecraft crews."[84]

Nevertheless, it is fair to consider the Tektite II all-woman mission a precursor to NASA's eventual acceptance of women as astronauts. Earle's team showed they could do the job without causing problems. The women themselves believed that their mission was a great step toward acceptance of women as scientists and especially as members of research expeditions, from which they had often been excluded. They keenly saw their experience as breaking down career barriers for women in science and exploration.[85] The idea of women in space was catching on, as evidenced by a long feature article in the *New York Times* during the lead-up to the 1973 Skylab missions: "Will man ever *live* in space? (If so, woman will live there too)."[86]

If this crew's performance didn't persuade NASA to admit women into the astronaut corps or to select a woman for the 1973–1974 Skylab missions, it did demonstrate that women could live and work in an extreme and isolated environment as effectively as men. As an analog to a space mission, the all-woman undersea mission was as productive and uneventful as anyone could hope. Astronaut Scott Carpenter, a veteran of the Navy's underwater Sealab program, was sent to observe Tektite from the control center during the women's on-site training mission, and he went diving with Earle to see some of their equipment.[87] NASA's attitude toward women in space seemed to be softening as it accepted and evaluated the all-women crew no differently from the male crews. Before their mission, NASA arranged for engineer Peggy Lucas to test equipment in the underwater facility at the Marshall Space Flight Center in Huntsville, Alabama. This opportunity made her the first woman allowed to work in the Neutral Buoyancy Simulator used for astronaut training.[88]

After all the planned missions were completed, NASA's Marshall Space Flight Center hosted a two-day symposium, "Tektite II: Men Undersea."[89] This event in 1971 brought together many of the participants from various agencies to assess the underwater facility and research program in sessions organized around topics in engineering, operations, and human behavior. The prelude to the program explicitly recognized that the Tektite II missions were analogous to manned spaceflight and the habitat was analogous to a manned spacecraft. On a panel that included NASA astronauts and aquanaut-engineers, Peggy Lucas and Renate True spoke about the effects of isolation and confinement on the job to be done. As the only two women among twenty-five men scheduled to speak, they surely were noticeable. Their appearance beside astronauts and other renowned researchers visibly elevated them as peers and signaled that their role was as relevant and acceptable as anyone else's.

The Tektite legacy affected the course of marine research and also spaceflight, demonstrating how productively scientists could live and work while isolated in an extreme environment. It led to new generations of undersea habitats, and lessons learned from Tektite informed NASA's Skylab space station. Skylab lessons then influenced Spacelab and International Space Station design and operations. Thirty years later, NASA again began to use another underwater habitat, Aquarius Reef in the Florida Keys, to prepare astronauts for International Space Station missions. Most important, Tektite erased doubts and modeled how women could have the same kinds of productive roles in spaceflight as men; the all-woman mission produced no physical or behavioral evidence for exclusion. This mission may have helped to advance NASA's position on training women as astronauts, given the similarities between underwater and in-space operations. That idea dawned on someone, because Sylvia Earle and Peggy Lucas both received inquiries from NASA inviting them to consider becoming astronauts.[90] Had they been less devoted to the sea, these aquanauts might have been America's first women astronauts.

Practicing for Science Missions

In December 1974, four women researchers participated in a five-day simulation as the crew of a space laboratory mission. All were employed at NASA's Marshall Space Flight Center, where the simulation occurred. It isn't clear whether they volunteered or were assigned to this session in a series of concept verification tests (CVT), but they were eager for the experience and would have welcomed being on an actual mission. Their job was to conduct eleven materials science experiments in a space-like laboratory environment while acting under mission-like communications and operations protocols.[91]

They were the only women to participate in the series of ten tests, each one focusing on a different set of science experiments, whether biomedical, physics, atmospheric, or astronomy. Three of the four, all in their thirties, were materials science specialists from the Materials and Processes Laboratory and were responsible for conducting the experiments: astronautical engineer Carolyn S. Griner, metallurgist Mary Helen Johnston, and physicist Ann F. Whitaker. Engineer Doris Chandler joined them as crew chief to handle their communications and schedule their daily workload. They worked in a

Women participants in a simulated space laboratory mission on November 13, 1974: (*left to right*) Mary Helen Johnston, Ann Whitaker, Doris Chandler, Carolyn Griner. *NASA*

general-purpose laboratory module, a big horizontal cannister, for eight hours a day, returned home for the night, and resumed the "mission" the next morning. Although they were not confined overnight in a crew quarters area as part of the simulation, in the spirit of a real mission they wore identical coveralls while on duty.

The purpose of this series of tests was to evaluate whether the kinds of experiments normally conducted in labs on Earth could be performed effectively in a space laboratory equipped to accommodate any discipline. NASA and the European Space Agency were developing such a laboratory, Spacelab, to be installed in the payload bay of the shuttle and connected to the crew cabin by a tunnel, so scientist-astronauts could move freely back and forth in the pressurized environment. Spacelab and the shuttle would provide power, water, air, vacuum, controlled temperature and humidity, communications, and computer and data services to permit 24/7 scientific research for up to two weeks. Spacelab would turn the shuttle into an intermittent space station. In these early years before the first Spacelab mission flew in 1983, the concept verification tests would help scientists and instrument developers understand how their equipment should be designed to fit into and operate within Spacelab.

NASA was also trying to verify what the roles of scientist-astronauts should be in the coming space shuttle era. To that point, only four scientists had served on mission crews: Harrison Schmitt on Apollo 17, and Joseph Kerwin, Owen Garriott, and Edward Gibson, each on one of the three Skylab missions. How adept might scientist-astronauts be in conducting other scientists' research? Should the astronauts be generalists or specialists? For this test (CVT-4), the women were not only materials science specialists but had also developed the experiments being tested, and they were among the few women holding such technical positions at the Marshall Center. The mission was simulated again the next year with a generalist crew of men who were not experts in materials science but received intensive hands-on training with the experiment hardware in advance of their test (CVT-4A).

Griner, Johnston, and Whitaker coauthored the report of these two successful tests with a full description of the laboratory, the materials science experiments, how well the experiment hardware and the lab interfaced, and the effects on delicate experiments of vibrations caused by the crew's movements in the lab. The crew noted that thoroughly understanding the equipment and the experiments allowed them to refine procedures and explore results during the mission. In fact, they reported that at least two experiments would have failed entirely without the crew's intimate knowledge of the hardware and science. The major difference between the two tests was that the second one required much more communication between the crew and a specialist "on the ground" nearby, which pointed to the need for more training and simulation for generalists.[92]

In interviews before the simulated Spacelab mission, these women expressed confidence that "a woman could do any job a man could do on a space flight" and that there were "plenty of qualified women available for scientific-type missions."[93] They also admitted their interest in going on a space mission themselves. During the next two years, Griner, Johnston, and Whitaker took steps to prepare for spaceflight. They received neutral buoyancy training (simulated weightlessness) in the deep-water tank used for astronaut training at Marshall and gained zero-gravity weightlessness experience on parabolic flights on NASA's KC-135 aircraft. They took flying lessons, and they may also have received familiarization training in Apollo spacesuits.[94]

Soon Griner and Johnston applied to become astronauts, and Whitaker was a finalist for a payload specialist position as a guest crewmember on the first Spacelab mission, STS-9 in 1983. Whitaker's selection for a possible Spacelab flight was announced *before* NASA announced any women astronaut candidates; it was reported that she "could become the first American woman in space."[95] Although none of the Marshall women were selected for the astronaut corps, all stayed involved with the early Spacelab missions. Johnston was named an alternate payload specialist for the Spacelab-3 mission of 1985; she trained with that crew and was on duty in the payload control center during the mission. She left NASA after Spacelab-3 to become a university professor and inventor with twelve patents. Whitaker's experiments flew on several shuttle missions, and Griner held technical management roles for science payloads. Griner eventually became deputy director and acting director of the Marshall Space Flight Center, the first woman to reach that level. Meanwhile, Whitaker earned a PhD and rose through the ranks to head the MSFC Science Directorate.

Like the Tektite story, the Spacelab simulation shows that NASA was on the way to bringing women into spaceflight in the years before it actually happened. In the mid-1970s NASA was wrestling with equal opportunity and affirmative action planning with a view toward admitting women into the astronaut corps. Management at the Marshall Center clearly was preparing its group of outstanding women scientists to qualify as astronaut candidates; otherwise, they would have had no access to the neutral buoyancy simulator, aircraft, or pressure suits. The Spacelab simulation is little more than a footnote in NASA history, but it is a significant story.

Occasionally photos of this group of women pop up during Women's History Month, with greater appreciation now than then. They were on the leading edge of women assuming prominent positions in the space agency. They proved their abilities as scientists and their readiness to become astronauts in the coming space shuttle era precisely when NASA was nearing the decision to recruit women for spaceflight. Their experience made it evident that the sex of a scientist or engineer was immaterial to the conduct of science, and women could do the kind of scientific experiments planned for Spacelab.[96] It was disappointing that they were not selected to fly, despite their timely demonstration of readiness.

Another Spacelab mission simulation occurred in 1977 at the Johnson Space Center, this one focused on developing life sciences missions.[97] Called Spacelab Mission Development Test III, it was a collaboration of JSC and NASA's Ames Research Center, the home of much of the agency's life sciences research focused on humans and animals. Like the materials science lab simulation, this one aimed to assess the feasibility of different types of experiments and equipment that could be supported in Spacelab.

One of NASA's longest-tenured African American scientists, Patricia S. Cowings at Ames, was selected to train and act as a mission specialist for this weeklong simulation and for all the preliminary planning and development activity. She had established a specialty in psychophysiology and focused much of her research on biofeedback—self-regulating techniques to avoid space sickness, control heart rate and blood pressure, and manage other bodily responses in microgravity and returning to Earth. Her research aimed to help astronauts adjust better to space and back.

Cowings claims to be the first woman scientist trained to be an astronaut.[98] Her role in the simulated mission was immersion in the science—operating and evaluating the experiments, procedures, equipment, and provisions of the lab to guide development of the same for life sciences research in Spacelab on the shuttle. Her work later extended to life sciences research on the space station. Her key role in this effort, like that of the Marshall women materials scientists, indicated that NASA was beginning to see women as potential astronauts. In fact, this simulation occurred in the same year as the initial recruitment of space shuttle astronauts and NASA's explicit effort to find highly qualified women and people of color to apply.

Simulating Weightlessness in Bed

Between the occasions when the Tektite and Marshall women scientists were demonstrating their suitability for spaceflight, NASA tested a group of women for physiological fitness. In 1973, twelve nurses participated in a study of the female body's adaptation to simulated weightlessness. Prolonged bed rest is an analog for the reduced use of muscles in space, where it is possible to "float" weightless without using one's legs to move about. Male astronauts showed evidence of general weakening without the normal stress on muscles, bones, and the heart, but there was no data yet from females.

The biomedical research team at NASA's Ames Research Center in California conducted a five-week experiment with twelve Air Force nurses selected for their medical and flight training.[99] During the first two weeks, all were evaluated physically and medically. In the second two weeks, eight of the women were confined to bed rest and were not allowed to get up or use their muscles except to raise themselves up on one elbow to eat. The other four women served as controls, confined to the same facility but able to move around as they wished. The last week was a recovery period. All were subjected to the same battery of tests before, during, and after the bed rest period.

The bed-rested women experienced similar deconditioning to men and also coped well with simulated weightlessness, thus clearing one of the last hurdles to considering women suitable for spaceflight. This may have been the first official US investigation into female fitness for space travel, although it echoes in part the testing of women at the Lovelace Clinic a decade earlier. If women were to be considered for spaceflight, science-based evidence of their fitness was essential.

Ending Racial and Sex Discrimination

As women struggled to become military pilots, found places in the science and engineering professions, and demonstrated that being female did not hinder their abilities to do what was considered "men's work," another barrier remained to their eligibility to become astronauts. That was the lack of legal protection from discrimination rooted in social attitudes and traditions. Concurrently with changes in the military services, educational institutions, and workplaces—and entwined with them—social and legal actions on racism and sexism arose on many fronts during the 1960s and 1970s. A period of extraordinary social and political activism challenged the status quo through efforts to replace bias with equality in education, hiring, and advancement.[100] The overarching goal was to ensure equal opportunities for all to live their lives without restrictions based on bias. Brains and abilities, not genitalia and skin color, would determine opportunity.

Legislation and court decisions during two decades finally removed the barriers to women's eligibility to become astronauts.

The civil rights movement led the way. Emboldened by integration of the military services and postwar prosperity in the 1950s and desperate to overcome social injustices, African Americans organized to make their claims for educational and economic equality. Their attack on segregated education and vindication by the Supreme Court in the 1954 ruling in *Brown v. Board of Education* foretold continuing pressure to remove other racial barriers. The Civil Rights Act of 1957, signed by President Dwight Eisenhower, was the first federal civil rights law passed in almost a century. It held some promise for protection of voting rights by establishing a federal Commission on Civil Rights and a Civil Rights Division within the Department of Justice and was strengthened in 1960 with a more effective version.

President John Kennedy and his administration committed to solving the race problem and quelling racial violence and pushed for a comprehensive Civil Rights Act to dissolve lingering discriminatory practices. His 1963 *Report to the American People on Civil Rights* urged comprehensive legislation for equal treatment of all citizens. After contentious debate and filibuster, Congress passed the act and President Lyndon Johnson signed it into law in 1964 after Kennedy's death. This Civil Rights Act, like others before it, outlawed discrimination on the basis of race, color, religion, national origin, and for the first time included sex in its Title VII prohibition of discrimination in employment. Its passage also reinvigorated the long-ignored Fourteenth Amendment to the Constitution, passed almost one hundred years prior to guarantee equal protection of the laws to all citizens. By executive order, President Johnson bolstered the Civil Rights Act by requiring employers to take "affirmative action" in hiring racial minorities and women. None of this happened quickly or easily.

As women began to organize anew in the 1960s, giving rise to "second wave feminism," they developed an evolving agenda of specific demands.[101] Winning rights to equal education and equal employment was just the beginning. Women wanted to end any legal and societal inequalities that subordinated, patronized, or restricted women in fulfilling their human potential. They wanted to disrupt stereotyping of women and their treatment as sex objects in advertising and entertainment. They sought an end to help wanted ads that separately listed some jobs for men and others for women. They demanded equal pay for equal work, without salary and promotion penalties for married women assumed to be supported by their husbands. In order to work, women wanted childcare centers, tax deductions for childcare expenses, paid maternity leave, and guaranteed return to the job held before childbirth. They demanded full responsibility for their own bodies and reproductive choices, including rights to contraception, abortion, and comprehensive health care. As they encountered conflict between responsibilities at home and at work, they argued for more family-friendly policies, such as flex time and sick leave for childcare. They sought reforms in divorce and alimony laws. Feminist Gloria Steinem, in her 1971 *Address to the Women of America*, characterized the women's movement as nothing short of social revolution.

Activists in this burgeoning women's movement saw clear parallels between racism and sexism. They regarded the social tactics and political strategies of the African American campaign for civil rights as useful models for a similar campaign for women's rights. Women, taking some cues from the civil rights movement, set out to develop a broad constituency, powerful advocates, social activism, legal challenges, and organizations with political clout to effect legislation and social change for women.[102] With passage of the Equal Pay Act in 1963 and the Civil Rights Act in 1964, which established the Equal Employment Opportunity Commission to investigate complaints, women seemed to have a package deal to abolish discrimination.

But desired change took another ten years of struggle after the Civil Rights Act to gain legal equality in the workplace and education. The National Organization of Women (NOW) formed in frustration at sluggish enforcement of the law and vigorously advocated for it; the Equal Employment Opportunity Commission had such a backlog of complaints from women and African Americans that civil rights protections were basically hamstrung. Women and Blacks were learning the difference between laws prohibiting discrimination and laws enforcing nondiscrimination. Legal remedies were useless without enforcement. NOW became the point of the spear for the women's movement and represented women's interests much like the National Association for the Advancement of Colored People (NAACP) advanced interests of the African American community. Other organizations, such as the Women's Equity Action League and the National Women's Political Caucus, joined the effort. By the end of the 1970s national and world conferences addressed women's rights and formed action plans to effect broad social and political change.

The Equal Employment Opportunity Act of 1972 extended the Title VII protections of the 1964 Civil Rights Act against sex-based discrimination to educational institutions and to state and local governments and smaller private employers. Similarly, the Education Amendments Act (Title IX) of 1972 and the Women's Educational Equity Act of 1974 were additional legal remedies to protect equal rights for women, with provisions for enforcement. The Equal Credit Opportunity Act of 1974 and the 1978 Pregnancy Discrimination Act prohibited more specific kinds of discrimination. The real culmination of the effort to abolish sex-based discrimination was the Equal Rights Amendment (ERA), which finally passed Congress in 1972 and was sent to the states for ratification. The time seemed ripe, at last, for this simple statement to be incorporated into the Constitution: "Equality of rights under the law shall not be denied or abridged by the United States or by any State on account of sex."

NOW and other organizations worked hard for ratification of the ERA, lobbying across the nation, holding marches and rallies, and keeping the quest alive in the media. As victory seemed almost certain, a STOP ERA backlash stalled the effort by raising fears that women would lose more than they would gain, such as safety rules on the job, the right to maternity leave and child support payments, separate restrooms, and the specter that women would become subject to the draft and forced into combat. The measure fell three states short of the three-fourths majority required by the 1982 deadline and to this day has not been officially ratified.

The two decades of activism and foment summarized here amounted to a revolution in social institutions. By the late 1970s, most American women benefited from more freedoms and greater confidence in their ability to shape their own lives as they chose. Besides legal rights, they had acquired a vocabulary for the times: male chauvinism, sexism, women's lib, patriarchy, misogyny, gender norms, gender-neutral language, gender gap, and gender pay gap entered the lexicon, as did consciousness-raising, glass ceiling, sisterhood, no-fault divorce, domestic violence, date rape, freedom of choice, right to life, and women's studies courses. Women were free to delay marriage and childbearing to launch a career. They had options to keep their birth surnames after marriage or to hyphenate their last name with their husband's or jointly share a new name. They could be addressed as Ms. rather than Miss or Mrs. without revealing their marital status.

A new archetype emerged as the Superwoman who now had education, career, marriage, and family if she chose them, confidence to speak her mind and use her power, and independence to make her way in the world.[103] For some, women's liberation brought pressure to "have it all." Some women traded their previous discontent with limitations for anxiety about managing all their new options and achieving enough. The bar for success was complicated by unanticipated stress for women striving to balance their ambitions in both the workplace and home. Having it all meant making adjustments, being resilient, and redefining marriage as a partnership of equals with shared responsibilities for home and family life.

The social revolution powered by the civil rights and women's rights movements continued into the new century as laws extended protections to groups still suffering from discrimination and additional rights were codified. Yet opposition to those rights and protections has not disappeared over the decades; it continues in attempts to repeal established laws. Most recently, the landmark 1973 *Roe v. Wade* Supreme Court decision protecting a woman's right to abortion via a constitutionally implied right to privacy and an explicit right to equal treatment was overturned by a more conservative Supreme Court in 2022.

Despite the failure of the Equal Rights Amendment, enough equal rights and opportunities laws passed in the 1960s and 1970s so that young women of these decades could prepare to qualify to become astronauts. They still had no women astronauts as role models, but they now had legal remedies for societal inequalities, and they could chart the course for their ambition. This is the broader social milieu in which the pilots of the Lovelace Women in Space program, the aquanauts of the Tektite program, the scientists of the Spacelab simulations, and aspiring younger women lived, on the cusp of change. The military services began opening the academies and pilot training programs to women, universities were becoming more open to admitting women into science and engineering, business and industry were making space for women in technical careers, and society in general was shifting toward acceptance of equality of the sexes. Men and women were beginning to share workspaces as peers. The next step was for NASA to understand that it was in its interest, and the nation's interest, to hire the most capable people from a more diverse populace.

Redefining Astronauts for the Space Shuttle Era

While men were walking on the Moon, NASA was already planning its next giant leap: a reusable spaceplane that would provide routine access to space by shuttling astronauts to and from Earth orbit and an eventual space station. The space shuttle would be roomy enough for larger crews and tons of cargo—satellites, a laboratory, observatories, scientific instruments, even construction elements ready for assembly into a space station. It would also carry something else, as President Richard Nixon noted in the January 1972 announcement approving the space shuttle program: "men and women with work to do in space can 'commute' aloft."[104] In principle, the decision to include women in spaceflight had been made. The question was how to implement it. The next five years included discussion, debate, planning, recruitment, and myriad decisions before women were finally admitted to the threshold of space.

NASA faced three major challenges in the early to mid-1970s: declining funds and public support, difficult and costly development of a technically innovative and complex space transportation system, and harsh criticism as an organization out of step with the times, still too white and male. Each of these politically charged challenges kept NASA under hostile scrutiny that it had rarely experienced while accomplishing spaceflight feats during the 1960s. Having banked on political and popular acclaim to burnish its image, the agency was slow to respond to these new pressures and risked losing its luster. Of these challenges, the third is most relevant to this story. The Civil Rights Act was almost ten years old, but the state of NASA employment was stagnant. NASA ranked among agencies with the lowest ratio of minority employees—barely 5 percent compared to an average 20 percent government-wide—and almost 88 percent of its female workforce was clustered in the lowest pay grades.[105]

NASA was trailing the rest of the government in compliance with new federal equal opportunity and antidiscrimination statutes and not keeping pace with society's expectations. Its status quo was under pressure from without and within.[106] The NAACP and NOW questioned the agency's lack of visible progress in hiring a more diverse workforce, including minority and women astronauts. Representative Charles Rangel, a founder of the Congressional Black Caucus, called for a Civil Rights Commission investigation to determine why there were no Black, Hispanic, or women astronauts and what plans NASA had to rectify their absence. Internal pressure came from the agency's Equal Opportunity office headed by Ruth Bates Harris, a Black woman executive, who prodded NASA leadership to be more aggressive. In 1973 she produced an internal report for the NASA administrator charging that the agency's equal opportunity program was an abject failure and that NASA's management was uncommitted and insensitive to civil rights.[107] She noted particularly that NASA had existed for a generation without training one woman or one minority astronaut: "The only three females so far sent into space by NASA were two spiders and a monkey."

Harris was summarily fired as too disruptive and abrasive, an event that misfired and became a political and legal scandal calling attention to NASA's shortcomings. NASA took a beating in the press and felt the fury of civil rights and women's organizations; the NAACP Legal Defense Fund and Civil Service Commission became involved in her case as an illegal dismissal, and Congress launched hearings. During one, Harris testified that NASA executives were taking credit for her proposals to recruit women and minorities into the astronaut corps yet had not made a step in that direction. With no evident improvement in hiring compliance per laws passed in the 1960s, Senator William Proxmire, a vocal critic of NASA, held hearings to investigate its EEO office and address the continued absence of women and minorities in significant positions. The Harris imbroglio demonstrated that NASA was ignoring the law and equal opportunity requirements in hiring, and worse, it appeared to be institutionally racist and sexist. This trouble lasted for two years until NASA agreed to rehire Bates into another position.

The first glimmer of hope that NASA was waking to the need for action came in a talk by administrator James Fletcher to an EEO conference in early 1972. He said that NASA was "working on plans to get members of minority groups into space."[108] That wasn't quite true, however. Six months later, at a retreat with the NASA center directors, Fletcher called for a plan to be developed by early 1973 for the next selection of astronauts. The group agreed that the Manned Spacecraft Center in Houston (soon to be renamed the Johnson Space Center) would take the lead on this plan. It was charged to project the number of astronauts needed for shuttle missions, develop a schedule for recruitment and training, and define selection criteria. The task proved harder than expected, and after various iterations, the plan was delivered in 1976, three years after its original due date.

Meanwhile, reacting to Congressional pressure, the NASA Public Affairs Office in 1974 prepared guidance on how to talk about the status of women and minorities at the agency. Struggling to figure out how to visibly include women and minorities, the agency seemed more concerned with improving its public image and support than acknowledging women and minorities as worthy participants in spaceflight. The chief of public affairs wrote to Fletcher that "it may be time for us to go out and get ourselves a black astronaut. . . . From a public affairs standpoint [this] would by far be the most visible means of demonstrating NASA's commitment to EEO." He noted further that it would win favor among minorities to create a space hero for them like Jackie Robinson, who broke the color barrier in baseball. This portrayal of a historically significant action as merely a symbolic gesture seems more transactional than principled.[109]

The coming space shuttle era was a perfect opportunity to recreate an astronaut corps more representative of America's population, because both the design and the purpose of the spacecraft could accommodate crews of seven or eight people and more varied missions than in the past. Instead of a cramped canister seating two or three astronauts shoulder-to-shoulder without a modicum of privacy, the airplane-like shuttle offered a two-deck crew cabin with a curtained-off toilet and personal sleeping bags or bunks. Workspace could be expanded into a laboratory module anchored in the payload bay. The design and amenities of the shuttle reflected a new philosophy of spaceflight as living and working in space for practical purposes.[110] Astronauts would be workers conducting research, deploying satellites, starting planetary probes on their journeys elsewhere in the solar system, and eventually assembling a space station. This multipurpose agenda called for a different kind of astronaut, often referred to as a "new breed," who did not need to be a test pilot but was an engineer or scientist instead. While still daring, going to work in space by commuting in a reusable spaceplane would make spaceflight more routine.

Plans for the recruitment and selection of shuttle astronauts envisioned two groups who would train and work together. Pilots certainly: the shuttle was configured for two pilots in the cockpit, and NASA still put a premium on test flight experience for this new vehicle. Pilot astronauts would be responsible for flying and maneuvering from launch through landing. The other group, called mission specialists, were scientists, engineers, and physicians, who would be responsible for operating the spacecraft systems and payloads, conducting research, performing spacewalks, keeping the shuttle humming, and doing the non-piloting work to meet each mission's objectives. Each crew would include two to five mission specialists. The resident astronaut corps by the mid-1970s was a mix of thirty-one pilots and nine scientists, most of whom had joined NASA too late for Apollo missions to the Moon or had not been chosen for Skylab and had not yet flown. These men would fly on the first shuttle missions while new astronauts were training. NASA's leadership was especially interested in the mission specialist category, because that is where women and minorities would more likely fit. To meet the need for mission specialists, NASA could recruit from a much larger population of scientists and engineers (including its own workforce), find the best-qualified in that talent pool, and then give them specific technical training and work experience to ready them for spaceflight.

This plan was generally satisfactory except that it didn't address how to ensure applications by women and minorities. In the past, NASA had simply put out a press release announcing that it was seeking a new group of astronauts and listing the requirements for applicants. Almost a decade had elapsed since the last recruitment, and NASA worried whether enough people, particularly women and minorities, would notice and apply this time. A more aggressive recruitment strategy of targeted outreach would be necessary, and that became part of the plan, too. NASA outlined a massive campaign to attract applicants, using its personnel and EEO staff to reach out to every organization that might harbor or communicate with qualified women and minorities—graduate schools, professional societies, scientific and technical organizations, other agencies, conferences, commissions, magazines, and newspapers. NASA officials would also make public appearances around the country and directly contact known prospects.

Finally, the long-planned day arrived, and on July 8, 1976, the space agency issued a news release that carried the announcement "NASA to Recruit Space Shuttle Astronauts."[111] The release described how the space shuttle would be used and defined the education, experience, and physical qualifications for pilots and mission specialists, with instructions how to apply. The pay and benefits were the same as in federal civil service or military service (likely to be less remunerative than those in industry or medicine, but attractive to academics). This recruitment was expected to double the size of the corps by adding at least fifteen pilots and fifteen mission specialist candidates. The announcement prominently featured this statement: "NASA is committed to an affirmative action program with a goal of having qualified minorities and women among the newly selected astronaut candidates. Therefore, minority and women candidates are encouraged to apply."

NASA announced the new astronaut recruitment in July 1976, with screening of applicants to begin the next July and selection to be finalized by the end of 1977. This longer than usual process allowed plenty of time for word to spread and for repeated announcements if necessary to attract enough candidates. NASA felt confident that if the number of applicants was high enough it would find highly qualified women and minorities among the white men who always applied. By the end of 1976, however, the numbers were not high enough, and NASA leadership asked for a demographic analysis. Not enough women and minorities were responding to meet the goal of a diverse applicant pool.

It happened that one of NASA's top officials who often spoke at Star Trek conventions had become friendly with some of the stars of the television program. He invited Nichelle Nichols, famous for her portrayal of the series' barrier-breaking Black woman Lieutenant Uhura, to a conversation with NASA's heads of human spaceflight and EEO about the astronaut recruitment effort.[112] They discussed why these population groups were less enthusiastic about NASA and were not applying in droves. Nichols recalled replying frankly that NASA had been sending them a message for two decades: "Don't bother. You aren't needed here." NASA would have to build trust that this recruitment effort was sincere. Nichols suggested a nationwide media blitz of speeches, articles, public service announcements, commercials, and talk show appearances led by someone with great visibility and credibility—perhaps John Denver, Bill Cosby, or Coretta Scott King—to convince the public that this was indeed a new era for a new kind of astronaut. The spaceflight chief asked, "Why not you?"

At first Nichols thought that using a Hollywood actor might be dismissed as a publicity stunt and no one would take her seriously. However, she had already done some work with NASA and visited some of its facilities, and she could communicate her own passion for space exploration. The conversation evolved into a plan that NASA would hire her as its ambassador for four months to add credibility to the recruitment campaign. Nichols's condition was that if she rallied highly qualified applicants and NASA did not select them, she would file a class action suit against the agency. She took on this mission with tremendous energy. After some exposure to astronaut training and briefings about the space shuttle, she received a NASA blue astronaut jumpsuit and traveled widely to spread the messages that "Space Is For Everyone" and "NASA Wants You." She made appearances in major cities, on university campuses, and at conferences and corporations, where audiences were thrilled to see her. National newspapers, *People* and *Newsweek* magazines, *Good Morning America*, and other media gave her favorable coverage as a surrogate astronaut. The astronaut application count rose dramatically during her tour, with those from women increasing from less than one hundred to more than one thousand. Nichelle Nichols's effort in the recruitment campaign was deemed a major success and an experience that she found personally rewarding to use her celebrity to further the goal of a more diverse population in space.[113] She found no reason to sue NASA as threatened.

When the July 1977 deadline arrived, NASA had received 8,037 applications (soon revised to 8,079), 6,735 of them for the new mission specialist positions. Of the total number, 1,142 were from women.[114] There was no reliable way to identify exactly how many applicants were minorities, but as the process proceeded, they surfaced in reasonable numbers. Screening so many applications was a massive effort; each application included biography, academic transcripts, work history, medical history, references, sample publications, military records, and other material to be reviewed by the personnel officers to determine if the applicant met minimum requirements. The applicants were winnowed down to 208 finalists, who were invited to come to the Johnson Space Center for a week to meet with the Astronaut Selection Board and undergo a thorough medical and psychological/psychiatric evaluation. At that time, NASA anticipated selecting thirty to forty candidates.

Of those evaluated, 80 were pilots and 128 were mission specialist applicants.[115] Twenty-one, or 10 percent, of the finalists were women, all potential mission specialists; several women who applied to be pilots met the minimum qualifications but were not finalists.[116] Ten groups of invitees were scheduled at the Johnson Space Center from August through November. The groups were generally organized into either pilot or mission specialist applicants and housed as roommates in a nearby hotel for efficiency and camaraderie. There were only two instances of a lone woman in a group of men; two of the mission specialist groups each included eight women. At the center, these applicants underwent a thorough physical examination and battery of tests, sessions with psychologists and psychiatrists, and an interview with the eight-to-ten-member Astronaut Selection Board composed of NASA managers and astronauts, including for the first time a woman and an African American. They wrote and discussed an essay about why they wanted to be an astronaut and what they thought they could contribute to the space shuttle program, and they answered personal and technical questions posed by the board. They had opportunities to tour the facilities, use the gym, talk with current astronauts, and socialize informally with NASA personnel and others in their group to gain a sense of what working there as an astronaut might be like. They knew that every move and conversation was being evaluated, so the entire week was really a test. Some reported later that they didn't realize how much they wanted to be an astronaut until they survived this week of assessment; others withdrew after realizing that the career paths they were already on were more appealing; yet others were betrayed by their bodies, which harbored some previously unknown defect that caused them to fail the physical exam.[117] No one was immediately selected or rejected; prospects had to wait weeks for a notifying phone call.

All the pieces were now in place for women to become astronauts. Those interested in science and engineering were earning technical degrees and landing technical jobs. Those avid about aviation were making their way into the military services and advancing as opportunities opened, or they were becoming private pilots for their own pleasure and as an added credential. Women were becoming accepted in previously male workplaces, disproving myths about female inferiority, and earning grudging respect. The newly defined mission specialist position offered a new and wider path to space without being a test pilot, and the space shuttle held promise that more people would be flying more often in space. Women born in the 1940s and 1950s were the right age to reap the first benefits of the civil rights and women's movements and to enter the talent pool where NASA was now looking for potential astronauts. With the many legal, social, and political changes since 1960, women were finally qualified and eligible. At last, their time had come: women were poised to be a new version of "the right stuff."

As 1977 drew to a close, NASA leadership was reviewing the list of selections from this prolonged and methodical recruitment, with a view to making the results public in January. The mystery to be answered remained the number of new astronauts chosen and how many of them were women or minority group members. The new names and faces of NASA for the space shuttle era would soon be revealed.

CHAPTER 2

The Trailblazing Generation: Astronauts for Space Shuttle Missions, 1978–1990

After nearly five years of planning, recruiting, evaluating, and interviewing applicants, NASA announced the selection of new astronaut candidates at a press conference in Washington, DC, in January 1978. Members of the media peppered administrator Robert Frosch with questions and sought assurances about the selection process, the number of women and people of color, and the number of military and civilian pilots selected. Chris Kraft, director of the Johnson Space Center, fielded questions and explained the experience-based filters and rating process for competitive selection. He was satisfied that the men and women selected "represent the most competent, talented, and experienced people available to us today."[1]

The main press conference to introduce the new astronaut candidates to the public occurred on January 31. The selectees convened at the Johnson Space Center (JSC) in Houston to meet one another and the press. The NASA auditorium was filled to capacity with reporters, camera crews, VIPs, veteran astronauts, and anyone else who could squeeze in to see them for the first time. After opening remarks Kraft called the thirty-five men and women to the stage. NASA had chosen fifteen pilots and twenty mission specialists. Six of the mission specialists were white women: Anna Fisher, Shannon Lucid, Judith Resnik, Sally Ride, Rhea Seddon, and Kathryn Sullivan. Fisher and Seddon were medical doctors; the other four held PhDs in different fields. The two youngest, Ride and Sullivan at age twenty-six, had just completed graduate school; this would be their first professional job. Lucid, the eldest at thirty-five, and the other three were a few years into research and medical careers. The group also included three African American men (one of them a pilot) and an Asian American man. Of the total, twenty-one were military officers and fourteen were civilians. After training, they would join the twenty-seven older men in the active-duty astronaut corps and double the size of the corps to sixty-two astronauts. The new position of mission specialist made it possible for women and minorities to become astronauts as scientists and engineers without having to be high-performance jet pilots, the main reason the first recruitment to diversify the astronaut corps succeeded.

After the introductions and applause, the formal program ended, and media representatives were invited to meet the astronauts-to-be for interviews. As Kathy Sullivan remembered it, most of the media flocked to the women and the minority men—the "ten interesting ones"—as the big news story. The "standard white guys" were able to depart quickly, while media kept the other newcomers engaged for hours. Most of the astronaut candidates had never been exposed to such a media gaggle and were either bemused or exhausted by the experience. Sullivan recalled with gratitude the helpful role played by the only woman who had served on the Astronaut Selection Board, Carolyn Huntoon, and another prominent NASA woman, engineer Ivy Hooks. They advised the women not to worry about their wardrobe or hairdo but to focus on what they did and did not want to say for publication. Their conversations with the media would make a greater impression than their appearance. As the media started to create a sort of composite persona to explain to the public who the women astronauts were, they questioned the women about both their common and distinctive traits. Huntoon basically advised them to be smart, confident, and poised, and cautioned that they take some care in what they revealed.[2]

Journalists could not glean by observation alone that two of the women were medical doctors, three scientists, one an electrical engineer, one a newlywed, two divorced, one a married mother of three, one a classical pianist, another a tennis champion, one drove a Corvette, another owned her own airplane, and one was Jewish. Those were the kinds of personal details that reporters were eager to discover. A lot rested on how the first six women fielded such questions. They needed to be mindful of their privacy as they moved into new public roles and stay alert to comments that might stereotype them. Between interviews, the six spontaneously convened in the ladies' restroom to compare notes on questions and answers and start considering potential boundaries.[3]

Twenty-five years later, NASA historian Jennifer Ross-Nazzal surveyed the resultant media coverage from that day and the next few years as the women completed training and began to fly in space. Many articles, she noted, reported on their physical features, size, weight, and marital status. Such information was rarely included in stories about male astronauts. Women's magazines took a great interest in their fashion choices, style, beauty tips, hobbies, homes, and domestic pursuits. The petite women garnered more attention than the larger-framed "robust" ones, although all were specimens of good health and fitness, evidence that some media attention was biased toward the stereotypically feminine type of woman rather than the athletic or tomboy type.[4]

The total group of astronaut candidates selected into the 1978 class called themselves the Thirty-Five New Guys (TFNG), a sanitized variant of the military phrase, "the F'ing New Guys." They outnumbered the twenty-seven members of the astronaut corps held over from the 1960s, many of whom had not yet flown and were skeptical of the new-comers, especially the civilian scientists and engineers who had no military flying experience. With a larger astronaut corps, the "old guys" feared losing their opportunities after waiting more than a decade to fly and initially kept their distance. They had a grace period of two years while the TFNG went through their core training and evaluation, and they had been promised the first shuttle flights. But the new candidates proved so adept in training that they completed it in one year while the first shuttle launch schedule slipped from 1979 to 1981. Those changes left ample room for the "old guys" to feel uneasy. They carried forward the astronaut culture of the 1960s and early 1970s, in which military pilots reigned over second-class civilians and scientists.

Officially Group 8 in the history of astronaut selections, the TFNG class acquitted themselves well. They vindicated the capability and value of mission specialists and eased the acceptance of women and minorities in future classes by deflating concerns about sex and race. They gradually won the respect of the veteran astronauts. The women and the men of color in the class became instant role models for young people and symbols of achievement for everyone. By the year 2000, members of this class had completed their flights in space, five of them—including Shannon Lucid—having flown five times each. Altogether they served on fifty shuttle mission crews and cumulatively logged almost a thousand days in space, demonstrating everything the space shuttle was designed to do—an illustrious record for the first class of shuttle-era astronauts.

The Trailblazing Generation

After selecting the 1978 group of astronauts, NASA held additional recruitments at approximately two-year intervals to increase the size of the astronaut corps. During the period of 1978 through 1990, fourteen more women were selected to be astronauts. These first twenty, plus a few of the other women chosen after 1990, were part of the post–World War II baby boom. Most of the women selected by 1990 were midcentury girls, all born within a few years on either side of 1950, most of them starting school in the relatively placid 1950s and coming of age in the social and cultural milieu of the tumultuous 1960s. If they were old enough to notice, the signal events of their youth included the beginning of the space age with the launches of Sputnik, Yuri Gagarin, Alan Shepard, John Glenn, and Valentina Tereshkova in the period between 1958–1963. All were old enough to witness on television Neil Armstrong's and Buzz Aldrin's landing on the Moon in 1969. For some, the space age spelled their destiny; they cite one or more of those early missions as the pivotal event in shaping their aspirations.

These women generally were products of middle-class families, white neighborhoods, and public schools. As noticeably intelligent students who flourished in academics, they also spent their youth in such normal pursuits as Girl Scouts, sports, music and dance lessons, and school activities. Rock and roll and folk music were the sound tracks of their youth. *Star Trek* debuted on television in 1966 while they were teenagers. They might have seen *2001: A Space Odyssey* when it was released in 1968 and watched family-oriented shows on television and at the movies.

For young girls in the 1950s and 1960s, there were not many serious role models in popular culture, where the standard for celebrity was beauty, or in public life, where notable women were rare. Instead, they looked for encouragement and support from their parents and teachers, both men and women—people in their personal lives rather than the public sphere.[5] Although some encountered active academic bias against studying physics, engineering, and medicine, most were able to find an open-minded mentor whose guidance primed them for success. The deciding factor, they said, was proving their determination and capability. Once that was established, lab positions, research projects, funding, awards, and jobs tended to follow. Unsurprisingly, the aspiring scientists and engineers who faced early resistance often graduated with honors. Their tenacity was a harbinger of success.

Despite their achievements, the astronauts had faced discrimination in the world outside of NASA. Rhea Seddon recounted that she was not allowed to take out a loan for the condo she was buying in Texas, despite being a medical doctor with three years of experience and a newly named astronaut; at the age of thirty, she had to ask her father to cosign the loan to guarantee that he would pay the mortgage.[6] She had already experienced barriers as the only woman surgical intern when she was denied access to the doctors' lounge in the hospital and had to take her breaks in the nurses' restroom between surgeries.[7] Anna Fisher applied for a surgical residency only to be told by the department chair that women didn't belong in surgery, so she went into emergency medicine instead.[8]

The young women who became scientists, engineers, medical doctors, and astronauts shared a commitment to graduate school in preparation for a career. In an era when educated women who worked were mostly confined to teaching or nursing, and those without bachelor's degrees or licenses were secretaries, clerks, or retail workers, these women were determined to chart different paths for themselves. Some were the first in their families to attend college. They headed into graduate school in the late 1960s and early 1970s on the cusp of major changes in the educational and social environment. They confidently pushed forward on their journeys to become scientists, engineers, and physicians. They were often the first or only woman—or one of very few—in their advanced science, math, and engineering courses at both the undergraduate and graduate levels, and in medical school. Some were the first women ever granted a degree in a particular field at their university. As they moved onto a career path, they were often the first or only woman in a medical residency or a corporate office or at a field site. Most were aware that they could not fail, for that would only reinforce bias and make it harder for the next woman.

Shannon Lucid probably had the hardest time launching her career. She was a few years older than the other women astronauts, just enough so that she was truly an anomaly as a female chemist and biochemist at the University of Oklahoma. She spent ten frustrating years being rejected for positions for which she was fully capable and working through a series of assistantships before finally breaking through to what she considered a real adult job. Among the trailblazing women, those who had long aspired to spaceflight learned to fly airplanes in hopes of improving their chance to become an astronaut, and it was not uncommon for them to come to NASA with a pilot's license in hand. Lucid, passionate about flying since childhood, held multiple flight ratings and owned her own plane; she could have become a commercial pilot. Similarly, Marsha Ivins started flying at age fifteen to prepare to become an engineer and astronaut, Mary Cleave earned her pilot's license at age seventeen and hoped to fly jets at NASA, and Rhea Seddon took flying lessons after graduating from medical school, a gift from her father to celebrate her success and a conscious decision to improve her prospects whenever NASA decided to hire women as astronauts. More than half of the first women to become astronauts took flying lessons or earned pilot's licenses, although they were not required to do so.

Some of the women in this generation believed that eventually women would be astronauts. Others never aspired to be astronauts because it seemed impossible; they saw only a male space agency and male astronauts when they were young. Their interest sparked when they learned of NASA's intention to recruit and select women, and not until they applied and interviewed did they realize how much they truly wanted it. Sally Ride read NASA's announcement in the Stanford University student newspaper. Had she not picked up that particular issue on that particular day, she might never have known or applied. Kathy Sullivan's brother, a professional pilot, was applying and challenged her to apply, too. A budding oceanographer bent on exploring the seafloor, she didn't take him seriously until she read the announcement and realized that exploration in sea and space had much in common. Rhea Seddon, Anna Fisher, and several others were encouraged by professors or coworkers to apply, people who saw their potential before they themselves recognized it.

This first wave of twenty women astronauts selected through 1990 proved more than capable in training and in flight. They carried out every orbital task such as tending to spacecraft systems, science experiments, spacewalking, robotic operations, and more—all except actually flying the shuttle because there were no qualified female military test pilots in the early selection rounds. Eileen Collins broke through that barrier when she was selected into the 1990 astronaut class.

Each woman has a compelling origin story arising from or intersecting with other American stories and themes. They have become admired role models and influenced the paths of younger women who have joined the space program as astronauts and in many other roles. Almost all women astronauts have been first at something, usually without aiming to be, as their individual stories reveal. But, to paraphrase Judy Resnik, Kathy Sullivan, and others, being first isn't the point. If something is worth doing, it doesn't matter whether one is the first, fifth, or one hundredth. It is the doing that counts.

The Class of 1978

NASA received 8,079 applications during the first recruitment for space shuttle astronauts, 1,142 of which came from women. Through the winnowing process, 208 applicants (21 of them women, including Carolyn Griner and Mary Helen Johnston from the 1974 Spacelab simulation at Marshall Space Flight Center) were invited for a week of interviews and physical evaluations at the Johnson Space Center. In the end, NASA selected six impressively qualified women.[9]

These first six found NASA partially ready for them, welcoming them into the "NASA family." But not everything was fully prepared. For example, the astronauts' gym did not have a women's locker room or even a women's restroom, and no one thought to remodel it until after the women arrived. NASA issued workout clothes to the astronauts, including athletic supports for the men, but no one thought of sports bras for the women. Airfields where the astronauts flew while in T-38 training had locker rooms and restrooms only for men. Some NASA "old-timers" kept an eye on what the women wore to work and protested if their attire was "too casual," despite the men forgoing ties or wearing jeans. Carolyn Huntoon fielded some of these complaints by pointing out that

The first six women in newly issued, incompletely adorned astronaut jumpsuits, 1978: (*front, left to right*) Sally Ride, Rhea Seddon; (*rear*) Kathy Sullivan, Shannon Lucid, Anna Fisher, Judy Resnik.

there was no dress code for men, so why should there be one for women? Such issues gradually resolved as the workforce came to terms with treating women equally.[10]

This first group of women wanted to blend in, to be one of the guys and not call attention to themselves as women. Anna Fisher noted, "It never occurred to any of us to ask for special accommodations for anything."[11] Sally Ride and Anna Fisher shopped for khaki pants and polo shirts to match the men's office attire. They wore the same NASA blue flight suits and black leather aviator boots for flying and public appearances. Some wore little or no makeup to work. The women had fun together with the men playing in the employee volleyball and softball leagues, running and working out, competing in the annual chili cook-off, going to happy hours, and socializing with spouses and families.[12] All of this was consistent with their desire then (and still) to be regarded as astronauts, not women astronauts, and not a separate subset of the astronaut corps.

They knew, however, where there was resistance or bias to overcome. Some of the older male astronauts initially had trouble accepting women and scientists into their ranks. So, too, did some of their classmates.[13] Although the TFNG generally bonded and mutually supported one another, it didn't happen immediately. One of the men in the 1978 class, Mike Mullane, was honest enough to admit his initial doubts about women and scientists who hadn't been tested and seasoned the way military pilots had been. "I felt a subtle hostility toward the civilian candidates . . . [they] hadn't paid their dues," he confessed. "I was in another galaxy when it came to working with women. I saw women only as sex objects. . . . We were flying blind when it came to working with professional women."[14] The military men had never worked with women as peers; they related to women as secretaries, girlfriends or wives, or family members. Nor had they worked with scientists in their military realm of aerospace engineers and aviators. For some, it was a difficult transition to become more open-minded and tolerant of those who were not the stereotypical military fighter pilots.

The most senior woman at JSC at the time led the way in bringing women into the astronaut corps and addressing problems that arose. Carolyn Huntoon, who managed the research labs in the biomedical division and had been studying astronaut health since the Apollo era, was the only woman on the selection panel that conducted interviews for the 1978 class (and all the subsequent astronaut classes until she became JSC director in 1994). She was a strong and respected advocate for scientific and technical women in the space program and was herself a prominent scientist in her field of human physiology. She took the women astronaut candidates under her wing as an informal den mother for advice and counsel.[15] She became their hero. Some still remember her advice to always use good judgment, which helped them navigate whatever circumstances they found themselves in. Huntoon observed that as the shuttle era began, NASA didn't discriminate, but individuals did. Some gave women a hard time, and some simply couldn't get over the fact that women are not exactly like men. She had to explain that doing something differently doesn't mean one isn't up for the job, or that not acting like "one of the guys" wasn't a problem. She thought that the biggest challenge during these first years was to change attitudes, and she used her influence for that purpose.[16]

The trailblazing generation of women astronauts found a welcoming environment at JSC and believed that NASA had committed to their success.[17] Despite occasional hiccups, they felt equally supported by their trainers, flight surgeons, and other personnel. They have described the Astronaut Office as both a cooperative and competitive workplace. For some time, though, they sensed a not-so-subtle hierarchy that determined who could be a leader and who would be assigned to flights. At a macro level, the office sorted into pilots and mission specialists. Within that scheme, the hierarchy was finer grained. Military pilots were at the top, with more numerous Navy pilots outranking Air Force pilots. Civilian pilots ranked below military pilots. Next down the ladder were mission specialists, with military again ranked above civilian, and women ranked below men. This wasn't explicit or codified in writing; rather, it was deduced from the types of assignments one received and how long one waited for a mission assignment.

Another sign that all was not perfectly equitable was a sexist joke that circulated within the 1978 class, but not to everyone. The hand-drawn flowchart parodied astronaut selection criteria for women.[18] Among the qualifications shown were IQ, academic degree, breast size, weight-to-height ratio, and stance on the Equal Rights Amendment (ERA). The only path for a woman to qualify as astronaut material according to this chart was to be a US citizen, not a genius, have a PhD degree, be slim, have a thirty-eight-inch or larger bust, and be against the ERA. On the chart, pilot candidates had a straight path to selection simply by meeting all advertised requirements, and male mission specialists could breeze through despite meeting no requirements at all.

Some of the new male astronaut candidates saw the women as prospects for dating and vice versa, which was not unreasonable given shared interests in aviation and spaceflight. Most were in their late twenties and early thirties and not yet married. Over time, in fact, several astronaut couples paired up or married. At the beginning, though, women sometimes felt that some of the men acted like fraternity brothers, sizing them up as dating material rather than coworkers. Some weren't comfortable with sexual banter, raunchy jokes, and quips about who was sleeping with whom.[19]

The six women settled into training and then work assignments, proving their abilities and dedication. Seddon recalled that the women knew they were being watched for any signs of weakness, and no one wanted to "look like a wimp." Yet some tasks and equipment were sized for larger, heavier bodies, which made it hard for smaller women to perform well. In her case, it was the parachute size used in survival training. It was so large and she was so lightweight that she kept ascending when she was supposed to descend. Veteran Apollo astronaut Alan Bean, training supervisor for the TFNG class, reported that the women performed very well. "I always thought we were letting women do what was instinctively a man's job. I don't think that anymore," he said.[20]

The women, however, were not intimidated or discouraged, and they called out what they saw as misguided and inappropriate. The same male astronaut who admitted that he didn't know how to deal with women professionally said that he quickly learned to be careful about what he said and how he acted because a woman astronaut would give him an icy stare or insult him right back. Perhaps speaking for all of them, Seddon remarked, "We really were blazing a trail in defining what we could do in this world, what we were willing to do, and what we wanted to do."[21] One of Sally Ride's T-shirts said it all: "Women's Place is Now in Space."[22]

THE FIRST SIX
||||||||||||||||||||||||||||||||||

Anna Lee Fisher

When NASA entered her life, Anna Lee Fisher had just reached a critical juncture.[23] A 1976 medical school graduate, she was finishing an internship in family medicine as Anna Sims when she heard about the astronaut recruitment announcement shortly before the deadline for applications. She and her fiancé, Bill Fisher, a physician resident, were interested in becoming astronauts, so both applied. Rather than commit to a medical residency, they decided to pause and work in emergency rooms for a while to keep the astronaut option open should NASA be interested in either or both of them. They submitted their applications in June 1977, and in August NASA called Anna to come to Houston for a week of interviews and evaluation. That was the very week she and Bill had scheduled their wedding. In an early test of their resilience, they scrubbed their wedding plans and travel, quickly arranged a small, local, weekday wedding with an overnight honeymoon, and three days later she flew to Houston. Bill was called for an interview a few weeks later.

Sims, who hadn't had time to notify NASA of her name change, interviewed in the same group of twenty mission specialist applicants as Rhea Seddon and Shannon Lucid. During her interview with the selection board, she volunteered that she wanted to have children, knowing that it might well prompt them to reject her. She made the cut, and Anna Fisher was introduced as one of the thirty-five new astronaut candidates. Bill Fisher was not selected but reapplied and was accepted in 1980. For a time, theirs was the first and only two-astronaut marriage, one that lasted for twenty-three years.[24] She was featured on the cover of the September 1, 1982, issue of the *Saturday Evening Post* for its story "Make Way for Ladies in Space."

NASA ensured that all candidates were broadly trained beyond their academic specialties, in Anna's case chemistry and medicine. In the few years after her core training, Fisher worked on mastering the Remote Manipulator System (RMS) robotic arm, contingency spacewalk procedures, the Extravehicular Mobility Unit (EMU) spacesuit, techniques for repairing the shuttle's thermal tiles and blankets, and flight software. She worked on preparations for the second, third, and fourth shuttle missions, and was a Cape Crusader at the Kennedy Space Center on a team of astronauts who oversaw vehicle and payload readiness for each shuttle mission on behalf of the assigned crews. She flew as an emergency physician in rescue helicopters for launch and landing and served as CAPCOM, or capsule communicator, in the Mission Control Center.[25]

Fisher gave birth to her first child in 1983. While pregnant, she was assigned to mission STS 51-A, which then flew in November 1984 after a year of crew training and the birth of her baby. She was the fourth US woman in space, following Sally Ride, Judy Resnik, and Kathy Sullivan. The crew deployed two communications satellites and retrieved two other satellites stranded in orbit during a previous mission. Fisher served as flight engineer for launch and entry, assisted the commander during rendezvous with the stalled satellites, and operated the RMS arm to seat the retrieved satellites back into the payload bay for return to Earth. The retrieval didn't go exactly as planned, so Fisher had to improvise maneuvering the satellites on the robotic arm in concert with her two EVA crewmates. This challenging flight was hailed as the first space salvage mission. It demonstrated a valuable capability of the shuttle; returning the failed satellites for refurbishment and relaunch saved their owners and insurers from an expensive loss. The crew proudly displayed signs for "Ace Repo Company" upon completing their unprecedented task.

The mission also received media notice as the flight of the "first mom in space," overlooking the fact that many of the male astronauts were fathers.[26] Fisher was assigned immediately to a second mission, but flights were suspended for eighteen months after the 1986 *Challenger* tragedy, so she took on a variety of other responsibilities, including service on the Astronaut Selection Board in 1987. Upon the birth of her second child in 1989, when she was thirty-eight years old, Fisher took leave and part-time work from the Astronaut Office until her daughters were in school and then returned to work full-time as a management astronaut, one not on active flight status, from 1996 until 2017, staying on duty longer than any of the other first women astronauts and retiring as the last of the 1978 TFNG group.

Upon returning from leave, Fisher found that the Astronaut Office had changed markedly.[27] Most of her class had moved on, and she didn't know the younger astronauts. Everyone had desktop computers to use email and make spreadsheets and Power-Point briefings, and the pace of office work had increased noticeably. She had to learn quickly to catch up to these innovations. She soon became chief of the space station branch in the Astronaut Office and then head of International Space Station (ISS) crew operations, working almost exclusively on ISS projects—early operations planning, coordinating with international partners, supervising assigned crews, and serving as CAPCOM for ISS expeditions. After the loss of *Columbia* and the STS-107 crew in 2003, Fisher had a key role in the investigation of NASA managers' decision making and the possibility of a viable rescue. Her last project was working on flight instrument displays for the new *Orion* spacecraft for return to the Moon.

Like most retired astronauts, Fisher remains active, working as a consultant, instructor, and public speaker, sharing knowledge and wisdom gleaned from her almost forty years at NASA. Looking back on her career, Fisher says, "I knew that I wanted to be part of something bigger than myself—not just a job. I wanted to have my life make a difference. I never let the fact that I was a woman keep me from doing what I wanted to do. I just went ahead and did it."[28]

Flights: **1** on space shuttle • **4th** US woman in space • **6th** woman worldwide • Time in space: **8** days (**192** hours)

Shannon Wells Lucid

Shannon Lucid stood out from the other five women arriving at NASA to start training as astronauts. She was the tallest and, at age thirty-five when inducted into the astronaut corps, the oldest, with more work and life experience. She had been married for more than ten years, and with three young children was the only mother. She loved aviation since childhood and was inspired by science fiction to be a space explorer before she knew the word "astronaut." If the women in the 1978 class had organized themselves by seniority as the military men did, she would have been their de facto leader.

Lucid earned a chemistry degree in 1963 before being warned by a professor that no one would hire her in her field because she was a woman.[29] After three years of rejections and make-do jobs, she finally landed a laboratory technician position at Kerr-McGee oil company, where she met and married Mike Lucid and started their family. Fired for being pregnant, she decided to return to school for master's and doctoral degrees in biochemistry after her first two children were born. She then had another child (hiding her pregnancy under baggy lab coats to avoid being fired again) and began to work in health science and medical research at a level commensurate with her education and abilities. During those years, she also earned more aviation credentials and applied for many piloting jobs, to no avail—the airlines simply were not hiring women.[30]

Lucid was the only woman mentioned by name in the initial press conference announcing the selection of the first group of shuttle astronauts, and she attracted special media attention again when the class of 1978 was introduced in public.[31] In both instances, her distinction was motherhood. At least one reporter asked whether NASA had taken into account her responsibilities to her young children in selecting her for the astronaut program. "Not at all," replied JSC Center Director Chris Kraft, and commented instead on her eagerness to fly the T-38 aircraft. (She was already an experienced pilot, with a thousand flight hours and multiple ratings.)

By 1985, all six women of 1978 had been assigned to specific missions and flown. Lucid happened to be the last to launch on her first mission, but she soon became the first woman to fly four and then five times. She had four missions on the shuttle from 1985 through 1993, followed by a six-month stay on the Russian space station *Mir* in 1996. Her first three shuttle missions were satellite deployments, for which she operated the robotic arm and also conducted science experiments. On STS 51-G (1985) to deliver three communications satellites into orbit, she released a small SPARTAN satellite for astronomical observations. On her second mission, STS-34 (1989), Lucid and Ellen Baker were two of the five-member crew responsible for deploying the *Galileo* spacecraft for its journey to Jupiter and conducting various experiments. In 1991 on STS-43, the crew deployed a Tracking and Data Relay Satellite (TDRS), one of four positioned around Earth to keep NASA and the shuttle in continuous contact and also to track, command, and receive data from other US satellites. As always, the crew had research to accomplish, and they sent the first email message from space.

Lucid's fourth flight was best matched to her background. The second life and materials science laboratory mission, Spacelab LMS-2, flew as STS-58 (1993), with Lucid and Seddon as crewmates. Both were in their element on the two-week mission conducting more than thirty physiological research experiments to investigate the effects of spaceflight on the body. On this mission Lucid set records as the first woman to fly four missions and to have the most cumulative time in space, a few minutes shy of 839 hours (35 days).

Her fifth and final mission occurred in 1996 as a crewmember on *Mir*. Lucid had always dreamed of having her own laboratory but hadn't imagined it being in space. On *Mir*, she had that opportunity and more: she was the laboratory director and could schedule her work as she desired. While her two Russian crewmates tended to their duties, Lucid had full responsibility for carrying out a varied program of experiments provided by researchers through NASA. She returned with some eight hundred pounds of experiment samples and data from this research.[32] She published her memoir of this mission in 2020 as *Tumbleweed: Six Months Living on Mir*.

A primary goal of putting astronauts on *Mir* was to learn more about the human body's adaptation to long-duration spaceflight and its recovery back on Earth. The longer Apollo lunar missions had lasted two weeks, and Skylab crews had spent twenty-eight, fifty-nine, and eighty-four days in orbit, but fewer than twenty astronauts had that much time in space. Shuttle missions typically lasted seven to ten days, with the longest being fourteen to seventeen days, so there wasn't much data about effects of long-duration spaceflight. That knowledge was essential for a permanent human presence on the International Space Station and eventual crews on long missions to the Moon or Mars. Lucid found that her fifty-three-year-old body adapted well; she was disciplined about daily exercise in orbit, and her recovery proceeded well upon return. She astonished everyone by her determination to walk off the shuttle unassisted, however wobbly her gait, and after a few days resting at home she tackled rehab to regain muscle strength and resumed working.

Lucid was one of only two women to reside on *Mir*, the other being cosmonaut Yelena Kondakova, and also was the only woman among the seven US astronauts on *Mir* crews. While there, she set US records for the longest single spaceflight (188 days), the most time in space by any woman (223 days), and the most time in space by a US

Shannon Lucid on the *Mir* treadmill, 1996. *NASA*

astronaut to that point. Her records were not surpassed until the International Space Station became operational for long-duration stays.[33] After her return, IMAX released *Mission to Mir*, a large-screen documentary film made from footage shot by astronauts on the shuttle and *Mir*. It featured Lucid demonstrating her activities and narrating her thoughts as a long-term resident in space.[34]

Known for her determination to do hard things, Lucid accepted the *Mir* assignment with enthusiasm, although it kept her away from home and family for two years of training plus the long stay in space. She diligently did the job she signed up for, but she set boundaries for extra tasks, such as interviews and public appearances. Prioritizing time with her family, she often declined to participate in ancillary astronaut activities. She guarded her family's privacy and declined for them to be interviewed. Opposed to any form of discrimination, she refused to do any single-sex events.[35] NASA's public affairs office was flooded with requests for interviews and appearances but sought to coordinate such postflight activity with respect for her privacy.[36]

Lucid is diffident about awards and celebrity; to her, doing her job well was expected, not exceptional. Like many other astronauts, she is not concerned about setting women's spaceflight records, because spaceflight is not inherently gendered.[37] Although she prefers to keep a low profile, President Bill Clinton persuaded her to come to the White

House to receive the Congressional Space Medal of Honor after her pioneering stay on *Mir*. She is the first and thus far only living woman to have received this high award.[38]

Lucid stayed in the astronaut corps for thirty-four years, from 1978 until 2012, when the space shuttle program ended. She had served often, perhaps more than anyone, in her favorite role as CAPCOM for shuttle and space station missions, and she was on console as CAPCOM for the final shuttle mission.[39] Having turned eighty years old in 2023, Shannon Lucid is still the most senior woman among the astronauts.

Flights: **4** on space shuttle, **1** *Mir* mission • **6th** U.S. woman in space • **8th** woman worldwide • Time in space: **223** days (**5,355** hours)

Judith Arlene Resnik

Judy Resnik already had three engineering degrees, including a PhD, and six years of engineering work experience when NASA selected her as an astronaut candidate. An electrical engineer by training, she worked on radar controls and telemetry for rockets and missiles at RCA, biomedical engineering at the National Institutes of Health, and commercial product development at Xerox. She had experience in research, design, testing, and project management. Resnik had never planned to be an astronaut, but when she learned that NASA was seeking qualified women, she "decided to take a chance."[40] She put herself on a physical fitness regimen and started flying lessons to improve her chance, but her academic and professional experience already placed her among the most qualified applicants reviewed by NASA.

Known from childhood as a brilliant student who started elementary school a year early, Resnik was a math whiz, made a perfect score on the math Scholastic Aptitude Test (SAT), and was the valedictorian of her high school graduating class. Resnik was also an accomplished classical pianist. Her family of Ukrainian Jewish immigrants valued academic achievement and their heritage. Resnik attended Hebrew School, celebrated her bat mitzvah, and married in her hometown synagogue, but as an adult did not identify herself as Jewish. Nevertheless, the Jewish community and press claimed her as the first Jewish American astronaut, and she was honored by the Jewish American Hall of Fame, although she preferred to be known simply as "an astronaut."[41] She was skeptical about the value of being first, saying "Firsts are only the means to the end of full equality, not the end itself."[42] She was confident that she was chosen for flight, not on the basis of her sex or religion, but simply on her considerable abilities.

Upon arrival at Johnson Space Center, Resnik quickly became integrated into the 1978 class. She and her first officemate, Jim Buchli, jointly designed a graphic logo for the TFNGs and had it silkscreened onto T-shirts for every member.[43] The cartoonish image was a shuttle in space with thirty-five astronauts clinging to or working on it, dangling from edges, riding on the robotic arm, doing spacewalks, and evidently enjoying this swarm of activity. She was soon known as "JR" among her classmates, who found her to be a disciplined perfectionist but with a great sense of humor.[44]

Resnik trained intensely on the Remote Manipulator System (RMS) arm and worked closely with its Canadian manufacturer to finesse its design and operability. The arm was a complex mechanical device with electronically operated joints, motion control, and cameras operated from a workstation inside the orbiter. It was an electrical engineer's dream. Resnik had excellent skills, and some colleagues thought she or Sally Ride, another master of the robotic arm, might be selected as the first woman to fly. Ride was assigned to the seventh shuttle mission (STS-7 in 1983) and was the first US woman in space, and Resnik followed as the second American woman on the twelfth mission (STS 41-D in 1984), the first flight of the *Discovery* orbiter. When asked by a reporter if she would have preferred to be the first instead of the second American woman in space, she gave a perfect astronaut's answer, saying, "I would like to be any person at any time in space."[45]

As RMS operator, Resnik was responsible for executing the signature task of the mission—extending and retracting a pleated 102-foot lightweight solar array wing, the tallest structure ever handled in space—and assisting with the deployment of three communications satellites and conducting onboard experiments. She and her crewmates also used the IMAX camera to film shots for *The Dream Is Alive*, the first of several documentary films shot on location in space for viewing on gigantic commercial theater screens. Especially close to her father, Resnik caused some smiles by displaying a "Hi Dad" sign during the mission as well as an "I ♥ Tom Selleck" sticker for her Hollywood crush. Her mission commander gave her excellent marks for her performance, saying that she "fit right in" with the five-man crew, "carries her share of the load," and "usually outdoes the rest of us."[46]

Upon completing the 1984 mission, Resnik was assigned to another satellite deployment and science mission, STS 51-L, launched in January 1986. This became the ill-fated tenth flight of *Challenger* and twenty-fifth shuttle mission overall. Barely a minute past liftoff, one of the solid rocket boosters failed catastrophically, causing the boosters, external tank, and orbiter to break apart under aerodynamic stress as if in an explosion. As millions watched in person and on television, the launch ended horrifically in contrails as pieces of the vehicle plummeted into the Atlantic Ocean off the coast of Florida. The crew never made it to space; weeks later, their wrecked cabin was found deep underwater with the astronauts still strapped into their seats.

This tragedy stunned the nation and the world by demonstrating that the vaunted shuttle for routine spaceflight was not as safe as advertised. It also shocked NASA, which had grown complacent about the shuttle as a safe operational vehicle. Only after the accident investigation would NASA admit that the space shuttle was still an experimental craft and take steps to remedy safety issues. The loss of the crew was widely mourned by a public introduced to them through an unusual amount of prelaunch publicity highlighting the most diverse crew yet, composed of an African American, a Japanese American, two women—one of them a teacher flying as a private citizen—a corporate guest astronaut, and two military pilots from the Air Force and Navy. Four of them were TFNG classmates.

Judy Resnik's home state US senator, John Glenn, speaking at her memorial service in Akron, Ohio, placed her and her crewmates squarely in the American tradition of curiosity, exploration, and discovery and urged the nation to "fix this and get on with it" in their honor.[47] Only thirty-six years old when she died, Resnik surely would have had an acclaimed future.

Many tributes and honors named for her followed her death—buildings, schools, scholarships, awards, a star, an asteroid, a crater on the Moon, and more. The most poignant may be a burgundy recliner chair that Resnik had managed to move into her NASA office when she arrived. After *Challenger*, Anna Fisher made sure that it stayed somewhere on the Astronaut Office floor as a reminder of Judy's spirit for as long as she remained at NASA.[48] When Fisher retired in 2017, she was the only TFNG left who had worked with Resnik, and the chair was still there.

Flights: **2** on space shuttle (**1** completed) • **2nd** US woman in space • **4th** woman worldwide • Time in space: **6** days (**144** hours, **57** minutes)

Sally Kristen Ride

The first US woman in space, Sally Ride, has been written about more than any of her fellow women astronauts, yet she was an intensely private person who guarded herself against the intrusions of fame. Her name is recognized by millions and will linger in history for ages, but even some of her close friends say they never fully knew her. She was an enigma to many.[49]

As an undergraduate at Stanford University, she had double majors in physics and English, the latter fueling her interests in language, poetry, drama, and the arts of self-expression. Her studies in physics led her to the rigorous and innovative field of astrophysics for a doctoral degree. She was enchanted with space and cowrote children's books about it, as well as technical papers for scientific journals. As a high school and collegiate tennis player, she honed the disciplines of fitness, teamwork, and competition, which served her well in the astronaut corps, yet she had a likeable, relaxed manner that belied her fierce intelligence and determination. She had never planned to be an astronaut, but upon seeing NASA's announcement seeking women for space shuttle missions, she applied, and once selected she gave it her best efforts.

No one knows with certainty why she was chosen to be the first of the six women astronauts to fly on a space shuttle mission and break the decades-old sex barrier to spaceflight. Any of the six could have been first and flirted with wanting to be first. All six flew within a two-year period from 1983 to 1985. Astronauts found the crew selection process to be so opaque that they never had a clue how those decisions were made, and those who made the decisions never filled in the blank. There was only surprise, suspicion, and speculation about why one person rather than another was assigned. This caused a fair amount of anxiety in the astronaut corps, especially among the military men who wondered how they ranked and when they would get their chance. The women, being fewer, generally were more philosophical about it, assuming that everyone would fly in due course.

Some have remarked that Ride was masterful in operating the robotic arm, scheduled for use on the first mission to which she was assigned, STS-7 (1983). She may have been selected for that technical skill. Others speculate that as a long-time competitive athlete she was calm and focused under pressure and strategic in dealing with problems. Still others think that her inherent reserve and poise might have been advantageous for surviving the expected frenzy of media attention that would descend on the first woman to fly, whoever she might be. During the press conference for this mission, Ride deflected questions about being first by saying, "I didn't come into this program to be the first woman in space; I came in to get a chance to fly in space."[50] Nevertheless, she and others knew how important her performance would be for the future of women in space. "The World is Watching," warned *Ms.* Magazine beside the cover story portrait of Ride before her historic flight.[51]

Ride flew twice before all of the 1978 women flew once. On her first mission in a crew of five, she was the primary RMS operator and used it to deploy and retrieve a small co-orbiting scientific satellite. She and the crew also deployed two large communications satellites and conducted early versions of materials science and biomedical experiments that would evolve on later missions. Ride, at age thirty-two, was the youngest astronaut to fly on the shuttle. She again was the lead RMS operator on her second mission, STS 41-G (1984), with crewmate Kathy Sullivan becoming the first US woman to do a spacewalk. This was the first spaceflight with two women aboard.

Ride was assigned to a third mission that was canceled after the 1986 *Challenger* catastrophe, after which she took on other duties. She was the only woman and astronaut to serve on the presidentially appointed commission to investigate the *Challenger* accident, and years later she served again on the investigation board for the 2003 loss of *Columbia* and crew. She worked in NASA's new Office of Exploration at headquarters and led a study team to make recommendations for America's future in space. The "Ride Report" outlined the types of missions and capabilities the nation needed to pursue to maintain leadership in space exploration.[52]

Ride left NASA in 1987 after nine years and spent the next twenty-four years as a professor of physics at the University of California in San Diego and as cofounder and chief executive of Sally Ride Science, an organization aiming to close the gender gap in scientific and technical careers by cultivating young girls' interests in those fields. Sally Ride Science developed curriculum materials, published books, trained teachers, held workshops, and staged science festivals around the country where students could meet scientists, engineers, and astronauts and engage in technical activities. Ride saw herself as primarily a physicist, but her fame as an astronaut gave her a platform to advocate for STEM education long before the acronym existed. Being an astronaut was actually the shortest period of her career.

The announcement of Ride's death from cancer at the age of sixty-one in 2012, made on the Sally Ride Science website, contained a bit of new information that caused another flurry of celebrity.[53] It named a woman, Tam O'Shaughnessy, as her life partner. O'Shaughnessy wrote the death notice with this revelation. Few people knew that Ride was ill, and only their families and closest friends knew that she and O'Shaughnessy were more than best friends and business partners. Few beyond her peers in the astronaut corps knew that she had been married during the 1980s to fellow astronaut Steven Hawley. The LGBTQ+ community immediately celebrated Ride as "the first gay astronaut." Since then, almost everything written or presented about Ride repeats this title, one that she never considered for herself (and one that may not be true). She has been appropriated in death as an icon for a cause and community that she didn't choose to represent.[54] As social attitudes about sexual identity evolved, Ride might have eventually revealed more about herself—or not—but that is unknowable.

Ride thought titles and labels were irrelevant. It was enough that she saw herself as a physicist, athlete, and astronaut. She could not avoid the one label that was factually undeniable—America's First Woman in Space—but she wore it lightly with a sense of humor.[55] She preferred to be quietly influential rather than in the spotlight and was vigilant about not lending her name to commercial enterprises or becoming a spokesperson for causes other than science education. She was known to decline requests and invitations that did not align with her values, and she refused to profit commercially from her inescapable fame.

Sally Ride was awarded the Presidential Medal of Freedom posthumously by President Barack Obama in 2013.

Flights: **2** on space shuttle • **1st** US woman in space • **3rd** woman worldwide • Time in space: **14** days, **7** hours (**343** hours)

Margaret Rhea Seddon

Rhea Seddon recalls that she "was raised to be a Southern lady. I was a little girl growing up in a small Southern town, taking piano lessons and ballet lessons, and assuming that I would be like my mother and be a nice wife in a nice home someday."[56] But Seddon began to envision a different future at age ten when she watched for Sputnik to pass overhead and wondered whether people might go into space. She discovered the sciences in her teen years, especially biology, and started thinking about a career in health care. She went to the University of California, Berkeley for its renowned life sciences program, then returned to Tennessee for medical school. Seddon also went to flight school, earned her pilot's license, and began to log flying hours. When she applied to be an astronaut, she was completing her residency in surgery. She also was an experienced emergency room physician.

During the 1960s, Seddon followed the news about spaceflight and found it fascinating. Her interests in space, life sciences, and medicine converged when she applied for and was selected in the first class of astronauts for the space shuttle. She hoped to use her medical training and do biomedical research in space—exactly what happened after she joined NASA. Along the way, though, she and all the other astronaut candidates had to learn a lot of engineering to understand the space shuttle and all the equipment and systems that flew on it.

Seddon flew on three space shuttle missions, one to deploy two communications satellites and two to conduct research. On all three missions, she served as the crew medical officer responsible for the well-being of her crewmates. She was the second medical doctor and the first woman physician to fly on a NASA mission.[57] On her first flight, STS 51-D (1985), she operated the robotic Remote Manipulator System arm for the EVA crew and for a challenging, unrehearsed attempt to activate a stranded satellite. She and the crew also conducted several medical experiments.

Her next mission was STS-40 (1991), designated Spacelab Life Sciences-1, which she said was "the one flight on the manifest that I had been waiting my whole career to fly."[58] Spacelab was a large, well-equipped, reusable laboratory carried in the payload bay of the shuttle; it was connected by a pressurized tunnel to the crew cabin so the crew could go back and forth to do their work. Investigators from around the world competed to have their research projects selected for flight and carried out by the mission and payload specialists. The primary focus of this mission was to study how humans, plants, animals, and cells responded to microgravity and readapted to Earth's gravity upon return. It happened to be the first mission with three women in its seven-member crew— Seddon, Tamara Jernigan, and payload specialist Millie Hughes-Fulford.

Two years later, Seddon flew on a similar all-research mission, Spacelab Life Sciences-2 (STS-58, 1993). This time she was the payload commander in charge of experiment operations and the scientific crew, which included classmate Shannon Lucid. A fourteen-day mission with round-the-clock work in two shifts, SLS-2 was deemed the most successful Spacelab mission to that point.

Before and between missions, Seddon completed a variety of medical and technical assignments at JSC, supported the STS-6 crew during their training, served as CAPCOM, and gained management experience as assistant to the director of flight crew operations. An assignment she especially enjoyed was serving as a physician on the rescue helicopter team deployed for shuttle mission launches and landings. She also kept up her medical skills by taking weekend shifts at nearby hospitals throughout her time at NASA.

Seddon and her pilot astronaut husband, Robert L. "Hoot" Gibson, gained some fame as the first astronaut pair to marry after joining NASA. Not yet assigned to a mission, Seddon was the first astronaut to become pregnant and give birth. The media hailed their son, born in 1982, as the first "astrotot." Their second child was born during the long delay after the *Challenger* tragedy, a window of opportunity for another pregnancy before her second flight. She managed a third pregnancy and birth after her last mission. Seddon and classmate Anna Fisher, also an MD who had two children about the same time, normalized pregnancy within the astronaut corps. It was important to them both to disprove the old adage that "women just get pregnant and leave." Both succeeded in keeping their careers on track while starting their families, and NASA learned that it was not a problem for women astronauts, almost all of childbearing age, to experience this natural process.

During her nineteen-year career at NASA, Seddon oversaw some of the earliest and most sophisticated life sciences research in space. After her three missions and Gibson's five, they left NASA in 1996. She went first to Vanderbilt University to work with researchers preparing experiments to fly on the final Spacelab mission, Neurolab, in 1998. She then returned to the medical community as an administrator in the Vanderbilt Medical Group. There she developed an aviation-based crew resource management approach to patient safety, quality of care, and teamwork and taught this concept at healthcare institutions around the country. Seddon shared her astronaut experiences in many public appearances, posted a lively blog on her website, and published her memoir, *Go For Orbit: One of America's First Women Astronauts Finds Her Space*, in 2015.

Flights: **3** on space shuttle • **5th** US woman in space • **7th** woman worldwide • Time in space: **30** days, **2** hours (**722** hours)

Kathryn Dwyer Sullivan

Kathy Sullivan must have been born with explorer genes in her DNA. She loved maps and was so interested in faraway places that she plotted routes for family road trips.[59] She was fascinated with Jacques Cousteau and other explorers. She also loved languages and mastered them easily, leading her to envision a career in the State Department, traveling around the world as a member of the Foreign Service working in embassies and consulates. But her life took a different turn when she was required to take a science course as a university freshman, and she discovered geology and the Earth sciences. She came to realize that working in those fields offered travel opportunities to exotic places and the chance to make discoveries. The explorer in her made a new plan that culminated in oceanographic expeditions to study undersea geology when she was in graduate school.

Sullivan was firmly on that path when her older brother, a pilot, suggested that she apply to NASA to become an astronaut, as he was doing. She was not yet aware that NASA was recruiting women scientists to be mission specialists on the space shuttle, but as she learned more about the role, she realized that she was well prepared for spaceflight. Research expeditions at sea operate much like missions in space, and through her PhD program Sullivan was already well experienced in the focus, discipline, planning, and teamwork required to conduct research—and survive—in isolated and hazardous circumstances. A submersible operating from a surface ship wasn't too different from a spacecraft, and dive gear had much in common with spacesuits as life

Kathy Sullivan suited up as the first American woman to do an EVA, 1984. *NASA*

support systems. When she realized that astronauts and oceanographers had similar jobs in lethal environments, she decided to apply. She was summoned to be interviewed and evaluated in Houston. Her brother, who wasn't invited, cheered her on.

As Sullivan completed initial training, including scuba diving, and worked on various technical assignments, she took the initiative to express interest in extravehicular activity. She was comfortable in the underwater neutral buoyancy facility used for EVA training, and she was already certified to wear a pressure suit through an assignment on NASA's high-altitude research aircraft. She also realized that her height and body frame were a good fit for the spacesuit, which some petite women could not wear comfortably or safely. Pursuing EVA as her path to spaceflight while others were focusing on the robotic arm as their likely ticket to space was rewarded with her first flight assignment. In 1984 Sullivan became the first American woman to do a spacewalk.

The agenda for that mission, STS 41-G, was primarily Earth observations. The crew deployed the Earth Radiation Budget Satellite and used an imaging radar system, a large format camera, and other instruments to study the Earth via remote sensing and imaging from space. The highlight of the mission was the 3.5-hour EVA by Sullivan and crewmate David Leestma to demonstrate the feasibility of refueling a satellite in orbit with toxic hydrazine propellant. This shuttle mission had the largest crew launched on a single spacecraft to that point—seven—and was the first time two women served on a crew, with Sally Ride on her second flight as the robotic arm operator and Sullivan as an EVA specialist. Sullivan just missed being the first woman ever to do an EVA. Learning of NASA's plan for her mission, the Soviets hurried their second woman cosmonaut, Svetlana Savitskaya, into space in 1983 to claim that record and also make her the first woman to fly in space twice, snagging that distinction from Ride.

Sullivan spent the next six years serving as CAPCOM and Cape Crusader for other shuttle missions and preparing for one of NASA's most heralded missions, deployment of the Hubble Space Telescope on STS-31 in 1990. She was again assigned to the EVA role, and with her crew-partner Bruce McCandless, spent years developing and testing the tools and procedures for future crews to be able to service and upgrade the telescope in orbit. She and McCandless trained underwater for any contingency that might arise during the deployment, such as the failure of a solar array to deploy automatically, that would require them to go outside and manually solve the problem. During the deployment sequence in orbit the pair were suited up in the airlock, ready to go outside without delay. It proved unnecessary, and confined there, they missed seeing the release of the observatory that had long occupied their attention; they had to review it later in IMAX and other footage. Sullivan did, however, earn bragging rights for being on the highest-altitude shuttle mission, a record-setting 380 miles.

After the 1986 *Challenger* tragedy that grounded NASA missions, Sullivan joined the US Navy Reserve as an officer to exercise her oceanographic and leadership skills. She commanded a unit based at the Naval Air Station in Dallas, Texas, that supported the meteorology and oceanography forecasting mission of the Oceanography Command Center in Guam. While training for her third shuttle mission, she was deployed on cruises and activated during the Persian Gulf war operations in 1991.[60]

On Sullivan's third flight, STS-45 in 1992, she was payload commander in charge of all the research comprising the Spacelab Atmospheric Laboratory for Applications and Science (ATLAS-1), the first of several high-priority Missions to Planet Earth to study the chemical and dynamic nature of the atmosphere. Perhaps the most exciting experiment involved firing an electron beam into the atmosphere to cause an artificial aurora, a beautiful instance of atomic excitation of atmospheric gases. Sullivan was the first astronaut to fly as payload commander in an important acknowledgment of the leadership potential of mission specialists. Several others held that role on subsequent missions.

Having completed three missions and not seeing more flight or leadership opportunities ahead, Sullivan left NASA in 1993 and embarked on a long post-astronaut career as a scientist-executive. She accepted public service appointments under four presidents of both political parties, first as chief scientist of the National Oceanic and Atmospheric Administration and on advisory panels and later as NOAA administrator. Between those two federal positions, Sullivan spent a decade as president and CEO of the Center of Science and Industry in Columbus, Ohio, guiding its physical and programmatic expansion for STEM education. From there, she moved to Ohio State University to become the initial director of the Batelle Center for Mathematics and Science Education Policy. After her second stint at NOAA, Sullivan was the Lindbergh Fellow at the Smithsonian's National Air and Space Museum, where she wrote *Handprints on Hubble*, a behind-the-scenes account of inventing the philosophy, design, and practical tools and procedures for in-orbit servicing of this complex observatory.

Sullivan then returned to private life but hardly retired, becoming active as a speaker, launching a podcast *Kathy Sullivan Explores*, writing a book on spacewalking, and traveling to exotic places. In 2020, she made history again as the first woman to descend to the deepest known point on Earth, the Challenger Deep in the Mariana Trench of the Pacific Ocean. This feat landed her in Guinness World Records, with two additional records as the first person to travel both to space and the deepest point on Earth and as the individual who has traveled the greatest vertical distance.[61] She had reached an unusually high orbital altitude on the 1990 Hubble Space Telescope deployment mission; that altitude combined with the ocean depth gave her a vertical range of 386.5 miles (622.1 km). As of 2023, Sullivan resumed public service upon appointment to the President's Council of Advisors on Science and Technology. Her career has been distinctive and distinguished as a high-level federal executive and a leader in institutions devoted to science, technology, and math education and scientific research.

Reflecting on her astronaut career in a 2019 interview, Sullivan noted that she had not suffered harassment as a woman, but she had observed that women working in most professions typically encounter more obstacles to their advancement than men.[62] After watching NASA women struggle to earn the ability to be flight controllers and flight directors, she was pleased to see how many women have moved into all positions that are integral to spaceflight. As a trailblazer in EVA, as a science payload commander, and in almost every position in her post-astronaut career, Sullivan has stayed alert to the potential of women and always mentored them toward success.

Flights: **3** on space shuttle • **3rd** US woman in space • **5th** woman worldwide • Time in space: **22** days, **4** hours (**532** hours)

The Class of 1980

The second shuttle astronaut recruitment began as the first group of candidates completed their initial year of training. In August 1979 NASA announced that it would accept astronaut applications annually, with the next group of candidates to be selected in mid-1980. NASA specifically urged qualified Hispanics to apply, citing the acceptance of women and other minorities in the previous group.[63]

This round resulted in about 3,200 applications, less than half as many as the first recruitment, with 390 women applying. To improve their chances, 341 people (mostly men) applied for both pilot and mission specialist positions.[64] The selection of nineteen new candidates announced in May 1980 included twelve who had interviewed in the first round: one Hispanic, one African American, two women—Mary Cleave and Bonnie Dunbar—and Anna Fisher's husband, William F. Fisher. Eleven of those selected, including both women, were identified as mission specialists.[65]

Mary Louise Cleave

Thirty-three-year-old Mary Cleave already had nine years' experience as a research scientist and engineer with the Ecology Center and Utah Water Research Laboratory at Utah State University. She specialized in microbial ecology and civil and environmental engineering. She credited Title IX of the Education Amendments Act of 1972, which banned discrimination in education, for making her career possible. "Affirmative Action made a huge difference," she says, "and I was right on the leading edge of it."[66]

Mary Cleave's first passion was flying. At age fourteen she started lessons and earned her private pilot's license at seventeen before she could drive. But her dream of a career in aviation collapsed when she learned, upon applying, that commercial airlines weren't yet hiring women as pilots, and at 5 feet, 2 inches she was too short to be a pilot or stewardess. She turned instead to biological sciences for her undergraduate and master's degrees and then, with her aptitude for problem-solving, became the first woman to earn a doctorate in engineering at Utah State University.

Cleave applied to NASA for the first shuttle astronaut recruitment after finding the notice tacked on a post office bulletin board. She was still completing her PhD coursework and writing her dissertation, and she was not invited to Houston for interviews and testing. She applied again after completing her graduate work and taking a research position and was selected for the 1980 astronaut class. A pilot at heart, she especially looked forward to flying in the T-38 jets used for astronaut training, saying "This was my only opportunity to fly in high performance jets. The space-flights were a bonus."[67]

Cleave flew on two space shuttle missions, both on *Atlantis*. On the first, STS 61-B in late 1985, she had primary duties as flight engineer to support the commander and pilot during ascent and entry, and to operate the remote manipulator system. She guided this robotic arm to lift and deploy three communications satellites from the payload bay. Then she maneuvered her two EVA crewmates as they demonstrated space construction techniques—the EASE/ACCESS experiment—in the payload bay. She also was responsible for a set of 3M crystal growth experiments. In 1989 Cleave flew again on the STS-30 mission to deploy the *Magellan* spacecraft to Venus, the first planetary probe to be launched from the shuttle. This time she was responsible for in-flight maintenance, photography, and various experiments and was jumpmaster in case of an emergency bailout.

Seeing the Earth from space and noticing its visible changes in the short interval between her flights triggered a desire to work directly on environmental issues. In 1991 Cleave moved to NASA's Goddard Space Flight Center in Maryland to manage an Earth-observing satellite project. Thirteen years later, she took senior leadership positions in the Office of Earth Science and the Science Directorate at NASA headquarters, overseeing research and exploration programs that studied Earth, space weather, the solar system, and the universe, becoming the first woman to rank as associate administrator for science. She retired from NASA in 2007 after eleven years as an astronaut and sixteen years in science and engineering management and policymaking. Elected to the Sigma Xi scientific research honor society during her university years, Cleave was elected to the American Academy of Arts and Sciences during her career. She received many awards and honors, including NASA Engineer of the Year in 1998.

As one of the early women in the astronaut corps who came from an academic research environment, Cleave felt welcome and had good assignments, but she recalled that the prevailing military culture required adjustments by everyone. It was challenging not to take things personally but accept that they were all in a new situation and try to make it work. In the run-up to the first space shuttle missions, which began in 1981, the Astronaut Office was understaffed and eager for more help to handle the workload. Cleave said that the older astronauts gave the new ones unfamiliar tasks without instruction, challenging them to figure things out. She was the first in her class to get a specific task: fix the space shuttle toilet, not an especially enviable job but one suitable for a civil engineer experienced in sanitation systems. This effort earned her the appreciative nicknames "sanitary fairy" and "first space plumber" and a sign stating, "The best man for a job may be a woman."

Reflecting on her time as an astronaut, engineer Mary Cleave thought her problem-solving contributions to life support systems and remote manipulator arm operations were her most significant accomplishments. Deploying the *Magellan* spacecraft to Venus was a highlight of her career. The most satisfying aspect of being an astronaut was working with smart, focused people, all committed to shared goals. "It was fabulous!" Cleave said. "I was really lucky to be there."[68]

Mary Cleave died of natural causes in late 2023, at age seventy-six.

Flights: **2** on space shuttle • **8th** US woman in space • **10th** woman worldwide • Time in space: **10** days, **22** hours (**262** hours)

Bonnie Jeanne Dunbar

Bonnie Dunbar watched *Flash Gordon* on television and spotted Sputnik overhead in the clear night sky of rural Washington, where her family farmed and raised cattle. She decided at the age of eight that she wanted to be an astronaut or build spaceships. Rising from a rural elementary school and small-town high school, she became the first in her family to attend college and earned engineering degrees from two universities. That trajectory led to her being the first woman to enroll in the ceramic engineering program and, in 1975, the first to earn a graduate degree in ceramic engineering from the University of Washington.[69]

While growing up in the 1960s, Dunbar followed the space program and was convinced that eventually women would become astronauts, too. She read everything she could find in the library and encyclopedia about space, enjoyed Jules Verne and space science fiction writers, and carefully planned her studies to take all available mathematics, chemistry, and physics courses. Nevertheless, she was discouraged from going to college by a high school guidance counselor, was seated at the back of an engineering classroom so she wouldn't be a distraction to the boys, and was unable to attend Caltech or join ROTC because they didn't admit women. One of her favorite professors betrayed her by voting against her admission to graduate school because he said he wasn't sure that she could meet the challenges. Dunbar credits her parents for instilling her strong work ethic and confidence, and she credits her successful career to those mentors who recognized her potential without gender bias at a time when a woman was rare in their professions. In math and engineering problems, the solution is either right or wrong, and being right settles any doubts about one's ability. She recalls, "My response to any negative attitudes was to ace the class."[70]

Steered toward ceramic engineering research by one academic mentor at the university, she did advanced lab work on thermal protection materials for the space shuttle as an undergraduate. That led to an engineering position with the space shuttle prime contractor, Rockwell International, where she was the youngest engineer but the only one with ceramic experience working on production of the shuttle's tiles.

While at Rockwell, Dunbar applied to NASA's first call for space shuttle astronauts. She interviewed in Houston as one of 208 finalists but was not selected in the 1978 class. Instead, NASA encouraged her to come to work at the Johnson Space Center, where she was hired as a payload officer and flight controller. She quickly gained mission

Bonnie Dunbar in the launch tower white room in final preparation to enter the shuttle on the last of her five missions, 1998. *NASA*

experience as a member of the Skylab deorbit guidance and navigation control team and as a coordinator of science and engineering payloads for early shuttle missions. Realizing that all the new astronauts except pilots had doctoral degrees, she started PhD coursework in mechanical/biomedical engineering while on the job.

Dunbar reapplied to be an astronaut and was selected in the 1980 class, the first woman to be drawn from the NASA workforce. She continued working on her PhD research while in astronaut training and earned her doctorate in 1983. She held a variety of scientific, engineering, and management assignments during her twenty-seven-year career at NASA, ultimately ranking as one of the most experienced and senior members of the astronaut corps.

Dunbar closely followed Shannon Lucid in becoming the second woman to fly on five missions, which both accomplished less than fifteen years after their first flights.[71] (At that time, only legendary astronaut John Young had flown in space six times, a record set again at seven flights by Jerry Ross in 1995 and tied by Franklin Chang-Diaz in

2002.) Dunbar achieved that status in 1998 after flying on a science mission sponsored by Germany, a satellite deployment and retrieval mission, a microgravity science lab mission, and two Shuttle-*Mir* docking missions. As a crewmember on these international missions, she lived and worked abroad for months at a time and developed a strong network of space scientists, engineers, and managers in partner space agencies. In the mid-1990s she completed cosmonaut training in Russia, the first American woman to do so. This breadth of experience prepared her for leadership positions within NASA.

Reflecting on her five missions, Dunbar clearly prioritized scientific research. The space shuttle always carried experiments in the middeck lockers and often carried a well-equipped laboratory module called Spacelab or an extended middeck called Spacehab in the payload bay. These facilities effectively made the shuttle into a temporary space station stocked with experiments developed by scientists around the world who had trained the crews to conduct the research on their behalf. After a mission, instruments and data were returned to the investigators, who could then refine their designs and procedures and perhaps fly the experiments again. This iterative, lessons-learned process was crucial in preparing for longer-term scientific research aboard the planned space station. Dunbar was instrumental in doing these experiments, refining the techniques, and defining future capabilities for microgravity research in space without the constraints of gravity.

Dunbar always had a key role in operating Spacelab systems and carrying out research, whether the experiments were in biology, physiology, materials science, fluid physics, crystal growth, technology, or another field. The German space agency organized and managed her first mission, Spacelab D-1 (Deutsche-1) flown as STS 61-A in 1985 with more than seventy-five onboard experiments. She was the only woman in the eight-member crew, the largest crew to date on a single spacecraft from launch to landing. Her second mission, STS-32 in 1990, deployed a communications satellite and retrieved a technology satellite—the Long Duration Exposure Facility (LDEF)—that was covered with material samples to investigate how well they held up in the space environment after five years in orbit. Dunbar captured the LDEF using the remote manipulator arm and nestled it back into the payload bay for return to Earth. That mission also included a variety of experiments and first-time astronaut Marsha Ivins. As payload commander on STS-50, the US Microgravity Lab-1 mission in 1992, Dunbar led four payload crewmembers, including Ellen Baker, in around-the-clock laboratory work to complete more than thirty experiments in thirteen days, the longest shuttle mission to that point.

Dunbar's last two missions were the first and eighth docking missions with Russia's *Mir* space station, a program that involved crew exchanges, cargo transfers, and scientific research. Although she was certified to fly as a cosmonaut on Russian spacecraft, she flew as a mission specialist on the shuttle side of these missions. STS-71 in 1995 carried a Spacelab module with equipment largely dedicated to medical evaluations of the two returning *Mir* crewmembers. On her last flight, STS-89 in 1998, she was payload commander again and was responsible for all research and logistics activities, including twenty-three experiments plus cargo transfers to and from *Mir*.

Between and after her flights, Dunbar filled assignments at Johnson Space Center and NASA headquarters, in Germany, and in Russia. As her portfolio of responsibilities expanded, she rose into managerial and senior leadership positions in life and microgravity sciences, risk management, and research partnerships with universities. She believed that everything she was called upon to do, and everything she initiated, had a foundation in engineering and problem-solving. One of the problems she tackled and insisted on righting was the design of some of the biomedical devices used in Spacelab research. They were sized for male bodies and did not fit female crewmembers who also needed to use them. She found the same problem with the pressurized Advanced Crew Escape Suits (ACES) developed in the 1990s and had to fight for them to be modified because they were unsafe for smaller astronauts to wear. In fact, she refused to wear one in flight to force its redesign.

Mission specialists came into the astronaut corps as experts in one field of science and were unfamiliar with some of the others. Assuming that knowledge breeds interest and commitment, Dunbar established a science support group to familiarize astronauts with the kinds of research being done on the shuttle before they were assigned to such missions. She had other roles that involved planning and explaining the science agenda for shuttle missions, particularly in microgravity science, her area of expertise. She spent three months in Germany as NASA liaison to the Spacelab D-1 mission science teams and thirteen months in Star City, Russia, training as a backup for a stay on *Mir*. Dunbar also served as assistant director of Mission Operations and acting director of Flight Crew Operations in preparation for International Space Station activity, then moved to the Center Director's Office to build up research partnerships with universities. Upon retiring from NASA in 2005, she led the Museum of Flight in Seattle, taught at the University of Houston, and joined the engineering faculty at Texas A&M University, where she directed the Institute for Engineering Education and Innovation.

Despite an enviable record of five missions totaling seven weeks in space, the only thing Bonnie Dunbar never accomplished was a full orbit of ninety uninterrupted minutes at a window, just absorbing the incredibly beautiful scenery of the Earth. She still hopes to do that someday.

Flights: **5** on space shuttle • **7th** US woman in space • **9th** woman worldwide •
Time in space: **50** days, **8** hours (**1,208** hours)

The Class of 1984

NASA announced another astronaut recruitment in May 1983 and planned to start making annual selections in 1984. The number of candidates to be selected annually would "vary depending upon mission requirements and the rate of attrition in the existing astronaut corps." The announcement reminded applicants of NASA's affirmative action program and its aim to select qualified minorities and women. Openings for six pilots and six mission specialists were anticipated in this cycle, and there were no changes in the qualifications and job descriptions for either type of astronaut.[72] Almost five thousand applications arrived by February 1984 when screening of applicants began. From that crowd, NASA invited those most qualified to interviews and evaluation, and in May announced selection of seventeen new candidates.[73] Seven were pilots, and three of the ten mission specialists were women—Ellen Shulman (now Baker), Marsha Ivins, and Kathryn Thornton. This recruitment netted one Hispanic pilot and five current NASA employees. Members of this group completed their training in 1985, but their first missions were delayed by the 1986 *Challenger* catastrophe and subsequent grounding of the space shuttle fleet.

Ellen L. Shulman Baker

Among the first women to enter the astronaut corps from the NASA workforce, Ellen Shulman was a graduate of the Cornell University Medical College. She completed her residency in internal medicine and joined NASA as a physician in the Flight Medicine Clinic at JSC in 1981. She received a master's degree from the University of Texas School of Public Health in 1994.

Now known as Ellen Shulman Baker, she flew on three space shuttle missions at regular intervals from 1989 to 1995. The primary purpose of her first mission, STS-34 (1989), was to deploy the *Galileo* planetary probe on its journey to Jupiter. That accomplished, the crew turned their attention to various scientific and medical experiments during their five days in space. Baker and Shannon Lucid were crewmates on this mission. Baker was trained to perform a contingency EVA if necessary, operated several small experiments, and was a subject in a variety of life sciences studies.

She was assigned to fly next on a Spacelab mission, STS-50 (1992), designated US Microgravity Laboratory-1, the first extended duration flight of the space shuttle. This two-week research mission focused on biological, physical, and materials sciences, with the crew working around the clock in two shifts. She and Bonnie Dunbar were crewmates but on opposite shifts with different duties. Baker was the mission flight engineer and also conducted experiments on the middeck, handled much of the Earth observation photography, and served as the crew medical officer. As flight engineer, she assisted the commander and pilot during launch and return and managed the orbiter systems while they slept. She was prepared for a contingency EVA if required, and she held amateur (ham) radio sessions with students around the world.

Baker was then assigned to STS-71 (1995), the first space shuttle mission to dock and exchange crewmembers with the Russian space station *Mir*. The shuttle carried the Spacelab module stocked with biomedical experiments to be conducted during ten days in space. Payload commander Baker and Bonnie Dunbar were crewmates again on this Spacelab-*Mir* Life Science mission focused on joint activities with the *Mir* crew and studying the human body in microgravity. Baker was responsible for Spacelab and medical activities and was again prepared for a contingency EVA. This distinctive mission tallied as the one hundredth US human spaceflight.

When not in training for a spaceflight, Baker held various jobs in the Astronaut Office, primarily related to crew health and safety. She led the education/medical branch in the Astronaut Office and helped to develop group training activities for crews assigned to long-term spaceflights. After her three flights, she continued to serve as the astronaut representative for many crew health and safety activities and served as an advisor to the space medical research community.

After retirement from NASA with thirty years of service, Baker joined the MD Anderson Cancer Center in Houston. There she worked with the global oncology team to provide support and technical assistance to low- and middle-income countries seeking to develop cancer prevention programs and improve treatment services. She also directed a program to improve access to cancer prevention and treatment services in underserved communities in Texas. Baker was an editor for the second and third editions of *Principles of Clinical Medicine for Space Flight*, for which she wrote chapters on human responses to spaceflight and gynecological and reproductive considerations.

Like most of the first generation of women astronauts, Baker did not imagine becoming an astronaut when she was young. "They didn't let girls go into space when I was little," she said. It began to seem a possible, but perhaps unattainable, goal when she saw a newspaper article about NASA urging women to apply just as she was finishing medical school in 1978, the year the first women were selected. Like other women astronauts, she believed, "I could do anything I set my mind to."[74] Her timing and experience were impeccable, and she was one of seventeen astronaut candidates selected in 1984. Apart from spaceflight itself, she has said, "What I like best are the people. I work with a wonderful group of people."[75]

Baker credited her parents, who instilled her work ethic, as "unconditionally supportive." They also set examples of achievement: like their father, she and her brother became physicians, and her mother was the first woman elected to be borough president of Queens, New York, a position she held for many years. As to being among the earliest women astronauts, Baker said, "I don't see women and men. I see astronauts."[76] Although there may be opinions and biases, she noted, "It's really the quality of the work that helps to make progress."

Flights: **3** on space shuttle, including a docking with *Mir* • **9th** US woman in space • **11th** woman worldwide • Time in space: **28** days, **14** hours (**686** hours)

Marsha Sue Ivins

An aerospace engineer, Marsha Ivins wrote to NASA as a university freshman and experienced pilot in 1970 to ask about the future for women astronauts in the space program and whether aerospace engineering would be the best preparation. She recalled being told as a child that she could not be an engineer or an astronaut because she was a girl, so she relied on herself for the courage to try.[77] She wanted to be an astronaut or work in some part of the space program since the first American was launched when she was ten years old. Concurrent with this interest, she took up flying airplanes, another uncommon pursuit for a teenage girl in the 1960s, soloed at age sixteen, and earned her private pilot's license at age seventeen. She became a flight instructor and in college taught some of the men in her engineering classes to fly. She holds multiple aviation licenses and ratings and has logged more than seven thousand hours in various aircraft.

Ivins received a mildly encouraging response to continue her studies and give it her best effort. NASA didn't envision needing more astronauts for some time, the letter said, but she should apply whenever NASA announced a new selection. Ivins followed the advice and, after graduating from the University of Colorado, went to work at Johnson Space Center as an engineer for the orbiter cockpit displays. There she became a flight engineer and pilot on NASA's Gulfstream aircraft. She applied in the 1978 recruitment and again in 1980, this time reaching the weeklong interview and evaluation stage. Her third application and more interviews led to selection in 1984, and she was a member of the astronaut corps until retiring in 2010. Her first flight into space was delayed by the *Challenger* catastrophe, but Ivins quickly had four missions during the 1990s and in 2001 completed her fifth shuttle mission as the most experienced member of that crew. She is one of only six women to date to have flown five times.

In 1990 Ivins served as flight engineer on STS-32, assisting the pilots during ascent and entry, and had a number of other responsibilities, including the IMAX camera and performing test echocardiographs on her crewmates. The crew deployed a large communications satellite and then retrieved the Long Duration Exposure Facility (LDEF), a satellite covered with samples of various candidate materials for use in space. It had been in orbit since 1984 as an experiment to learn how the materials would hold up to the extreme temperatures, unfiltered sunlight, and bombardment by radiation and orbital debris particles in space. Among her other tasks, Ivins did a thorough photographic survey of the satellite as it was brought into the payload bay.

Two years later, Ivins was on STS-46, again serving as flight engineer and an IMAX camera operator. She participated in a variety of research projects and helped to deploy a European scientific satellite. A highlight of this mission was an attempt to test the Tethered Satellite System, an experiment in dragging a long cable anchored to the shuttle at one end and a small satellite at the other through Earth's magnetic field to generate electricity. The cable jammed during reel-out, so the experiment could not be completed and was rescheduled for a future mission.

Ivins's third flight, STS-62 in 1994, was a Spacelab mission called US Microgravity Payload 2 to study the effects of microgravity on materials science experiments. These experiments and another set of engineering and technology experiments were mounted

Marsha Ivins wrangling cameras in orbit, 1994. *NASA*

in the payload bay and controlled by investigators on the ground. The crew engaged in biomedical and materials science experiments and photography projects inside the orbiter. Ivins also operated the RMS arm to test a new feature and take a closer look at the instruments in the payload bay.

In 1997, Ivins flew on the fifth Shuttle-*Mir* mission (STS-81) to exchange crewmembers, transfer supplies and experiments, and carry out scientific research in the onboard Spacehab module, a commercial version of Spacelab. She was responsible for most of the transfers and was the lead intravehicular crewmember (spacewalk choreographer) should an EVA become necessary.

Ivins's last mission, STS-98 in 2001, took her to the International Space Station to deliver the US *Destiny* laboratory module. Again she was flight engineer, loadmaster for equipment transfers to and from ISS, and an IMAX operator. She used the RMS arm to lift the *Destiny* module out of the payload bay for installation on the ISS and maneuvered the EVA team during spacewalks. She also helped activate *Destiny* after its installation.

Ivins was fortunate that her missions averaged ten and a half days, giving her a total of almost eight weeks in orbit. On each mission, every crewmember had a variety of primary and secondary jobs, and over the course of five missions Ivins did almost everything a mission specialist could be assigned: flight engineer, robotic arm operations, EVA choreography, scientific experiments, photography, logistics transfers, and troubleshooting or repairs as needed. As the photography lead on all five missions, she took the initiative to add new types of film, cameras, and camera technologies to the crews' repertoire.

Apart from flight, Ivins served as a Cape Crusader (astronaut support personnel) at Kennedy Space Center and CAPCOM, plus other technical assignments, during her time in the astronaut corps. She held two long-term assignments: reviewing all imagery from inside the shuttle and space station to select the best shots for public release and overseeing the packing and stowage of crew equipment to ensure that everything was packaged, labeled, and organized for maximum efficiency with no confusion.

After her last flight, Ivins continued supporting shuttle and ISS missions as flight crew equipment lead and in other roles. As the end of the shuttle program approached, she led the branch working on post-shuttle missions, what was called the Constellation program from 2004 to 2009. It featured a new Orion spacecraft, and Ivins was the primary Astronaut Office representative during its development.

Ivins believes that the most important skill she learned as an astronaut is one worth acquiring: systems engineering.[78] Understanding how all the technical systems and human organizations worked to support spaceflight became a useful thought process for problem-solving in general. Thinking about integrated systems that function together leads to asking better questions, seeing the bigger picture from all angles, and finding effective solutions. That, and leaving the planet five times, is what Marsha Ivins found most rewarding about spaceflight.

Flights: **5** on space shuttle, including missions to *Mir* and the International Space Station • **11th** US woman in space • **13th** woman worldwide • Time in space: **55** days, **21** hours (**1,341** hours)

Kathryn Cordell Thornton

The only girl in her high school physics class and one of only two women in her university-level advanced physics classes, Kathy Thornton went on to earn bachelor's, master's, and doctoral degrees in physics and a postdoctoral fellowship in nuclear physics. She worked as a civilian physicist for the US Army Foreign Science and Technology Center before heading to NASA. Thornton saw the 1983 astronaut recruitment announcement and thought it sounded like an interesting job but assumed she had "about a one in a million chance" to be selected. She applied anyway and was accepted.[79] She once said that her father wanted her to study something more useful than physics; he never imagined it would lead her into two productive careers.

Thornton, or "KT" as other astronauts called her, arrived at NASA married, with a two-year-old child. After training, she had typical technical assignments in flight software and vehicle integration testing, CAPCOM duty, and crew equipment responsibilities before and between her four shuttle missions. No one except her crewmates know exactly what she did on her first flight, a classified Department of Defense mission (STS-33 in 1989). Neither NASA nor the Department of Defense released information about the payload or purpose, but journalists speculated that it was to release an electronics intelligence spy satellite. Already holding a security clearance from her prior work with the Army, she was the only woman and one of only two civilians to fly on any of the classified DOD shuttle missions. STS-33 was the last shuttle mission of the 1980s.

Thornton had trained to be a contingency EVA spacewalker for that mission, which didn't happen, but she logged her first spacewalk soon after on the STS-49 (1992) mission to retrieve, repair, and release a stranded INTELSAT communications satellite that had not deployed properly. Capturing the satellite proved more difficult than expected and finally succeeded with the only three-person EVA in history. Thornton, the only woman in the seven-member crew, worked inside to support the three days of spacewalks by her crewmates, and then she and one of the men carried out a separate EVA to test methods for assembling and repairing space station elements. Those tests were originally scheduled for two days but were reduced to one after the time-consuming INTELSAT activities. Still, this was the first shuttle mission with four EVAs, and Thornton became the second US woman to spacewalk.

As things happened, Thornton participated in two of the Top Ten EVAs later ranked as spectacular.[80] Her third flight was on the first mission to service the Hubble Space Telescope, STS-61 (1993). Again, she was the only woman aboard with six men. For this high-stakes mission to correct the telescope's flawed vision and replace some of its components, two EVA pairs worked alternately for five days to complete the demanding workload. Thornton and her EVA partner worked outside on two days.

Thornton's duo drew two of the most dramatic tasks: installing a telephone booth–size box containing corrective optics (like contact lenses) for the instruments and replacing the two original solar arrays. For both tasks, Thornton was "hands-on," standing at the end of the fifty-foot-long RMS arm to guide the COSTAR corrective optics box straight onto guide rails inside the telescope. On an earlier EVA she perched on the arm to jettison a solar array too damaged to bring back. In what was called her "Atlas act," she held the forty-foot-long solar wing motionless at shoulder height for ten minutes while standing on the RMS arm high above the shuttle, awaiting the signal to release it. When she gently let it go and the shuttle backed away, the solar array began to flutter like a bird flying. When she was identified as the astronaut, she was compared to Atlas, a Greek mythology titan who bore the weight of the heavens on his shoulders. In reality, she said that she was "not exerting one ounce of effort" in holding it, only trying to hold it steady without imparting any momentum that might cause a collision.[81] With twenty-one hours and ten minutes of EVA, Thornton became the leading woman spacewalker, a record that held for almost fifteen years until Sunita Williams surpassed her in 2007.

Thornton's last flight was STS-73 (1995), the second US Microgravity Laboratory Spacelab mission. She was payload commander for a varied group of experiments in materials science, combustion science, fluid physics, biotechnology, and other fields. The crew worked 24/7 in two teams for sixteen days, one of the longest shuttle missions. This was the only time Thornton flew with a crew that included another woman, astronaut Catherine "Cady" Coleman, although they worked on opposite shifts and probably saw more of each other in training than in orbit. Thornton described her other flights like camping trips with brothers.[82] Later she, Coleman, and three of the men appeared in a 1996 episode of the television show *Home Improvement*, "Fear of Flying."

Thornton became familiar to the media and public at the time of the Hubble Space Telescope servicing mission, which drew intense interest. She was profiled as an "astronaut-mother-cookie baker" or an "astronaut, spacewalker, nuclear physicist, wife, and mom," what some might have called a superwoman.[83] At the time, she had three young children under the age of ten living with her in Texas and a commuting husband who was a professor at the University of Virginia. She said that upon leaving the office each day, her focus was on her family, but her 7:00 a.m. to 7:00 p.m. work schedule meant she had to do a lot of mom tasks, like baking cookies, at 2:00 in the morning. About being an astronaut and mother, she remarked, "Leaving home each time is the hardest thing of all. After you've done that, strapping on a rocket is no big deal."

After twelve years of NASA and a long-distance marriage, Thornton was ready for a more regular life and less travel. She left the agency in 1996 and moved with the children to Virginia to live with her husband. She joined the faculty of the University of Virginia as a professor in the School of Engineering and Applied Sciences and in time became an associate dean of the graduate school and director of the Center for Science Education. After twenty-three years, she retired from the university in 2019. Over time Thornton served on many advisory, review, and study committees for NASA and the National Research Council. Looking back on her dual career, she has said of being an astronaut, "I can't imagine any other job that would hold my interest would be that much easier. . . . I can't imagine any job that [would] be more fun."[84] Thornton would like for people to know this about women astronauts: "They are astronauts, just like the guys, only we eat less, weigh less, and take up less space."[85]

Flights: **4** on space shuttle • **10th** US woman in space • **12th** woman worldwide • **40** days, **15** hours (**975** hours) • EVA time: **21** hours, **41** minutes

The Class of 1985

In 1985, faced with starting a new recruitment so soon after the previous one, NASA reconsidered its plan for annual recruitments and shifted to an ongoing acceptance of civilian applications for pilots and mission specialists. Selections would occur each spring or as needed to meet mission requirements or replenish the astronaut corps as members left or retired.[86] The military services would follow their usual annual nomination process. Even if selections were made annually, this rolling approach to handling applications in a continuous flow might be more efficient than screening a flood of thousands of applications at once on a tight schedule through a process not yet automated. For this selection cycle, NASA did not announce a new recruitment but instead reconsidered qualified finalists from 1984, from which thirteen candidates were selected—six pilots and seven mission specialists, two of them women, Linda Godwin and Tamara Jernigan, who were scientists.[87]

Linda Maxine Godwin

Linda Godwin doesn't recall being either encouraged or discouraged by anyone about studying physics.[88] In 1980 she was only the second woman to earn a PhD in physics from the University of Missouri. She had become interested in science during the early space program and simply followed her interest, never thinking of becoming an astronaut at a time when there was no path to space for a woman. That changed when NASA began to recruit women for the astronaut corps while she was in graduate school.

Godwin followed advice that she often gives others—always look for a way forward—when she applied three times to become an astronaut. The first time she had not completed her graduate work in physics and was not selected into the 1978 class. She soon finished her degree and worked on earning a private pilot's license. The next time she made it to the interview week but was offered a different NASA job. After working a few years as a payloads officer and flight controller, she applied again and was selected in 1985. She stayed at NASA for thirty years, during which she flew on four space shuttle missions, including to *Mir* and the International Space Station. She found her way forward to spend cumulatively thirty-eight days in space with ten hours of spacewalking. "I really wanted to do this and was willing to keep trying."[89]

The main purpose of Godwin's first mission, STS-37 (1991), was to deploy the massive Gamma Ray Observatory, the second of NASA's Great Observatories after the Hubble Space Telescope that launched the previous year. Godwin was the prime RMS operator, tasked with lifting it out of the payload bay for deployment. When one of its solar arrays did not unfurl, she positioned the observatory so the EVA team could open it manually. She also operated the arm to support a planned EVA to test space station assembly techniques with large structures. In addition, she and the crew conducted physics and chemistry experiments and amateur radio communications.

Godwin was payload commander on STS-59 (1994), the first Space Radar Laboratory mission. Three large radars and a carbon monoxide sensor in the payload bay operated to study Earth's surface and atmosphere. The crew worked in two shifts around the clock to ensure that the radar could image all the targets, which required many changes in orbiter attitude and management of the high-data-rate tape recorders.

Linda Godwin during the *Mir* EVA, 1996. *NASA*

Members of the crew did extensive Earth photography and prepared written narratives to complement the instrument data.

The third Shuttle-*Mir* docking mission, STS-76 (1996), delivered Shannon Lucid for her long stay and more than two tons of equipment and supplies. While the two spacecraft were docked, Godwin completed a six-hour EVA to mount debris and contamination detectors on the docking adapter attached to *Mir*. She was the third American woman spacewalker, following Kathy Sullivan and Kathy Thornton. Working inside the shuttle, she conducted biology experiments in the Biorack mini-lab and handled KidSat, a digital camera programmed from classrooms for imaging studies of Earth that was one of Sally Ride's educational projects after she left NASA.

Godwin's last mission, STS-108 (2001), brought her to the International Space Station as the primary robotic arm operator, loadmaster, and spacewalker. This mission delivered almost 5,000 pounds of supplies, equipment, and experiments to enable the

ISS residents to start research. The goods were packed in a transportable stowage unit, the *Raffaello* logistics module, that Godwin attached to the ISS to be unloaded, and her RMS partner later detached and returned it to the shuttle, using the robotic arm. She also served as loadmaster for these transfers. With another crewmate, Godwin did a four-hour EVA to install thermal blankets on the solar array rotational mechanisms, and she was responsible for monitoring the shuttle's ISS docking system. The mission also dropped off a new crew and brought the previous crew home. Godwin was one of five US women to visit both *Mir* and the International Space Station.[90]

Godwin was among the first women to move into management positions during her time at NASA. After serving as branch chief for CAPCOM and other functions, she became the first woman deputy chief of the Astronaut Office, a position that she held from 1993 to 2000. She later became deputy director of Flight Crew Operations and finally was an Astronaut Office representative to the Constellation program to work on requirements, design, and budget issues.

After four missions and about fifteen years in management roles, Godwin retired in 2010, as did her husband, fellow astronaut Steven Nagel. Taking another path forward, they joined the University of Missouri as faculty members, she as a physics and astronomy professor and he as an engineering professor. She told the university community, "My main goal is to draw students into physics and present them with options related to science and the scientific process, no matter what they go on to do. . . . Learning physics helps you understand how the whole world works."[91]

Flights: **4** on space shuttle, including visits to *Mir* and the International Space Station • **12th** US woman in space • **14th** woman worldwide • Time in space: **38** days, **6** hours (**918** hours) • EVA time: **10** hours, **14** minutes

Tamara Elizabeth Jernigan

On the younger end of the age range of the trailblazing women astronauts, Tammy Jernigan was twenty-five years old when selected as an astronaut candidate, by which time the first six were approaching or passing age forty. Despite that range, they shared many similar experiences during the 1960s and 1970s when women's status was in flux.

Jernigan had watched the Apollo missions to the Moon and grown interested in space and science, but she didn't consider becoming an astronaut until she was a nineteen-year-old undergraduate studying physics and playing varsity volleyball at Stanford University while NASA was recruiting the first female astronaut candidates. Aware that Sally Ride, a recent Stanford graduate, had been selected, Jernigan began to think she might have a chance. After completing her bachelor's degree in physics, she earned two master's degrees in engineering science and astronomy and took flying lessons while working at NASA's Ames Research Center in California. Jernigan applied for the astronaut corps in 1984 and 1985, was selected the second time, and then completed a doctorate in space physics and astronomy before her first mission.[92]

Although there were few girls in physics classes in her high school and university in the 1970s, Jernigan remembers how many people encouraged and mentored her, starting with her mother. Her math and science teachers, her volleyball coaches, and NASA engineers and scientists were good influences. She excelled in both academics and athletics, earning honors and awards in both. Jernigan noticed some instances where women were less welcome, but she "tended to focus on positive interactions and ignored insensitive comments and bad advice."[93] She took it in stride when a laudatory article about her appeared in the *Los Angeles Times* with such personal details as her height, weight, figure, eye color, hair style, and that she rarely wore any makeup but mascara.[94]

During training she had technical assignments in software development and verification, payloads, and mission development. She also served as CAPCOM for five shuttle missions in 1989 and 1990. She enjoyed being CAPCOM "because I always had my finger on the pulse of the mission."[95] Later Jernigan was deputy chief of the Astronaut Office and became increasingly involved in preparing for space station assembly, operations, and external maintenance.

Jernigan was assigned to five quite different missions during the 1990s. Her first was STS-40 (1991), the second Spacelab Life Sciences mission, dedicated to studying how humans, animals, and cells react to microgravity and readapt upon return to Earth. Crewmembers, rodents, and jellyfish were test subjects for investigations of all major bodily systems. In addition, there were secondary experiments in materials science and plant biology. Four of the seven crewmembers had MD degrees. STS-40 was the first mission with three women aboard: payload commander Rhea Seddon, guest payload specialist Millie Hughes-Fulford, and Jernigan, who was the flight engineer and one of the two crewmates prepared for contingency EVA. The crew worked in two teams around the clock; Jernigan's responsibilities focused on many of the orbiter systems plus photography and video work, more than research in the lab.

Her second mission was STS-52 (1992), called the US Microgravity Payload-1. It was an eclectic payload that included a satellite to be launched, experiments mounted in the payload bay, and various experiments in the crew cabin. One of Jernigan's primary responsibilities was the LAGEOS satellite, a sphere covered with tiny prisms to reflect laser beams fired from Earth. This laser ranging technique was used to measure Earth's shape, rotation, tectonic plates, and other studies in geodesy. Again, Jernigan was one of two contingency EVA members in addition to other duties.

The mission closest to her scientific interests was STS-67 (1995), the second dedicated astronomy mission called ASTRO-2, for which she served as payload commander in charge of meeting all the research objectives. Jernigan was already the first female professional astronomer to fly in space, but on this mission she was flying with a suite of telescopes and other instruments for ultraviolet observations of hot stars, faint objects,

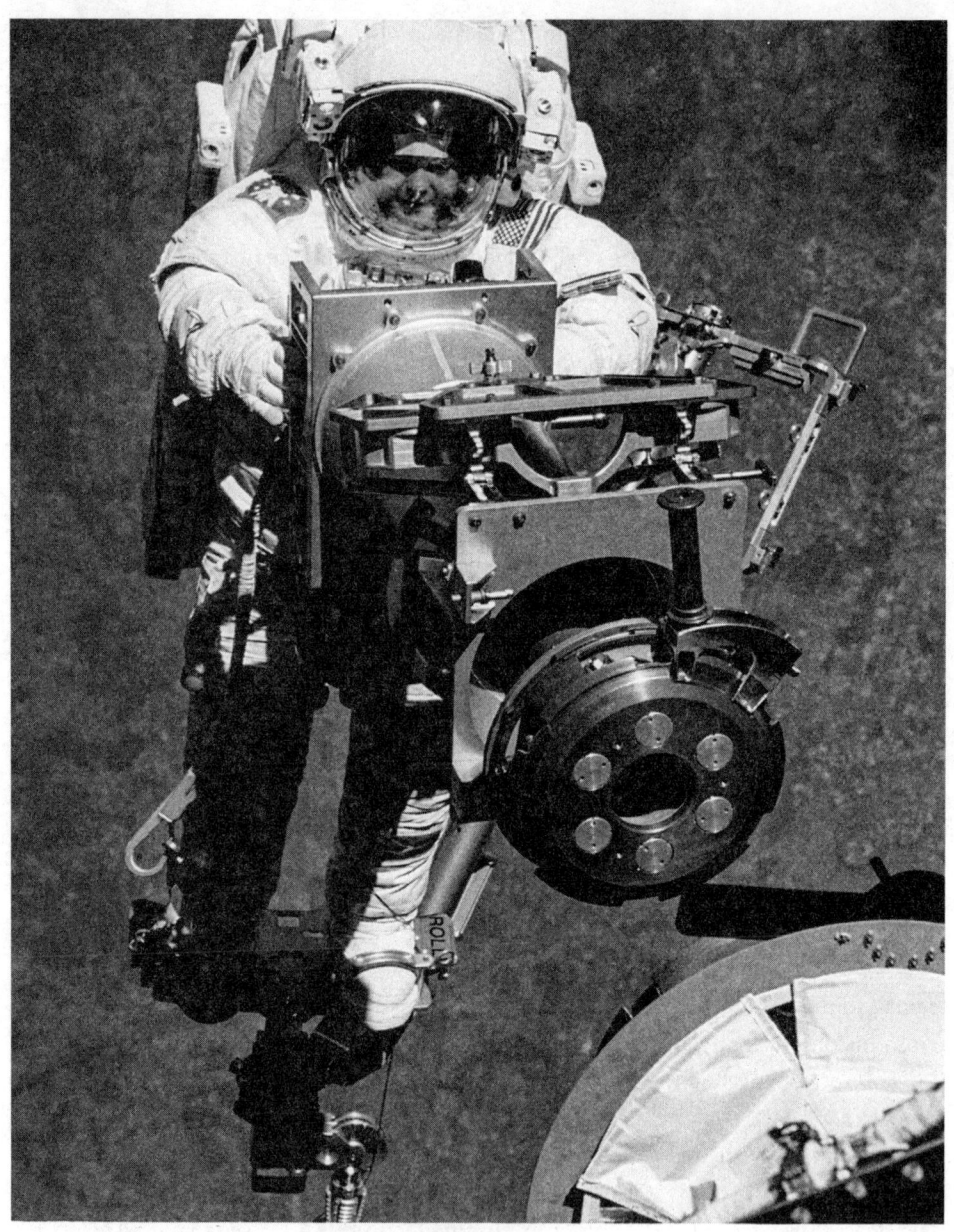

Tammy Jernigan in her almost eight-hour ISS EVA, 1999. *NASA*

and distant galaxies. She was ready for a contingency EVA and had other duties, but astronomy was the priority.

Jernigan's fourth mission, STS-80 (1996), set the record as the longest of the 135 shuttle missions: 17 days, 15 hours, 53 mins. The crew deployed and retrieved two free-flyer satellites that co-orbited with the shuttle. One, the Wake Shield Facility, was a physics experiment—a huge disc that produced a nearly perfect vacuum behind it as it moved through space. Experiments mounted on the backside of the shield tested the formation of thin-film materials for use in semiconductors, with a view toward their possible perfection. Jernigan had prime responsibility for the astronomical satellite ORFEUS. She and crewmate Tom Jones had been scheduled for two days of EVA to test equipment for space station tasks but were foiled by a jammed hatch on the airlock.

Jernigan's last mission, STS-96 (1999), was the second flight to the nascent International Space Station and the first docking. The ISS then consisted of two joined modules, one Russian and the other American. The next shuttle mission would bring the first resident crew, so the purpose of STS-96 was to get the place ready for occupancy. With Jernigan as payload commander and loadmaster, the seven-member crew delivered and stowed almost two tons of supplies and equipment—clothing, sleeping bags, water, spare parts, medical kits, anything the incoming ISS crew would need—and also configured the onboard systems for activation. After years of preparation to go outside, Jernigan finally had her chance to perform an EVA; she and Daniel Barry installed two cranes and positioned tool kits on the ISS exterior. For the second time, Jernigan was on a mission with two other women, now with Ellen Ochoa and Canadian astronaut Julie Payette. Ochoa operated the RMS arm while Payette was the EVA choreographer. Jernigan and Ochoa also worked closely together monitoring the trajectory, approach, and docking, and they were responsible for opening the ISS, checking the air pressure, and sampling the air before entering.

In 1999 Jernigan married fellow astronaut Peter J. K. "Jeff" Wisoff, also a physicist whom she had known since graduate school at Stanford. Both flew multiple missions between 1990 and 2000. They left NASA in 2001, she after twenty-one years, sixteen of them as an astronaut. They returned to California where both took positions at the Lawrence Livermore National Laboratory and expanded their family with two children.

Flights: **5** on space shuttle, including **1** to ISS • **13th** US woman in space • **16th** woman worldwide • Time in space: **63** days (**1,513** hours) • EVA time: **7** hours, **55** minutes

The Class of 1987

In line with its rolling application process for annual selections, NASA had expected to announce new astronaut candidates in early 1986, but the *Challenger* tragedy put that on hold; no one knew when flights of the grounded shuttle fleet might resume. Instead, NASA waited a year before adding more astronauts and announced this class in June 1987. Of almost 2,100 applicants, 117 were interviewed and fifteen were selected. Of twenty-three women interviewed, Jan Dozier (later Davis) and Mae Jemison were hired. The 1987 class of fifteen included seven pilots and eight mission specialists. Ten were military officers and five were civilians. Of those five, four were current NASA employees.[96]

A member of Congress noticed these demographics and questioned the selection criteria and process.[97] Selections were trending toward fewer candidates from universities and industry and more who were already working at NASA. The congressman, who served on the House committee that oversaw NASA, was concerned that the agency was encouraging applications broadly but selecting almost no one outside the military services and NASA. He and other critics thought that the selection process was becoming too narrow and inbred. The agency's administrator, James C. Fletcher, rebutted, "All qualified applicants are considered equally and those determined to be the best qualified are chosen." Still, the statistics were concerning enough that the House brought up the issue in the next round of budget legislation to alert NASA that it was under scrutiny to ensure a fair and balanced selection process.

Nancy Jan Smotherman Davis

Jan Davis started and ended her NASA career in Huntsville, Alabama, bracketing her time as an astronaut with engineering and management positions at NASA's Marshall Space Flight Center (MSFC). Her family moved there when she was a child in the early 1960s, during the booming years of the early space age amid testing of Saturn rockets and preparation for missions to the Moon. Davis never lost her early excitement about spaceflight and in 1979 returned to the rocket city to work as an aerospace engineer at MSFC. There she was involved in structural analysis of the Hubble Space Telescope and the telescope that became the Chandra X-ray Observatory. She earned her master's and doctoral degrees while working at NASA, and she became a scuba diver to work in the MSFC neutral buoyancy simulator with astronauts who were training for Hubble deployment and servicing EVAs. Along the way, she became a registered professional engineer and earned a patent.[98]

Although Davis experienced space history as it was happening during her youth in Huntsville, hearing the thunder of rocket test firings, going to school with children of the von Braun rocket team, and celebrating the Apollo 11 Moon landing in a community working to send astronauts into space, she did not aspire to be an astronaut until 1978. The first six women, born about the same time as she, were her inspiration; she carried a picture of them in her briefcase for years.[99] It wasn't just that women were entering the astronaut corps, but the new position of mission specialist beckoned her as an exciting opportunity. When she realized that she was on the right path for that job, she earned her pilot's license and completed a doctorate in mechanical engineering before applying in 1984 and again in 1985. After the solid rocket booster, a technology that MSFC managed, was determined to be the technical cause of the 1986 *Challenger* tragedy, Davis "didn't think there was any chance that anyone at Marshall would ever be selected" to be an astronaut.[100] She was working on redesign of a part of the booster for return-to-flight when she received the call to be interviewed in 1987. On her third attempt, she became the first Marshall employee selected for the astronaut corps.

Davis enjoyed everything about astronaut training and how the class drew together in friendship and camaraderie. She completed technical assignments in mission development, supported shuttle payloads, served as CAPCOM for seven missions, became the Astronaut Office representative for RMS robotics and then the chief of the payloads branch. She approached the robotic arm as if it were a flying skill requiring the same kind of spatial awareness and quick decision making as piloting.

In 1989, Davis was assigned to her first mission, STS-47 (1992). Another crewmember, Mark Lee, had been named the payload commander. After training together for more than a year, they married in 1991. The media misreported that their marriage was a secret, but Davis denies that, saying they notified the Astronaut Office in advance and that it was not a problem.[101] NASA allowed them to fly as planned; they were already assigned to opposite shifts on the 24/7 mission.[102] Davis and Lee became the first, and so far only, married couple in space.

Although it was designated STS-47, this was actually the fiftieth space shuttle mission flown since 1981. Called Spacelab J as a joint mission with Japan's space agency,

Jan Davis at the RMS control station in the shuttle, 1997. *NASA*

it was focused on biomedical research, life sciences, and materials processing in microgravity. The crew included a payload specialist from Japan and the first African American woman astronaut, Mae Jemison. Also aboard were frogs, fruit fly larvae, two fish, chicken embryos, and fungi. Davis was responsible for operating Spacelab systems and a variety of experiments. She also served as crew medical officer and a contingency EVA specialist on this and her next two missions. She noted that her small frame and short arms were not ideal for EVA, so she concentrated on robotics and payload work.[103]

Two years later Davis was on STS-60 (1994), the first of twelve Shuttle-*Mir* missions and the first time a Russian cosmonaut flew on a US spacecraft. This mission did not actually go to *Mir*; it was the preliminary flight to exercise joint US and Russian space operations. Davis was in charge of the RMS arm and planned to deploy and retrieve the Wake Shield Facility, a huge disk that created an ultra-high vacuum in its wake where perfect thin-film materials could be formed. However, the electronics on the shield misbehaved, preventing deployment, so Davis used the arm to maneuver its experiments around outside the payload bay. The crew carried out research in the middeck and the Spacehab experiments module, and they tested a technique for tracking orbital debris.

Davis was the payload commander on STS-85 (1997) and again the prime RMS operator. This time she deployed and retrieved a satellite for atmospheric study, and she also operated and evaluated a robotic arm that Japan's space agency was developing for use on the International Space Station. This twelve-day mission kept the crew focused on many experiments in various fields. "Looking at the Earth from space . . . was my favorite thing to do. But that was not our job," Davis recalled.[104]

After her last flight, Davis shifted into management. She went to NASA headquarters as the director of the Independent Assurance office for the Human Exploration and Development of Space. After that two-year stint, she returned to MSFC in 1999 as deputy director, and soon director, of Flight Projects, making her responsible for ISS elements and systems and the Payload Operations Center, a science-oriented version of the Mission Control Center. After the *Columbia* tragedy, she was named director of safety and mission assurance for all MSFC projects and personnel.

Upon retiring from NASA in 2005, Davis moved into private industry as an executive with Jacobs Engineering, a NASA contractor for engineering and other professional support services, and then Bastion Technologies, a NASA contractor for safety and mission assurance. After retiring a second time, she remains active in a number of organizations and boards, makes public appearances, consults, and in 2023 she published a book that is a dual memoir of her father's experiences as an aviator and hers as an astronaut.[105]

To encourage girls and young women to find their passion and to meet women leaders in technical careers, Davis and two astronaut partners, Susan Helms and Sandra Magnus, cofounded a nonprofit organization of role models, AstraFemina, in

2019. Davis is actively involved as an officer on the board and an inspirational STEM ambassador.

Flights: **3** on space shuttle • **14th** US woman astronaut in space • **19th** woman worldwide • Time in space: **28** days (**673** hours)

Mae Carol Jemison

Mae Jemison earned lifelong fame and a place in history as the first Black woman in space when she flew on the STS-47 Spacelab J mission in 1992. Jemison is one of several astronauts who cited Nichelle Nichols as an inspiration, proving that Nichols's work helping NASA recruit women and minorities for the 1978 astronaut selection paid off. During her flight, Jemison began her shifts by announcing, as Lieutenant Uhura would, "Huntsville, *Endeavour*. All hailing frequencies are open."[106]

Jemison's family moved from Alabama to Chicago in search of better opportunities when she was a young child. They were the first Black family in a suburban neighborhood, and she attended an integrated college-prep high school, where she excelled in science, math, and English. She always wanted to be a scientist and reach space someday. She also was passionate about many dance genres and later debated whether to become a scientist or professional dancer.

At Stanford University she majored in chemical engineering, where she says she was ignored as if there was no place for a Black woman in engineering. "I felt totally invisible," she remembered.[107] She was active in dance performance, choreography, and directing and had a second major in African and Afro-American Studies. After graduation Jemison earned her medical degree, became a general practitioner, joined the Peace Corps as a medical officer, served in Africa for two years managing a health care system and researching, and returned to California to practice medicine—all in the span of ten years.

On STS-47, Jemison had a new title as science mission specialist and served essentially as a payload specialist, which allowed her to concentrate on the heavily biological laboratory research without a load of general orbiter tasks. She and Jan Davis worked on the same shift conducting experiments on a collection of frogs, frog eggs, fish, chicken embryos, and fruit flies to evaluate how their systems or development reacted to microgravity and radiation. The crew themselves were subjects of several biomedical experiments, and there were also materials science experiments on board. Altogether the crew completed more than forty investigations during the mission.

Jemison flew only a single mission and resigned from NASA six months later in 1993, six years after her selection. She has not revealed much about her time there or what her experience in the astronaut culture was like, other than to thank NASA for opening many doors and giving her "the opportunity to make one of my dreams possible."[108] She had one of the shortest tenures as an astronaut, but being an astronaut was only part of her life. She had other goals in sight and wanted to apply her knowledge,

Mae Jemison suspended weightless inside Spacelab, 1992. *NASA*

skills, and energy in new ways.[109] Instead of returning to medicine, Jemison moved toward teaching, mentoring, and encouraging STEM education and wider involvement in space exploration. She established an educational foundation named for her school-teacher mother, worked on projects for underserved communities, traveled widely as a public speaker, and became involved in private space exploration initiatives.

In her memoir Jemison described herself as curious, intelligent, self-confident, assertive, outspoken, happy, willing to take stands and risks, competitive, hooked on science fiction, and always testing limits and challenging authority. No one was going to hold her back. Despite her disappointing experience as an engineering student, she said, "I knew who I was [and] who I intended to be. . . . I had as much a right as anyone to be there."[110] Jemison had always identified with science fiction's strong male characters rather than the supporting female characters until she discovered Madeleine L'Engle's women scientists and heroines and Octavia Butler's women of color. At home, she was immersed in Black history, music, and culture with pride in her heritage. Travel to and living in Africa made her realize how entwined history, religion, and culture are and how much needs to be done to address inequities, suffering, aggression, and violence.

Jemison coined a phrase in her memoir, "testimony to the possible," for all the inspirational influences on her life. She claimed that she always knew she would travel in space. Her story, like that of so many of the astronauts, is a testament to the power of personal ambition and perseverance to reach an almost impossible goal.

Flights: **1** on space shuttle • **15th** US woman astronaut in space • **20th** woman worldwide • Time in space: **7** days, **22** hours (**190** hours)

The Class of 1990

As NASA prepared for its sixth astronaut selection cycle of the space shuttle era, it was still trying to calibrate its handling of thousands of applications. After its decision in 1987 to accept applications continuously, the agency struggled with the seemingly endless workload. The manual process before computers took a full year. Training was another chokepoint; astronauts assigned to missions had priority use of simulators and facilities, which affected their availability for the stream of astronauts in training. There was too much pressure on limited resources.

As a result, in 1988, NASA announced that it was shifting from an annual to a biennial astronaut selection cycle, explaining that "the two-year process will moderate the demand on NASA resources required for candidate selection and training, while maintaining the manpower levels necessary to meet mission requirements."[111] The agency would continue to accept applications on an ongoing basis but go through the selection process every other year. As usual, the number of candidates selected would depend on flight rate, mission requirements, and astronaut attrition. The two-year selection cycle was the norm from 1990 to 2000.

The recruitment for 1990 was announced in late 1989.[112] For the first time, the announcement described the stages of the intensive review process and named some members of the Astronaut Selection Board to give applicants a better understanding of what they faced. The 1990 recruitment received nearly 2,500 applications; 1,945 met qualifications, 107 men and women were interviewed, and twenty-three candidates were selected.[113] Of these, seven were pilots and sixteen were mission specialists. Five women were selected, but something was noticeably different. For the first time, some of the women candidates came from the military services: Major Eileen Collins and Major Susan Helms from the US Air Force and Captain Nancy Sherlock (later Currie) from the US Army. For the first time as well, a woman—Collins—was selected to be a pilot astronaut. While other women astronauts had been private pilots, all were chosen as mission specialists. The last barrier to being a space pilot had just disappeared.

Astronaut candidates from the military services had a different process and status than civilians. They applied within their service, where they were vetted and rated. Their applications were then forwarded as nominations for NASA's consideration. If NASA selected them, they remained on active duty as military officers assigned to duty at NASA, as if they were "on loan" to the space agency, and they kept their military benefits. Their assignment might extend indefinitely, they might return to service voluntarily, or they might be pulled back by the service for a special need, such as a national emergency. In the early space shuttle era when the Department of Defense used the space shuttle for classified missions, only military astronauts flew as those crews.

The 1990 selection was a landmark. Women were now able to enter the astronaut corps on par with military officer men. The position of mission specialist had made it possible for women scientists and engineers to be selected competitively with men, but until 1990 all of the women were civilians. In contrast, many of the mission specialist men were military aviators and officers. Their rank had value in the military-model culture of the Astronaut Office, where, informally at least, military astronauts were of a higher caste—more often the leaders—than civilians and women. The arrival of Sherlock and Helms signaled a leveling in the mission specialist group. Both held officer rank, and their credentials were similar to those of the military men. Sherlock had taken the Army's path as a helicopter pilot, the highest a woman could advance in Army aviation at the time. Helms was a graduate of the Air Force Academy and Air Force test pilot school's flight test engineer program. For pilot astronauts, something else was needed to level the hierarchy: advancing in rank by flying the right kinds of aircraft, graduating from test pilot school, and having significant leadership experience. Collins was the first woman astronaut candidate to meet that standard, and Helms was very close. Collins, Helms, and Sherlock were the leading edge of a wave of military women aviators reaching elite levels, and they achieved that as soon as it was possible after the opportunities were opened to women. As these three women and others who followed entered the Astronaut Office, the division between military and civilian and the hierarchical tradition of military males at the top disappeared.

Whether or not NASA took into account the congressional inquiry about the 1987 selection, the 1990 class of astronauts was evenly balanced between twelve military and eleven civilian candidates. The other two women were civilians with PhDs—Ellen Ochoa, a scientist working for NASA, and Janice Voss, an engineer working in industry. Ochoa was the first woman of Hispanic descent to join the two Hispanic men already in the astronaut corps.[114]

Eileen Marie Collins

The first woman to pilot the space shuttle and the first woman to command space missions, Eileen Collins grew up in Elmira, New York, the "Soaring Capital of America," and loved to watch gliders and airplanes. From age nine, she devoured books about pilots and aviation. She wanted to fly and become an astronaut, having no doubts that a woman could do that job.[115] Her dream of flight was her secret refuge from a precarious family life shaped by alcoholism, parental separation, financial insecurity, food stamps, and public housing while undergoing speech therapy for stuttering, being bullied, and feeling that she didn't fit in. As she completed high school, Collins decided that she needed to take charge of her life and get away from home. She wanted to join the Air Force but realized she had been coasting academically and needed to challenge herself to do better and not be "that woman who failed."

After high school, Collins went first to the local community college and then to Syracuse University for its Air Force ROTC program. Upon graduating with a degree in mathematics and economics in 1978, she entered the Air Force and completed basic training. She was then selected into one of the first groups of women Air Force pilots for a test project to determine whether women could fly military aircraft and was assigned to Vance Air Force Base in Oklahoma, which had never trained "girl pilots." She completed that training successfully in 1979 and became a T-38 jet trainer instructor pilot, the second Air Force woman to do so and the only woman T-38 flight instructor at Vance for three years. Her next assignment was commander and instructor pilot for the C-141 transport aircraft, operating from Travis Air Force Base in California, for another three years. She had requested assignment to a fighter aircraft, but women were not yet allowed to fly combat planes, so she flew the big transport. During that time she flew troops and evacuated civilians during operations in Grenada in 1983, earning combat pay and a medal even though women were not yet officially allowed to fly in combat. She later said that flying and commanding the C-141 was great preparation for the shuttle; it too was a large, complex vehicle with a large crew, various kinds of payloads, and missions that demanded a high level of coordination.

Collins then spent three years teaching mathematics and being an instructor pilot at the US Air Force Academy before being admitted to Air Force Test Pilot School in 1989. A major at the time, she was the senior member, and thus the leader, of her class, the first woman ever in that role. Collins was the second woman test pilot to graduate from this school, and NASA immediately selected her for the astronaut corps. Her eleven-year trajectory as a military pilot was possible only because of policy changes in the 1970s that opened military pilot training, pilot assignments, and test pilot schools to women aviators. She came through the pipeline as soon as it was possible to do so.

After basic astronaut training, Collins had technical assignments in orbiter engineering support and as a Cape Crusader and CAPCOM. During her time at NASA, she also served the Astronaut Office as chief of two different branches and as its chief information officer. She flew on four space shuttle missions, twice as pilot and twice as commander. She was the first woman in both roles.

Collins's first two missions were part of the Shuttle-*Mir* series. STS-63 (1995) was a rendezvous-only, or near-*Mir*, mission to test the proximity maneuvers and communications systems for docking missions. The flight included Spacehab, packed with experiments, with Janice Voss in charge of the research agenda and a cosmonaut crewmember. Two years later, Collins piloted STS-84 (1997), a shuttle-*Mir* docking mission to transfer supplies and equipment and to exchange the US astronauts assigned to *Mir*. Cosmonaut Yelena Kondakova was a member of the shuttle crew.

Deployment of the Chandra X-ray Observatory was the focus of Collins's first flight as commander on STS-93 (1999). Cady Coleman was the crewmember

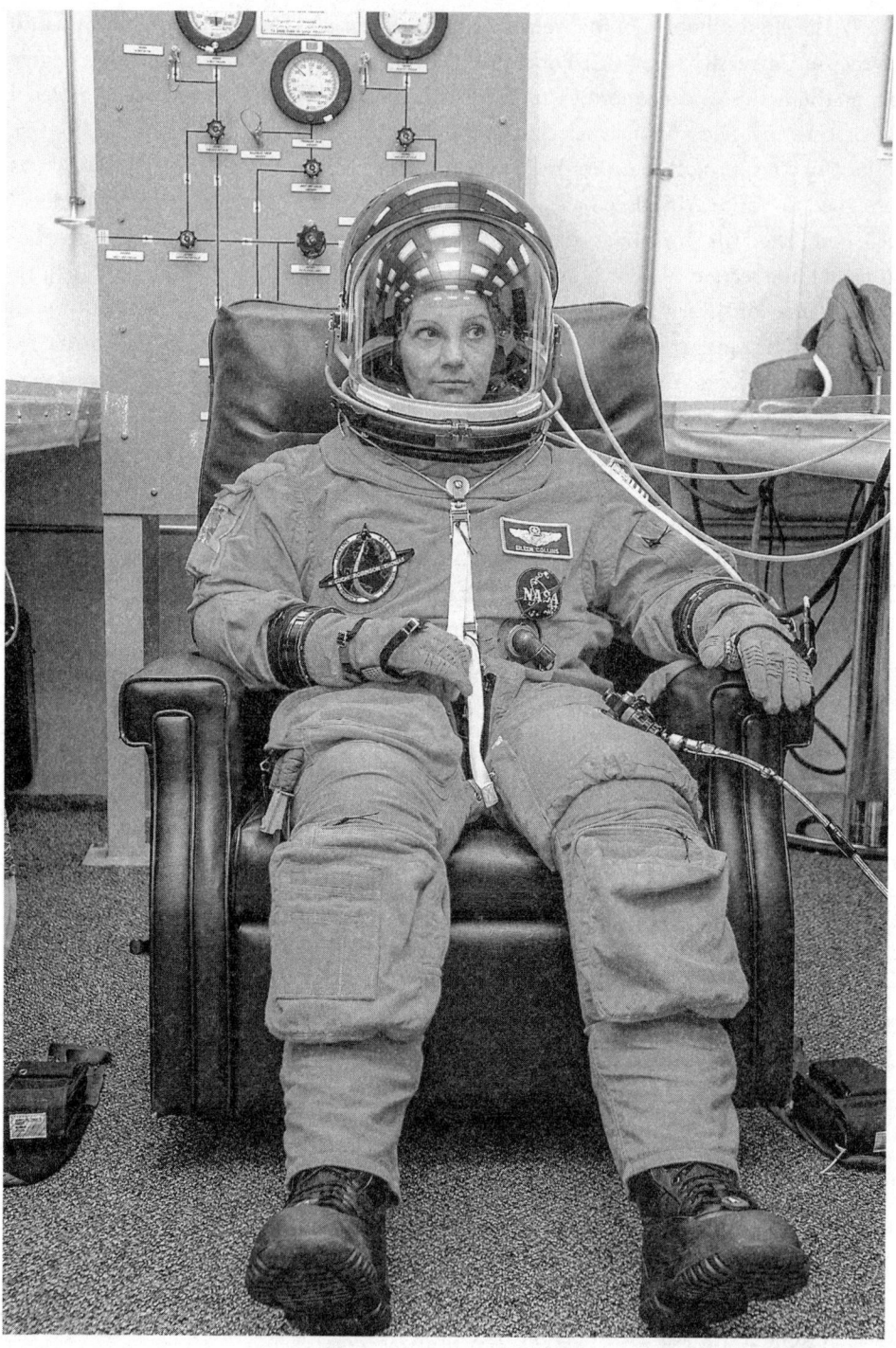

STS-114 mission commander Eileen Collins in the suit-up room before the space shuttle's return to flight, 2005. *NASA*

responsible for deployment. The shuttle also carried an interdisciplinary mix of life sciences, materials science, and remote-sensing experiments. The mission began dramatically with an electrical short and a hydrogen fuel leak in one of the main engines. Collins calmly maintained ascent and the shuttle safely reached orbit, albeit a few miles lower than intended.

Collins's next assignment, STS-114 (2005), was originally scheduled to follow the STS-107 *Columbia* mission in 2003, but it became the return-to-flight mission after *Columbia* disintegrated during its return. The shuttle fleet was grounded for two years to assess and remedy the causes of this tragedy. STS-114 would bring supplies and equipment to the International Space Station, and the crew would also test new safety enhancements for in-orbit inspection and repair of a damaged shuttle. Collins was the first person to perform a 360-degree rendezvous pitch maneuver, bringing the orbiter close to the International Space Station and doing a slow back flip to expose the belly and top of the vehicle to the ISS crew so they could observe and photograph any signs of damage. That inspection revealed two protruding insulation strips between tiles, which prompted an unprecedented EVA underneath the orbiter to remove them.

Collins's status as the first woman pilot and commander placed her in the spotlight for media attention and gave NASA opportunities to tout its progress in diversity. The agency treated her selection and flight as mission commander as high-profile events. First Lady Hillary Clinton and President Bill Clinton hosted the announcement at the White House during Women's History Month in 1998.[116] Both blended the themes of space exploration, education, equality, and inspiration and greeted Collins as an important role model. The president noted that although she didn't fit the exact mold of the original astronauts, "heroes come in every size and shape and gender . . . and she clearly embodies the essential qualities of all our astronauts, then and now." The presence of Sally Ride, Hillary Clinton, and other VIPs suggested that a symbolic torch was being passed from one "first woman" to another, and that Collins was now in an elite sorority.

Among the crowds at Collins's first launches as pilot and commander were women pilots from the early 1960s Lovelace Women in Space program. Collins had met them and learned their story shortly after becoming an astronaut. She found them inspirational and invited them to witness her launches. More than thirty years earlier they had hoped to fly in space; she finally achieved their piloting dream.

After four flights and sixteen years as an astronaut, Colonel Collins retired from the Air Force in 2005 and left NASA in 2006 to focus on family life with her airline pilot husband and their two children born while she was an astronaut.[117] Her crews had fondly nicknamed her Mom, and that is the job she wanted to do full-time at home. She continued to make public appearances and consult, and in 2021 she published her memoir, *Through the Glass Ceiling to the Stars: The Story of the First American Woman to Command a Space Mission.* Her book inspired a documentary film, *Spacewoman,* that premiered in November 2024.[118]

Always driven by her mantra "Focus, Focus, Focus," she now muses, "Someday I would like to go into space as a tourist and have the time to have fun."[119]

Flights: **4** on space shuttle, including **2** to *Mir* and **1** to ISS • **20th** US woman astronaut in space • **27th** woman worldwide • Time in space: **36** days, **8** hours (**872** hours)

Nancy Jane Currie

Nancy Currie flew her first mission as Nancy Sherlock. She flew her next three missions as Nancy Currie after her 1995 marriage to another Army aviator. She had the distinction of being the first woman from the US Army to become an astronaut, and she attained the rank of colonel. The daughter of a B-29 bombardier in the Army Air Corps during World War II, she wanted to fly for as long as she can remember. Her interest in flying grew from her father telling her older brothers about it. She recalls, "I was the one that kind of listened. I was the one that wore his flight jacket to school . . . and went to air shows with him."[120] By the time she was in high school, she was thinking about a career as a medevac helicopter pilot, although that wasn't yet an option for women.

As an undergraduate, Currie joined the Army ROTC and was commissioned upon graduation. She entered rotary wing pilot training and fell in love with helicopters. She then became an instructor pilot in the UH-1H "Huey" at the US Army Aviation Center and held various leadership positions at the platoon and brigade levels. At the same time that she was becoming a master army aviator, she earned a master's degree in safety engineering. Currie claims that some of her students asked her if she had considered applying to the astronaut program. The Army nominated her in 1985, but the 1986 selection was postponed. NASA invited her to interview for the 1987 class but then asked her to take another position. The Army cooperated and assigned her to the Johnson Space Center in 1987 as a flight simulation engineer working with the pilot astronauts and the shuttle training aircraft, a modified Gulfstream that handled like the orbiter. She reapplied to become an astronaut and was selected as a mission specialist in the next selection cycle.

Currie was the flight engineer and primary RMS operator on all of her four missions, the only woman and possibly the only astronaut with that consistent a record in two demanding positions. Her first mission was STS-57 (1993), to which she and classmate Janice Voss were both assigned. Currie operated the RMS arm to retrieve and stow for return a satellite covered with automated experiments and solar observing instruments, the European Retrievable Carrier (EURECA) that had been left in orbit several months prior. She also maneuvered the EVA crew through two spacewalks. The shuttle was outfitted with a new laboratory facility, Spacehab, that augmented the crew cabin with more workspace for research.

The STS-70 (1995) crew deployed the fifth of NASA's tracking and data relay satellites (TDRS). Positioned in geosynchronous orbit around the globe, these communications satellites kept NASA constantly in touch with the shuttle, the Hubble Space Telescope, and many other satellites. Through them it was always possible to send commands and receive continuous data. Currie spent much of the mission attending to orbiter tasks and also participating in biomedical and remote sensing experiments. Mary Ellen Weber was one of her crewmates.

As she was flying in the 1990s, Currie earned a doctorate in industrial engineering, worked as CAPCOM, was the astronauts' representative for safety and habitability equipment, and served as chief of both the robotics and payloads branches.

STS-88 (1998) made history as the first International Space Station assembly mission. The first two modules were connected to form the initial core of the ISS. On this mission, the crew delivered the US *Unity* node and mated it to the Russian *Zarya* module that was already waiting there. On later missions, a docking port and laboratory modules would be attached to the *Unity* node. Currie operated the RMS arm to mate *Unity* to *Zarya*, and she supported spacewalks to make electrical connections and position equipment needed for future EVAs. The mission crew entered the spartan space station and began activating its systems, readying it for occupation by its first crew. In retrospect, Currie felt that joining *Unity* and *Zarya* to create a space station is the most significant contribution she made to the space program.[121]

Currie's final mission, STS-109 (2002), was the fourth servicing visit to the Hubble Space Telescope. The crew had a heavy schedule of activities to upgrade the telescope with new components and install a new power system and an advanced camera the size of a telephone booth. Replacing the power system required that the telescope be turned off during the task, with the suspense of waiting to turn it back on and confirming that it worked properly. Again, Currie operated the arm to retrieve the telescope, mount it in the payload bay, maneuver the EVA crews around it for five days of work, and redeploy the telescope back into orbit. It was an intense mission for the RMS operator.

Not long after she returned home, the loss of *Columbia* in 2003 led to Currie's assignments in management. She was named at first to lead the space shuttle safety and mission assurance office, and then moved up into more senior leadership positions, culminating as deputy director of engineering at JSC and later as chief engineer of the NASA Engineering and Safety Center.

Currie retired from the Army in 2005 and from NASA in 2017 to join the faculty of Texas A&M University as a professor of engineering practice in industrial and systems engineering. There she teaches courses and publishes in human factors and ergonomics, systems and safety engineering, and human-robotic interaction, bringing her vast spaceflight experience to bear on educating the next generation of engineers. Reflecting on a career that she couldn't imagine as a girl, Currie says, "I was literally in the right place at the right time all the way through my career."[122] She also enjoys riding and competing on her horses and indulging other interests. Since 2015, she has been known as Nancy Currie-Gregg.

Flights: **4** on space shuttle • **19th** US woman astronaut in space • **24th** woman worldwide • Time in space: **48** days, **15** hours (**1,000** hours)

Susan Jane Helms

Susan Helms joined NASA as a captain, left as a colonel, returned to the Air Force, and retired as a lieutenant general. She always wanted to be in the Air Force like her pilot father, and she was a trailblazer in fulfilling that ambition.

Remarkably proficient in math, and envisioning a military career, Helms thanked her high school guidance counselor for suggesting she follow an engineering path.[123] She said she "locked in" on engineering in eighth grade, and by her senior year she had plans to attend a home-state university with strong engineering and ROTC programs so she could enter service upon graduation in 1976. At the same time, the all-male military service academies had just been ordered to admit women. Helms applied for, and was accepted into, the first US Air Force Academy class to include women cadets and was among the ninety-seven who graduated in 1980. She credited her father, who coached her through a military-style physical training regimen after she applied, for her ability to pass the rigorous entry exam. The timing was perfect to set her on course to achieve other firsts as a military woman. About that, she said, "I didn't set out trying to be first. . . . I just walked through those doors when the doors opened."[124]

Armed with an aeronautical engineering degree, Helms began her military career with a series of assignments during the 1980s, first with the Air Force Armament Laboratory in Florida to work on fighter weapon systems, then graduate school at Stanford University for a master's degree in aeronautics/astronautics, back at the academy as an assistant professor of aeronautics, on to the Air Force Test Pilot School in California for the flight test engineer program, and then in Canada as an exchange officer working as a flight test engineer in the Aerospace Engineering Test Establishment. One of the first women to complete the Air Force flight test engineer program, Helms was a distinguished graduate and was named the outstanding flight test engineer in her 1988 test pilot school class.

While she was in graduate school, Helms's ambitions began to shift after she attended a talk by Sally Ride that inspired her to consider becoming an astronaut. Seeing *The Dream Is Alive*, an IMAX documentary filmed in space by astronauts, convinced her that she wanted to be an astronaut. The test pilot school graduation speaker, an admired astronaut from the Air Force, hinted that she should apply. Her destiny beckoned. Selected as a mission specialist, she had various technical assignments after initial training, including in mission development and RMS/robotics.

Helms was the first of the three military women in the astronaut class of 1990 to fly, and she was assigned to five missions during her twelve years at NASA, four on the space shuttle and one on the International Space Station. She was the first woman to be an ISS resident, staying almost six months during Expedition 2 in 2001.

On her first flight, STS-54 (1993), the crew deployed another of NASA's tracking and communications satellites (TDRS). Helms was the crew lead for an X-ray spectro-

General Susan J. Helms, highest-ranking of the military women astronauts. *US Air Force*

meter operated from the payload bay to study the hot gas between stars and other astronomical sources of X-ray radiation. She worked with other payloads, served as crew medic, and managed photo and TV activities, and she coordinated her crewmates' spacewalk to test techniques for ISS EVA tasks. An accomplished pianist, she had approval to bring a small electronic keyboard on the flight for occasional musical interludes.[125] (When not in space, Helms played and was a lead vocalist in the all-astronaut rock band

Max-Q that performed for NASA events.) Upon returning from the mission, she served as CAPCOM while training for her next flight.

Helms was flight engineer and primary robotic arm operator on STS-64 (1994), a mission that included a space technology experiment with a LIDAR (laser imaging, detection, and ranging) system for studying the Earth's uppermost atmospheric regions. Helms deployed and retrieved a co-orbiting scientific satellite and participated in onboard experiments.

Her next flight, STS-78 (1996), was the Life and Microgravity Spacelab mission, lasting a then-record sixteen days while the crew carried out more than forty investigations in biomedical and materials sciences. The science payload represented all of NASA's international partner agencies as a model for the types of research and cooperation to be accomplished on the space station. Helms was both flight engineer and payload commander, as well as the contingency EVA specialist.

Helms then trained in Russia for an ISS expedition. Training focused on learning Russian language and the systems and hardware for the Russian sections of the space station (*Zarya* and *Zvezda*). In 2000, on the STS-101 mission, Helms and the two men assigned to the same expedition had a chance to become familiar with the ISS before staying there. The shuttle crew spent four days delivering and checking out ISS hardware. Her work was focused on the *Zarya* module that was outfitted with the utilities (power, air handling, attitude control, computer network, and other basic systems).

Before launching to the ISS for the six-month Expedition 2 in 2001, Helms shut down her life on Earth.[126] She gave away furnishings, put what she wanted to keep in storage, relinquished her apartment, froze her bank account, had only a post office box address, and entrusted friends to keep her car and jewelry safe. She moved to space and considered it home, as if she were on a military deployment. She felt free of all earthly responsibilities.

The three-person Expedition 2 crew, two Americans and one Russian, traveled to and from ISS on the space shuttle. Before they moved into the ISS, they and the shuttle crew completed installation tasks, adding a mating adapter for docking spacecraft and a robotic arm to the ISS. Helms coordinated one of the EVAs, and she and ISS partner James Voss set a spacewalk record duration of eight hours and fifty-six minutes from the shuttle. She did not do an EVA from the space station, but she tested the new robotic arm and used it to install the *Quest* airlock, performed maintenance tasks, and started science operations with medical and other experiments. The theme of their expedition was "Open for Business."[127]

In 2002, Helms decided to return to the Air Force and continue her military career. She had advanced in rank to colonel and was eligible for command and staff assignments. The first was as chief of space superiority at the US Air Force Space Command in Colorado. Steadily advancing in rank, she became a brigadier general and took command of the 45th Space Wing at Patrick Air Force Base in Florida, which manages the

Cape Canaveral launch range and supports all launches. Next, she became director of plans and policy at US Strategic Command at Offutt Air Force Base in Nebraska, as a major general. As a lieutenant general, she became the first woman commander at Vandenberg Air Force Base in California, where she commanded two organizations—the 14th Air Force in Air Force Space Command, and the Joint Command for Space in US Strategic Command. She was pleased that the Air Force assigned her to command roles that best put her spaceflight experience to use.[128]

Helms retired from the Air Force in 2014 after thirty-four years of service, twelve of them as an astronaut. She is one of six women who are veterans of five space missions, the most to date by women. Since then, she has been a consultant, public speaker, member of NASA's Aerospace Safety Advisory Panel, board member for other organizations, and cofounder, with astronaut colleagues Jan Davis and Sandra Magnus of AstraFemina, a nonprofit organization dedicated to encouraging girls' interest in STEM fields.

Of her experiences being a woman in new places and roles from the Air Force Academy to positions in military command, Helms learned what she calls "the credibility of competency."[129] Gender can't be used against you if you are as competent as, or more so than, your critics, who will usually desist when they see you can do the job well. "No one can argue that you shouldn't be there," she concluded. Her advice for women in any profession: cultivate competence to neutralize naysayers.

Flights: **4** on space shuttle, **1** on International Space Station • **16th** US woman astronaut in space • **21st** woman worldwide • Time in space: **210** days, **23** hours (**5,063** hours) • EVA time: **8** hours, **56** minutes

Ellen L. Ochoa

Ellen Ochoa and four men aboard the space shuttle orbiter *Discovery* were engaged in conducting experiments and tending to the vehicle when she took time to film a short segment for an educational video for young children. She lifted her silver flute to her lips, and melodies from Mozart and Vivaldi filled the crew cabin.[130] The astronaut, a classical flautist, brought her instrument to space and performed a miniconcert. These moments on her first mission suggest how Ochoa has balanced her two great interests—music and math—since she was young and why she is an ardent advocate of STEAM education: science, technology, engineering, the arts, and math.

Ochoa grew up near San Diego, California, not far from the border with Mexico, from where her father's parents came as immigrants. She recalls doing homework as a family while her mother took one college course at a time, modeling for her five children the importance and pleasure of learning. Ochoa graduated at the top of her class from high school and university and was inducted into two national honor societies, Phi Beta Kappa for the humanities and Sigma Xi for scientific research. Having performed in recitals, band, ensembles, and an orchestra, Ochoa briefly considered a career in music but decided on science instead, earning degrees in physics and electrical engineering.[131]

She chose to major in physics as an undergraduate because the advising professor was far more welcoming than the engineering professor who made it clear that he didn't want her in his department. "It's a really difficult course of study and I just don't think you'd be interested," he claimed. Little did he guess that years later she would share three patents for optical system inventions, receive national engineering awards, and lead one of the nation's premier engineering organizations. In the meantime, she made the highest grades in all eighteen physics classes. Ochoa remembered the lesson learned from this experience, "You don't want to listen to discouragement from someone who doesn't know you."[132]

Ochoa realized that astronaut might be a possible career when the first space shuttle mission occurred in 1981 while she was in graduate school. The prospect of flying in this new kind of spacecraft and doing research that could not be done on Earth enticed her. So, too, did Sally Ride's flight in 1983, and she decided to apply to be an astronaut upon completing her PhD in 1985.

When NASA reactivated the astronaut selection process in 1987, Ochoa was a finalist invited for the week of interviews and evaluations, but she was not selected. Instead of seeing this as a failure, she decided to improve her credentials by earning a pilot's license and working for NASA to support its mission. She took a new research position in intelligent systems technology at NASA's Ames Research Center and within six months became a manager. For the 1990 selection, NASA reviewed her updated application, and she was selected on her second attempt.

Ochoa flew on two pairs of missions between 1993 and 2002, during which she operated the RMS robotic arm, conducted science experiments, and twice was the flight engineer. The first two, STS-56 (1993) and STS-66 (1994), were Spacelab missions designated as ATLAS-2 and ATLAS-3, or Atmospheric Laboratory for Applications and Science. Both were dedicated to atmospheric and solar science investigations of the effect of both human and solar activity on Earth's atmosphere, specifically the creation and destruction of ozone. Among other duties, Ochoa deployed and retrieved a satellite that carried instruments to study solar wind, solar flares, and solar energy on the ATLAS-2 mission. On ATLAS-3 she retrieved another atmospheric research satellite. She was payload commander as well as primary RMS operator on the second flight.

Back in the Astronaut Office before her next pair of missions, Ochoa worked as CAPCOM and had a leading role in preparing for the International Space Station. She traveled to Russia and the other partner nations to negotiate how they would work together as an international team, how ISS expeditions would operate, how crews would be selected and trained, which languages would be used, and other issues. Negotiations with the Russians over levels of cooperation were challenging because, as NASA's international relations personnel put it, "We wanted to get married, and they wanted to live next door."[133] Reconciling such different versions of a relationship exercised everyone's

Ellen Ochoa floating upside down beside the Volatile Removal Assembly Flight Experiment located in the Spacehab module on *Discovery*, 1999. *NASA*

negotiation skills. Ochoa also led the effort to prepare the astronaut corps for continuous 24/7 support in the mission control center, as permanent ISS occupancy would require.

Ochoa's next two missions, STS-96 (1999) and STS-110 (2002), were to the International Space Station. She was flight engineer and prime RMS operator on both, as well as loadmaster on the first. STS-96 was the first docking with the vacant two-module ISS core. The shuttle brought tons of equipment and supplies for the first resident crew to set up housekeeping there, and the shuttle crew readied the ISS for their upcoming arrival. On STS-110, the crew delivered and installed the central section of the long truss system to which solar arrays, radiators, and other equipment would be attached. Ochoa shared space station robotic arm duty with other crewmates during four EVAs and equipment handling. This mission was the first time all EVAs were carried out from the ISS rather than the shuttle and the first time the ISS arm was used to maneuver the spacewalkers.

Having completed four successful flights in her twelve years as an astronaut, Ochoa moved into management, first as the deputy director of the Flight Crew Operations Directorate and then in 2006 becoming the first woman director of that unit, which oversees the Astronaut Office and all aircraft and spacecraft flight operations. As deputy, she was on duty in the mission control center when news of the destruction of *Columbia* came in and immediately set her part of the response in motion.

In 2007, Ochoa was promoted to deputy director of the Johnson Space Center as NASA was planning the transition from shuttle to post-shuttle programs. After six years as deputy, she became the JSC director, a position she held from 2013 until retiring in 2018. She was the first Hispanic and second woman (Carolyn Huntoon was the first) to lead the center. Her tenure spanned three presidencies with shifting priorities for the space agency, and she pursued an agenda for innovation, inclusion, and a more streamlined organization.[134]

Now retired after thirty years at NASA, Ochoa remains passionate about encouraging students, especially girls and children of color, in STEAM studies. Since retiring, she has published bilingual children's books on STEAM topics in a series titled *Dr. Ochoa's Stellar World*. Ochoa considers it a great honor that at least seven schools have been named for her. She credits the women's and civil rights movements for opening federal jobs to women and minorities; the development of the space shuttle that could carry more astronauts and support doing research in space; and NASA's decision to diversify the astronaut corps with women, people of color, and more scientists and engineers—itself a consequence of the first change.[135] It is not uncommon to see "trailblazer" in articles about her or in award citations.

Flights: **4** on space shuttle, **2** to International Space Station • **17th** US woman astronaut in space • **22nd** woman worldwide • Time in space: **40** days, **19** hours (**979** hours)

Janice Elaine Voss

According to one of her mission commanders, Janice Voss was "born to fly in space." He added, "She's a commander's dream, will take any job and do it well, do it outstandingly. She was always happy, always smiling . . . and always doing her job perfectly well."[136]

Voss applied four times to be an astronaut and proved herself as one of only six women to fly on five space shuttle missions, all between 1993 and 2000. She had decided to become a scientist after reading Madeleine L'Engle's *A Wrinkle in Time* while in elementary school.[137] This book triggered her appetite for science fiction and problem-solving and gave her a sense that women could do anything. Years later, Voss corresponded with L'Engle, read from a copy of the book in space, and sent it to the author.

An academic prodigy, Voss leaped from kindergarten to third grade, graduated from high school at fifteen, and enrolled in engineering science at Purdue University as a sixteen-year-old.[138] She started working for NASA at age sixteen, alternating a semester of classes with a semester of work to gain on-the-job experience as a young engineer. Although Voss was younger than the women selected in the 1978 astronaut class, she was already working at NASA five years before they arrived and was an instructor in guidance and navigation when they were in training. While applying for the astronaut corps herself, she also earned master's and doctoral degrees in engineering. She then spent almost three years working in flight operations for Orbital Sciences Corporation before being selected by NASA in 1990.

Voss flew at two-to-three-year intervals during the 1990s, starting with STS-57 (1993). She usually had primary responsibilities for science payloads and robotic arm operations. On this mission, she shared duties with fellow mission specialist, Nancy Currie, in capturing for return the European Retrievable Carrier (EURECA) scientific satellite and carrying out a variety of experiments in biomedical and material sciences in the commercial Spacehab laboratory unit.

Her next mission, STS-63 (1995), also carried Spacehab experiments for Voss's attention, and she deployed and retrieved a co-orbiting scientific satellite using the robotic arm. The main purpose of the mission was to rendezvous with *Mir* and practice procedures and maneuvers for the upcoming missions to dock and exchange crews with the Russian space station.

Voss had a unique opportunity to fly twice in 1997, on STS-83 and STS-94, known as the Spacelab Microgravity Science Laboratory. Three days after launch, the mission was cut short by a fuel cell power-generation problem and had to return early. NASA scheduled a reflight three months later, and the entire payload and crew returned to space—the only time such a repeat flight occurred. Susan Still was pilot, and Voss was payload commander in charge of the ambitious life and materials sciences payload. The lab included a greenhouse for plant studies and a chamber for setting small fires to study combustion in microgravity.

Her fifth and final flight—the first shuttle flight of the 2000s, STS-99 (2000)—was a radar mapping mission to record the topography of Earth's land mass, especially remote and cloudy areas that had not been well mapped. The Shuttle Radar Topography Mapping instrument extended on a two-hundred-foot mast from the payload bay and required two-shift operations to scan and record data for more than forty-seven million square miles in high-resolution 3D images. The work of payload commander Voss, Janet Kavandi, and other crewmembers led to the most accurate and comprehensive digital topographic map of planet Earth.

After this mission, Voss resumed technical assignments within the Astronaut Office, focusing on science payloads and robotics. From 2005 to 2007, she worked at NASA's Ames Research Center as science director for the exoplanet-hunting Kepler Space Observatory, launched in 2009. Upon returning to Johnson Space Center she became chief for ISS payloads and supervised training of ISS crews to conduct experiments in space. Her colleagues said she wrote the best procedures, so clear and precise that anyone could follow the directions, and she was determined to make everything better.[139]

Voss succumbed to cancer in 2012 at age fifty-five. In her honor and memory as the first Purdue woman to become an astronaut, an impressive sculptural scale model of the solar system—the VOSS Model—stands on the Purdue campus. She once noted that "in an era when no one knows who the astronauts are as individuals—unless I'm wearing my flight suit, most people don't recognize me—everyone still wants to hear our story."[140] Voss donated materials from her childhood through her career to the Purdue library and archives so that others might find inspiration, saying, "If I can do it, then so can they."

In 2014, Orbital Sciences Corporation sent a Cygnus cargo spacecraft named in her honor to the International Space Station.

Flights: **5** on space shuttle, including **1** to *Mir* • **18th** US woman astronaut in space • **23rd** woman worldwide • Time in space: **49** days, **4** hours (**1,180** hours)

Other Trailblazing Women in Space, 1978–1992

Two women who were not NASA career astronauts flew during the same period as the trailblazers. Opportunities were opening through other nations' space agencies and through a NASA guest astronaut position called payload specialist that enabled a representative of space agency partners and customers with payloads on the shuttle to participate in the flight and oversee their interests. Typically these were scientists and engineers who might have made the astronaut selection cut. Most payload specialists underwent some basic training with NASA, while more technical specialists trained intensely with the crew to coordinate everyone's research responsibilities and aim for maximum research success.[141]

Many shuttle missions had payload specialist crewmembers. Millie Elizabeth Hughes-Fulford was the first woman to serve in this position on the first Spacelab Life Sciences mission (STS-40) in 1991, as a representative of the Veterans Administration medical center where she was a medical researcher. That was the first mission with three women aboard: Hughes-Fulford, Rhea Seddon, and Tammy Jernigan. Hughes-Fulford was a finalist in the 1977–1978 astronaut recruitment, and she provided several experiments for shuttle and space station missions.

Another woman flew on the shuttle as a spaceflight participant, a unique designation. She was the "teacher-in-space," Sharon Christa McAuliffe, who was competitively chosen through a public project sponsored by NASA. McAuliffe was to conduct classroom lessons and keep a journal while in orbit, but she perished during the final ascent of *Challenger* and never reached space.

Three other women might have flown on the shuttle during this period or later in the 1990s had plans not changed. The Department of Defense organized a cadre of thirty-two Air Force officers as manned spaceflight engineers (MSE) to serve as payload specialists on classified military missions. Three of the MSEs named in 1985 were women: Maureen Cecil Gaudette LaComb, Katherine Eileen Sparks Roberts, and Theresa Mary Stevens Tittle. Only two men among the Defense department's intended astronauts ever flew.[142]

Among the trailblazing women astronauts, more than half expressed in their own words a version of what Ellen Ochoa said concisely: "I never considered being an astronaut as an option because when I was growing up there were no female astronauts. It wasn't until the first six female astronauts were selected in 1978 that women could even think of it as a possible career path."[143] The absence of women in the early astronaut corps stifled the imagination, self-image, and ambition of many girls and young women. In contrast, some of the young trailblazers thought it inevitable that women would go into space, so they deliberately prepared for that future. The journeys of this first generation of women who became astronauts are stories of conviction, resilience, and perseverance that inspired the next generation of girls and young women to plot their own paths to space.

The Inspired Generation: Astronauts for Space Shuttle and Space Station Missions, 1992–2000

After the twenty trailblazing women astronauts set many firsts, much still remained for the next generation of women to achieve. Women born in the 1960s and becoming adults in the 1980s knew that they could become astronauts, because they had seen women do the job. Born while the Baby Boomer generation was in high school and college, they were the leading edge of Generation X, the cohort with birth years from 1965 to 1979.

As they were maturing, education and career paths in science, engineering, and military aviation were open to women. Title IX of the Civil Rights Act barring discrimination in education, which effectively required equal athletic programs for women and men, was passed in 1972, and in 1973 the NCAA began to offer scholarships to women. Collegiate tennis star Sally Ride and other astronauts in the first generation had not received such support. Military academies began to admit women in 1976, and the first female academy graduate to become an astronaut was already at NASA. Women had ready access to birth control and, since 1973, abortion to manage their reproductive life. College enrollments soared as more women pursued their education, and more women and mothers were entering the workforce.

Knowing that it was possible to become an astronaut, this second wave of incoming women had the benefit of precedent and could tailor their preparatory experience more directly to that goal in ways that few of the first women applicants could have foreseen, giving some of them credentials from a longer work history. With thousands of applicants for tens of astronaut slots, the odds of selection remained daunting, but seeing the successes of the first women astronauts spurred many others to apply. Twenty-three more women were selected from 1992 to 2000, including women of African American and Hispanic heritage, bringing additional diversity into the astronaut corps.

Whether they were inspired by the Apollo astronauts or by the women astronauts who flew on space shuttle missions during the 1980s, by *Star Trek* or *The Right Stuff*, or by their own skills and experiences to think about a career in space, the women of the 1990s generation were on a mission to achieve their ambitions and met less resistance and fewer obstacles than some of their predecessors. Their confidence and abilities were evident at every step along the way. This group of women entered the astronaut corps with nothing to prove; they had already met the highest standards and established themselves in technical careers. They were ready to embark on spaceflight on the space shuttle and space station.

In the realm of public service, women who were elected or appointed to important positions in government became visible as role models of achievement. Among the noteworthy during the 1970s and 1980s were the first woman to be elected as a governor of a state; a number of women selected to cabinet level positions; and the first woman to serve as an associate justice of the Supreme Court. The year when women astronauts of the inspired generation were first selected—1992—was called "The Year of the Woman," when a record number of women won state and national elections.[1] More barrier-breaking women also became visible in elite athletics, business leadership, entertainment, and other fields.

Politically, this was the era of four presidencies: Nixon, Ford, Carter, and Reagan. The Soviet Union collapsed and the Cold War ended, bringing about a warming of relations with the new Russian Federation. Meanwhile, in the US military services, the first women rose in rank to rear admiral, brigadier general, and major general. Finally, the first woman pilot and captain were hired by US airlines.

And the US Treasury issued the first currency depicting a woman: the Susan B. Anthony one dollar coin honoring the famous women's suffrage leader. The reverse design depicted the American eagle landing on the Moon based on the Apollo 11 insignia, an odd mélange of imagery, since women's suffrage and space exploration had no connection—except, perhaps, that of the steadily growing number of women in the astronaut corps.

Spaceflight Evolution in the 1990s

The purpose of the US human spaceflight program in the 1980s was to make the space shuttle operational and demonstrate routine spaceflight for practical benefits and advancement of scientific research, commercial enterprise, national security, and technology development. Space was a new workplace for commuting astronauts. As the shuttle fleet grew to four orbiters, the flight rate increased from two to three, then four to five missions a year, peaking at eight missions in 1985 with an even more ambitious 1986 flight plan. Mission crew size also increased from two to four, then five and seven members, which meant multiple opportunities to fly. Some astronauts had already logged two missions before the 1986 *Challenger* tragedy grounded the shuttle.

When flights resumed in late 1988, priority went to the backlog of missions on hold—deploying communications satellites and time-critical planetary spacecraft and carrying out the remaining Department of Defense missions. In 1989, the flight rate was back up to five missions and in 1990 to six missions, and for the rest of the decade a brisk schedule held with a peak of eight missions in 1992 and 1997. During the busy 1990s, some astronauts flew every two to three years as missions launched almost every other month, typically with five or seven crewmembers each. Eventually most astronauts were not reassigned after flying four times to ensure that everyone in the expanding astronaut corps had opportunities to fly at least once or twice.

The purpose of the US human spaceflight program gradually shifted in emphasis during the 1990s to a more strategic series of missions related to a long-desired effort: a space station. As soon as the shuttle began flying, NASA launched a campaign to win White House and Congressional support to develop a space station as "the next logical step."[2] The shuttle needed a place to shuttle to and from, and research would benefit from longer durations in space than one-to-two-week flights. If longer-range goals of the space program were to expand human presence elsewhere in the solar system—the Moon and Mars, for starters—longer stays in space near Earth would yield the knowledge and techniques for maintaining astronaut health, well-being, productivity, and efficiency. A space station would enable a permanent human presence in space, something NASA considered to be "the key to the future."[3]

NASA succeeded in persuading President Ronald Reagan and enough of his advisors that a space station was crucial to US leadership in space and opening space as a place for commercial activity, such as research and manufacturing. In his 1984 State of the Union address, Reagan directed NASA to develop a permanently manned space station. Congress, the media, and pundits were unconvinced, but the process of developing a space station concept design began. As often happens, planners were reaching for the stars, and the concept grew larger and more complex in an attempt to accommodate all potential uses, and thus more expensive. By 1990 the concept then known as space station *Freedom* had become so politically and economically unwieldy that the program suffered widespread criticism, funding cuts and threats of cancellation. Congress was insistent that it be scaled back, and NASA was forced again and again to trim its vision to a more affordable design in response to more than twenty votes in the House and Senate between 1990 and 1993. When President Bill Clinton entered office, he ordered NASA to downsize the space station again, reconfigure it to include Russian elements in the architecture, and embrace the Russian Space Agency as a new partner in the program. Serial redesigning and cost escalation had to stop. Clinton asked NASA to develop three simpler concept options; he chose one, and the concept survived a congressional move to terminate the station by a single vote. Years of crisis ended on that note, and what came to be known as the International Space Station (ISS) dominated spaceflight planning and many of the shuttle missions for the rest of the 1990s.

Bringing the Russians into the space station program meant more than integrating their technologies into design and assembly. It also meant learning to operate together after decades of rivalry, suspicion, secrecy, and political and economic instabilities. Cooperation in space meant overcoming language barriers, different cultural traditions, and different styles of engineering and management. It meant sharing responsibilities and building trust relationships between cosmonauts and astronauts, NASA officials and their Russian counterparts. To build a foundation for this new cooperation, the United States and Russia and their space agencies agreed to a series of joint missions to be flown in the mid- to late 1990s. These would involve Soyuz and shuttle flights to the Russian *Mir* space station to work out and practice the myriad protocols of spaceflight—hardware and software design, training, in-orbit rendezvous and docking, mission control, decision making, crew exchanges, division of labor, and more.

These Shuttle-*Mir* missions, as they were called, were considered to be Phase I of the International Space Station program. Twelve such missions were flown from 1994 through 1998, all but three of them with docking for crew exchange and transfer of supplies and equipment. Seven US astronauts completed long-duration stays on *Mir* as the American member of a three-person crew, and eight cosmonauts were crewmembers on the space shuttle. One astronaut rode in Soyuz to *Mir*. The biggest impact on the astronauts was the requirement to study Russian language and train on Russian hardware and procedures at the cosmonaut training center in Star City near Moscow. Not every astronaut was able to make that lengthy commitment, which continues to this day. So much was learned through these sojourns and missions, sometimes with difficult negotiations and diplomacy, that the transition to operational partnership on the International Space Station was eased.

The expansion of NASA's human spaceflight program to include designing, developing, assembling, and staffing the International Space Station and executing the Shuttle-*Mir* missions prompted a steady growth in the astronaut corps to handle the looming workload foreseeable into the first decade of the new century. Throughout the 1990s, NASA held biennial recruitments and selections, and the size of the new classes topped out at thirty-five candidates in 1996. The astronaut corps reached its maximum size of 150 in 2000. In preparation for the ISS, astronaut recruitment in the 1990s gradually shifted toward an emphasis on people with the right skills to assemble the station and serve as long-duration crewmembers in the 2000s. The new astronaut classes included men and women with notable experience and leadership in the military services who were accustomed to deployments away from home, people who had participated in research expeditions, and people already working with NASA on projects related to spaceflight. The Astronaut Office itself underwent some changes as it grew.

New rules and policies emerged to manage the larger corps more efficiently. Although the chief remained a pilot, it became more common for mission specialists to serve as deputy to balance the management role.

By the time the fourteenth class of candidates was selected in 1992, every woman selected before the *Challenger* accident (classes eight through eleven) had flown on at least one mission; six had already flown twice, and Shannon Lucid and Kathy Sullivan flew for the third time in 1992. Eight women would fly in 1992, including women of the twelfth class, while members of the thirteenth class and others were training for missions that would fly in 1993 and 1994. With frequent flight opportunities, it was a good time to be a woman in the astronaut corps. After more than ten years of women in the Astronaut Office, men were accustomed to working with them as peers, and women were serving as branch chiefs and deputy to the chief astronaut.[4] Camaraderie was strong, and while there would always be competition for flight assignments, spacewalks, and soon for missions to the ISS, the newcomers from 1992 on found the Astronaut Office a friendly and exciting place to work.

The Class of 1992

In keeping with NASA's plan to seek astronaut candidates in two-year cycles, a new recruitment was announced in mid-1991 for a 1992 selection.[5] This announcement stated that "NASA is particularly interested in individuals with backgrounds in medical sciences research, microgravity research, and materials processing." This was a nod to upcoming Spacelab scientific research missions in those disciplines, which themselves were preludes to research -planned for the International Space Station. The Astronaut Office was staffing up to fly five or more missions per year. Although the announcement said that the number of selections would be limited to match projected requirements, NASA chose nineteen American candidates from 2,054 qualified applicants.[6] There were only four pilots but fifteen mission specialist candidates. Among them were three women, two of them military officers and one a civilian PhD: Catherine "Cady" Coleman, Wendy Lawrence, and Mary Ellen Weber.[7] Five astronauts from the international partner agencies joined this group, making a total class of twenty-four, one of the largest.

Catherine Grace Coleman

Cady Coleman keeps expanding her list of titles and roles. She has been a chemist, engineer, pilot, astronaut, wife, mother, daughter, sister, speaker, podcaster, space news analyst and commentator, consultant, advisor, a university's resident global explorer, STEM/STEAM advocate, musician, film star, and author. Having retired from NASA in 2016 after twenty-four years as an astronaut and from the Air Force as a colonel in 2009, she is committed to sharing her experiences and lessons learned from spaceflight with audiences around the world and in all media. "Energized" seems to be her middle name.

Coleman never thought of being an astronaut until she was a student at MIT in 1978–1983 and Sally Ride came to campus to speak. Coleman had never seen a woman astronaut in person. The two met, and Coleman called it her turning point, when she began to think, "Maybe that could be *me*."[8] She was already on a path toward earning her science credentials in chemistry, and she was enrolled in the Air Force ROTC. She also was a competitive athlete on the crew team. She decided then to try to become an astronaut.

Coleman grew up with exploration as part of her family's life, so it was natural for her to be drawn to adventure.[9] Her father was a US Navy deep sea diver who was involved in the Navy's SEALAB program in the 1960s, designing and establishing underwater stations where scientists and engineers lived and carried out research missions. Through his work, Coleman was familiar with hazards, life support technologies, biomedical fitness, isolation and remoteness, and curiosity as factors in exploration. Although she loved chemistry, space exploration beckoned.

After earning her bachelor's degree in 1983, Coleman went to graduate school to earn a doctoral degree in polymer science and engineering. She also entered active duty in the Air Force and served as a research chemist in the Materials Laboratory at Wright-Patterson Air Force Base in Ohio. During that time, she became involved in analyzing materials samples from NASA's Long-Duration Exposure Facility (LDEF) satellite that had been retrieved by the space shuttle after being stranded in space for five years. The goal was to study how materials withstood the harsh environment of space. She also set several endurance and tolerance records in the centrifuge at the Armstrong Aeromedical Laboratory as a medical subject for experiments in physiology and technology. Coleman's next step was to apply to NASA, almost ten years after first having thought to do so.

Selected on her first attempt, Coleman completed basic astronaut candidate training and then began a series of assignments through the Astronaut Office mission support branch, payloads and habitability branch, and communications liaison (CAPCOM) branch. She worked on habitability issues in space station design, served as the Astronaut Office branch chief for robotics overseeing hardware development and astronaut robotics training, and led supply ship operations integrating the activities of NASA's international and commercial partners. In her last position at NASA, she worked in the Office of the Chief Technologist at NASA headquarters, focusing on innovation in public-private partnerships.

Coleman's first of two space shuttle missions, STS-73 (1995), was the second US Microgravity Laboratory flight, a Spacelab precursor to scientific research on the space station. As a mission specialist, she was in her element working in a lab stocked with materials science, combustion science, crystal growth, fluid physics, and biotechnology experiments. She and payload commander Kathy Thornton shared science duties and responsibilities for the Spacelab systems on opposite shifts. Coleman was a contingency

EVA crewmember should a spacewalk have been necessary. Lasting almost sixteen days, this was the longest shuttle mission at the time. Her second mission was STS-93 (1999), commanded by Eileen Collins. As lead mission specialist, Coleman was in charge of deploying the Chandra X-Ray Observatory, which joined the Hubble Space Telescope and NASA's other "Great Observatories" in orbit to study the universe across the electromagnetic spectrum. She was ready again for a contingency spacewalk, but the need did not arise.

Coleman then trained as a backup crewmember for ISS Expeditions 19, 20, 21, 24, and 25 and completed the eleven-day NEEMO-7 undersea mission and a ten-week Antarctic meteorite-collecting expedition. Her third and final space mission was a six-month stay on the International Space Station (Expeditions 26/27) as a flight engineer in 2010–2011. She served as the lead robotics and science officer for experiment operations and space station maintenance, and she was the lead robotic arm operator for the second-ever capture and berthing of an automated supply ship (Japanese HTV2) upon its arrival. While she was on board, the ISS was a busy traffic hub, with five automated supply ships and two shuttle mission crews coming and going. She traveled to and from the ISS via Russian Soyuz flights and welcomed the last crews of *Discovery* and *Endeavour* (STS-133 and STS-134) to the station before the space shuttle program ended.

To be assigned to fly on the ISS, Coleman had to qualify in the medium-size EVA suit after the small size was discontinued.[10] She became the smallest person certified for extravehicular activity on the ISS, but no spacewalks were scheduled for Expedition 26/27. Instead, Coleman operated the robotic arm for shuttle crew spacewalks.

Apart from her leadership and technical work in space, Coleman became known for playing her flute aboard the ISS and during a live-broadcast National Public Radio interview. She had arranged to take four types of flutes on the expedition: a penny whistle and an antique Irish flute from the Irish folk band The Chieftains, a flute belonging to Ian Anderson of the Jethro Tull rock band, and her own concert flute.[11] During her time on the ISS, she and Anderson performed the first duet by a musician in space and another on Earth (in Russia), dedicating it to Yuri Gagarin on the fiftieth anniversary of his historic flight in space. Coleman performs as a member of a small astronaut band called Bandella, and she also plays with local groups and occasionally with Jethro Tull or the Chieftains. Coleman is one of four astronauts of Irish heritage featured in a set of stamps issued by the Irish postal service to celebrate the fiftieth anniversary of the Moon landing.[12]

Coleman is passionate about sharing space with as many people as possible, which she does through myriad public appearances on stage, radio, and video interviews. In 2024 she published her memoir *Sharing Space: An Astronaut's Guide to Mission, Wonder, and Making Change*, and in 2023 she had a role in the documentary film *Space: The Longest Goodbye*. She appeared as herself in two 2021 documentary films, *Zero Gravity*

Cady Coleman with musical instruments she brought for her stay on the ISS, 2011. *NASA*

and *The Wonderful: Stories from the Space Station,* and in episodes of the television series *Home Improvement* (1996) and *Makers* (2015). She also advised actor Sandra Bullock in preparation for her starring role in the 2013 film *Gravity*, providing firsthand information about the realities of movement, body position, and activities in space.[13]

As an ambassador for spaceflight, Coleman remembers how pivotal it was for her to see the first US woman astronaut. She hopes that her efforts to share space will have a similar effect on someone who will think, "Maybe that could be *me*."

Flights: **2** on space shuttle, **1** on the International Space Station via Soyuz •
23rd US woman astronaut in space • **30th** woman worldwide • Time in space:
180 days, **4** hours (**4,324** hours)

Wendy Barrien Lawrence

Wendy Lawrence was born in a hospital at a naval air station to a naval aviator father and a mother who was a naval officer's wife and the daughter of a naval aviator. Lawrence has said about the Navy, "That's what we do in our family," so it was a natural aspiration for her to be a Navy pilot.[14] Watching Neil Armstrong and Buzz Aldrin on the Moon during the Apollo 11 mission changed her priorities, though, and being a navy pilot became a path to being an astronaut.

Like her father and grandfather, Lawrence attended the US Naval Academy, entering in 1977 in the second class to include women. It was tough on several counts beyond the curriculum and rigorous training.[15] Some, not all, of the male midshipmen thought women did not belong there and could not meet the academic and physical standards; they went out of their way to harass and hinder the women. The female midshipmen, determined to prove themselves, felt constant pressure to excel. In the absence of women's sports, Lawrence and others formed a women's crew team and lobbied for a coach. Rowing offered a release from pressure and a friendly support group. Lawrence felt another, unique pressure; her father, then a vice admiral, was superintendent of the academy while she was enrolled. She had to avoid making a single misstep that would feed any notion that she was favored. Lawrence excelled in academics and leadership, and she graduated near the top of the class of 1981.

Lawrence earned her wings of gold as a naval aviator in 1982. Impressed by her first flight on the H-46, she chose helicopters as her aircraft, knowing that she could get more flying time in rotor craft than jets; at that time, helicopters seemed the best career path for women pilots.[16] Lawrence was sent to sea and became one of the first two women helicopter pilots assigned to a long-term deployment to the Indian Ocean in a carrier battle group. Her at-sea duty included supporting oceanographic surveys. Altogether she logged more than eight hundred shipboard landings and flew six different types of helicopters. She earned a master's degree in ocean engineering while in the Navy and then returned to the academy to teach and coach the women's crew team. Lawrence rose to the rank of captain during her military career.

By 1991, with a decade of piloting experience and a graduate degree, Lawrence applied to become an astronaut. NASA chose her as a mission specialist in 1992; she was the first active-duty female naval officer to be selected. Lawrence served until 2006, flying on four space shuttle missions, including two missions to the Russian space station *Mir*. In 1996 she went to Russia as NASA's director of operations for US astronauts training in Star City. Lawrence and astronaut Scott Parazynski volunteered to train as backup crewmembers for a four-month mission on *Mir*, but both were disqualified by the Russians; she was too small to fit properly in the Soyuz spacecraft and wear the Russian Orlan EVA spacesuit, and he was too large for Soyuz. In jest, they ordered special name tags, Too Short and Too Tall, to wear occasionally.[17] Despite that disappointment, both "Russian rejects," as they called themselves, were assigned together to the seventh shuttle mission to *Mir*. As visiting rather than resident crewmembers, they bypassed the Soyuz and Orlan size issues.

Lawrence was the first woman graduate of the Naval Academy to fly in space and the first Navy helicopter pilot to do so. Her first flight as an astronaut was on STS-67 (1995), the second mission to carry the ASTRO observatory, a set of three large telescopes in the payload bay. It was a sixteen-day, two-shift mission to study ultraviolet emissions from hot stars, distant galaxies, and faint astronomical objects. Two women

Wendy Lawrence, one of the women who faced spacesuit fit problems, climbing out of a T-38 jet, 1998. *NASA*

in the seven-person crew—Lawrence and Tammy Jernigan, a third-time flier—shared mission specialist research duties. Lawrence also was the flight engineer during ascent and entry and the pilot for the shift opposite the commander and pilot's shift.

Rather quickly, Lawrence flew again on STS-86 (1997) and STS-91 (1998), both shuttle missions to dock with *Mir*. The first of these included an exchange of the US *Mir* crewmember, a joint spacewalk by Parazynski and a cosmonaut, and the transfer of supplies and experiments between the shuttle and space station. The later mission was the ninth and final docking in the shuttle-*Mir* series. This crew picked up the last US astronaut to stay on *Mir*, transferred cargo, and carried out research. Lawrence was the flight engineer and loadmaster for the transfers and also shared responsibilities for the onboard experiments with crewmate Janet Kavandi.

After her 1998 flight, the Navy assigned Lawrence to a rotation at the National Reconnaissance Office in the Advanced Systems and Technology Directorate. The two organizations wanted to apply her experience to advanced technology initiatives and partnerships. She remained a member of the astronaut corps during this assignment and returned to duty at the Johnson Space Center two years later.[18]

On her last spaceflight, Lawrence was a member of the STS-114 (2005) return-to-flight mission commanded by Eileen Collins after the *Columbia* tragedy. This was a high-stakes mission to demonstrate that the shuttle could safely fly again and that astronauts were able to verify whether the orbiter had sustained damage and repair any damage detected. This mission to deliver supplies and equipment to the International Space Station introduced several techniques to visually inspect the entire vehicle exterior and repair any damage to its protective shield. Lawrence had several operator roles on the mission—using the laser ranging device during approach to the ISS, overseeing the orbiter docking system, and operating the station's robotic arm during vehicle inspection, cargo transfers, and spacewalks.

Upon returning from this mission and completing post-flight duties, in 2006 Lawrence retired from NASA and the Navy. She transitioned into working with aerospace companies, advocacy for STEM education, and service on boards. She also supports programs at Space Camp, the Kennedy Space Center Visitor Complex, and Higher Orbits, an organization that engages students in STEM by developing experiments to be flown on the ISS. Additionally, she is a founding member of AstraFemina, a nonprofit that partners with other organizations to inspire today's girls to be tomorrow's STEM stars.

In 2019, as Lawrence was honored as a US Naval Academy Alumni Association Distinguished Graduate, her wife was acknowledged as matter-of-factly as anyone's spouse.[19] This public occasion with no fanfare was the first reported disclosure of their marital status. Lawrence had served in the military under the "Don't Ask, Don't Tell" policy, when such disclosure would have ended her career.[20] After she retired, that policy no longer affected her and eventually was repealed in 2011. She married in 2013.

Lawrence has given interviews on the topic of LGBTQ and space and says that "it's not a big deal" in the mission-driven astronaut corps; what matters is what each person contributes to the success of the mission, and everyone has opportunities to demonstrate their abilities. She also has noted that there are other astronauts who self-identify as LGBTQ, but not publicly, "Just because you're not hearing about it doesn't mean it's not in existence."[21]

The life lesson Lawrence shares with young people is, "You have to be able to persist."[22] Reaching a goal or fulfilling a dream is a process, sometimes a long one, and commitment and persistence are key. Setbacks and failures are opportunities to learn and figure out how to move forward. Lawrence's career in both the Navy and NASA reflect her persistent pursuit of a lifelong goal.

Flights: **4** on space shuttle; **2** to *Mir*, **1** to the International Space Station • **21st** US woman astronaut in space • **28th** woman worldwide • Time in space: **51** days, **4** hours (**1,228** hours)

Mary Ellen Weber

With degrees in chemical engineering from Purdue University and physical chemistry from the University of California, Berkeley as her academic credentials, Mary Ellen Weber came to NASA as a research scientist, aviator, and avid competitive skydiver. Weber gained experience as an engineer while still a student by interning, and then, after earning her doctorate, by pushing the boundaries in chip manufacturing as a research chemist at Texas Instruments. Along the way, she earned a patent and published papers in technical journals.

Her interest in aviation was sparked while at Purdue, where she began skydiving in the early days of the sport with round parachutes, belly-mounted reserves, and combat boots.[23] Growing up in an era when women were barred from many careers—the first American woman in space finally flew during her senior year in college—Weber never even dreamed of becoming an astronaut until graduate school, when her passion for aviation matched her passion for science and, at last, women were recruited to be astronauts. She went on to become a medalist at the US skydiving national championships, an instrument-rated pilot, rock climber, scuba diver, and accomplished researcher in industry before submitting her astronaut application.

When she joined the NASA astronaut class of 1992 as a mission specialist, Weber at age twenty-nine was among the youngest candidates ever selected. Within three years, she was in space on her first mission, STS-70 (1995). Weber was responsible for the checkout and deployment of the large Tracking and Data Relay Satellite (TDRS), one of several in an orbital constellation that NASA used to monitor and communicate with the space shuttle, space observatories, and various other scientific satellites. She also was the crew medical officer, the operator of myriad scientific experiments, a member of the flight deck crew for entry and landing, and the primary contingency spacewalker if an EVA became necessary.

In addition to its primary objectives, the STS-70 mission became known for two coincidences. The first was that four of the five crewmembers were from Ohio, and after the governor declared the fifth an honorary Ohioan, they became the first and so far only all-Ohio crew, proudly celebrated throughout the state. The other coincidence led to STS-70 being called the "woodpecker mission" when nesting woodpeckers damaged the foam insulation on the shuttle's external propellant tank. The entire shuttle stack had to be rolled back from the launch pad for repair of numerous divots, delaying the flight. This serious technical problem prompted a lot of humorous commentary.

Weber's next mission was to the International Space Station, STS-101 (2000), so early in the development of the ISS that it was still rudimentary and vacant. The mission for which the crew trained was postponed several times by delays in launching the Russian module they were to outfit. Meanwhile, the vacant station's batteries were losing the ability to recharge and needed to be replaced. This put the entire ISS program at risk, because dead batteries meant the ISS could not receive commands and thus no vehicle could dock with it. Rather than wait for the Russian module, NASA split this mission into two and tasked half of the original crew with urgent repairs. The other half of the crew was reassigned to a later mission to activate the Russian module after its arrival. Such a crew change had never happened before and was stressful for the entire mission team who, after more than a year of training, had to perfect new tasks in a matter of ten weeks. They persevered, though, and this crew delivered more than a ton of equipment, replaced the batteries, installed critical components inside and out, completed necessary maintenance, and used the shuttle to push the ISS to a higher orbit. As flight engineer, Weber was a member of the flight deck crew for both launch and entry, controlled the docking module, operated the robotic arm during a spacewalk, and directed equipment transfers from the shuttle to the station. The future Expedition 2 crew, including Susan Helms, participated in this flight for a preview of their soon-to-be home on the ISS the next year.

In the interval between missions, Weber had leadership opportunities within and beyond the Astronaut Office. She went to NASA headquarters in Washington, DC, as the legislative affairs liaison to Congress and worked closely with the NASA administrator. There she also worked on a team in a fledging effort to commercialize space research operations, a precursor to NASA's transformation into commercial space partnerships spearheaded and brought to fruition a decade later by Deputy Administrator Lori Garver. Weber also served on a headquarters team to revamp the ISS research facilities and worked with international partner space agencies to develop ISS training protocols and experiment facilities. Her time working with Garver at NASA headquarters to launch the commercial space era strongly affected her outlook.[24]

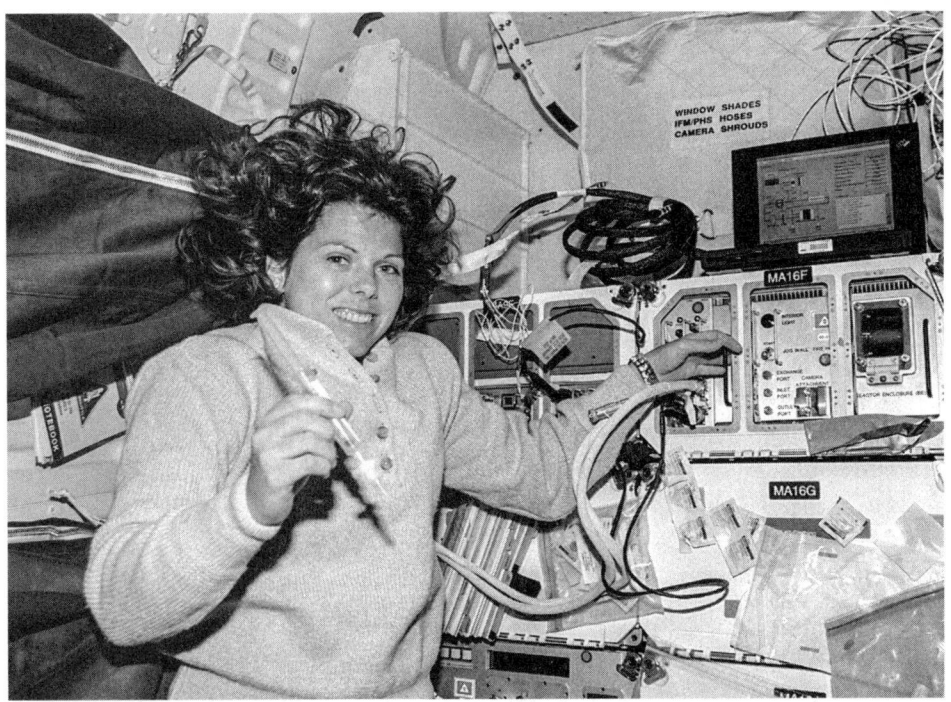
Mary Ellen Weber working with an experimental bioreactor on the shuttle, 1995. *NASA*

These activities and her master's degree in business administration likely influenced Weber's executive and entrepreneurial post-astronaut career. After ten years, she left NASA in 2002 and spent nine years as the vice president for government affairs and policy at the University of Texas Southwestern Medical Center, a world-renowned medical school and research powerhouse in Dallas. There she helped launch world-class research centers by working with the state and federal government and prominent scientists. Thereafter, she founded Stellar Strategies, a firm through which she maintains an active consulting practice and speaking schedule, providing guidance in leadership, strategy, and high-stakes decision making informed by experiences in spaceflight. Since 2012, Weber has been on the executive education innovation faculty at the Kellogg Graduate School of Management at Northwestern University. In addition, she was a member of the NASA Advisory Council committee for technology, innovation, and engineering for eleven years. She also serves on various boards and provides space analysis and commentary for news media. Weber has remained active as a skydiver for forty-plus years, having completed more than six thousand skydives and earned more than two dozen medals at the US National Skydiving Championships.

Weber notes that she has never thought of herself as a "first woman" or a female fill-in-the-blank, but always and only as a scientist, astronaut, or skydiver without a gendered modifier.[25] Reflecting on being an astronaut, she observes that many people still think of women astronauts as different from "real" astronauts, and it is important to dispel that myth.[26]

> Flights: **2** on space shuttle, **1** to the International Space Station • **22nd** US woman astronaut in space • **29th** woman worldwide • Time in space: **18** days, **18** hours (**450** hours)

The Class of 1995

The next biennial recruitment notice, although basically the same as the last, added a statement by the flight crew operations director, astronaut David Leestma: "We are looking for multi-faceted individuals who are not only outstanding in their chosen disciplines but who will be able to handle various technical assignments, maintain spacecraft systems and experiments, work well with others and have excellent communications skills. We also like to have a balanced skill mix and a culturally diverse group in the astronaut corps."[27] He was thinking ahead to staffing not just short shuttle missions but longer expeditions to the International Space Station. In the selection process, expedition skills, such as the ability to work and communicate well with crewmates from other cultures, began to gain importance. In this cycle, NASA selected nineteen US astronaut candidates from almost three thousand applications and welcomed four internationals to the group, for another large class of twenty-three.[28] Ten pilots and nine mission specialists were chosen, among them two women pilots, Pamela Melroy and Susan Still (later Kilrain), and three mission specialists: Kalpana Chawla, Kathryn "Kay" Hire, and Janet Kavandi. This group reported to NASA in 1995.

Kalpana Chawla

Kalpana Chawla was born into a Punjabi Hindu family in Karnal, India, and educated in India through her university-level studies. In 1982, at age twenty, she came to the United States for graduate school in aerospace engineering, earning a master's degree from the University of Texas at Arlington and a doctorate from the University of Colorado. She worked for several years at NASA's Ames Research Center in California in computational fluid dynamics applied to aerodynamics before applying to become an astronaut in 1993. Chawla became a naturalized US citizen in 1991.

A precocious, confident, even headstrong child, Chawla was so insistent on going to school that her family misreported her birthdate so she would be eligible to start sooner.[29] When she was three years old she chose her own formal name, Kalpana ("imagination"). She loved playing outdoors, sleeping under the stars, and going to the local airfield to watch planes. She rebelled against some of the expectations for girls' appearance, clothing, and activities, and her mother and older sisters supported her independence. Her more traditional father had doubts about her atypical academic and career ambitions but relented in the face of her determination. Chawla was the first female to earn a BS degree in aeronautical engineering from Punjab

Engineering College, where she was one of only seven women studying any engineering field.[30]

Upon arriving in Texas in 1982, Chawla met the man who would soon become her husband, a British-born American citizen. Both of them earned various flight ratings, and he became a professional pilot and flight instructor while she was in graduate school. Chawla eventually earned flight instructor and commercial pilot's licenses for single and multiengine aircraft, seaplanes, and gliders. She loved to fly.

When Chawla was selected into the 1994 astronaut class as a mission specialist, the couple returned to Texas to pursue their respective careers. She became known as "KC" in the astronaut corps, succeeded in her technical assignments in robotics and software testing, and was assigned to her first mission the year after she completed basic training. Usually photographed with a radiant smile, petite Chawla was well liked for her kindness, humor, positive outlook, and perfectionism.[31] Chawla was a committed vegetarian, and as an astronaut she worked with NASA's food science team to develop vegetarian options for shuttle and space station menus.

Chawla was the prime robotics operator for STS-87 (1997), a US Microgravity Payload mission that also included solar observations, deployment and retrieval of a scientific sat-

Laurel Clark and Kalpana Chawla (*center*), STS-107 crewmates, 2003. *NASA*

ellite, and testing of EVA tools and procedures for future space station assembly. The crew had some difficulty deploying and activating the satellite. They were unable to recover it with the robotic arm but finally succeeded in capturing it manually during an EVA.

After that flight, Chawla was assigned as the Astronaut Office representative for shuttle and station flight crew equipment and then became head of the crew systems and habitability branch. She flew again on *Columbia* in 2003 on the STS-107 mission, the last shuttle flight dedicated to scientific research. She operated experiments in alternating daily shifts during the sixteen-day, twenty-four-hour work cycle, and she served as flight engineer assisting the pilots during launch and entry. Chawla was seated on the flight deck when *Columbia* disintegrated over Texas just sixteen minutes before its scheduled landing. As in the *Challenger* tragedy of 1986, the entire *Columbia* crew perished.

According to her wishes, Chawla's remains were cremated and her ashes spread where she loved to hike in Utah. As the first woman of Indian origin to fly in space, she is revered in India by children and public figures alike and admired by Indian Americans in the United States and elsewhere. In the year after her death, her husband and Indian family traveled around India giving talks about her career and presenting mementos for display in schools and elsewhere.

In his grief, her husband challenged the romanticized view of astronaut deaths by writing, "Kalpana and her *Columbia* crewmates did not willingly sacrifice their lives, pay the ultimate price, or any other such nonsense. The *Columbia* crew lost their lives in an avoidable accident and had every intention of returning home safely."

Flights: **2** on space shuttle • **25th** US woman astronaut in space • **33rd** woman worldwide • Time in space: **31** days, **14** hours (**758** hours)

Kathryn Patricia Hire

Kay Hire achieved two careers in parallel. For thirty years she was a NASA engineer and astronaut, while for thirty-five years she was on active or reserve Navy duty as a naval flight officer, instructor, wartime command staff officer, science and technology program manager, or commanding officer. In both careers she rose into significant leadership positions.

Hire grew up in Mobile, Alabama, and while in high school, she noticed when the first Navy women entered flight school in nearby Pensacola, Florida. As she applied for college scholarships, the ROTC recruiter asked if she had considered applying to the US Naval Academy in Annapolis, Maryland. She was unaware of the military academies but learned that the previously all-male institutions would soon admit women. She remembered that the Navy had opportunities for women to fly, so she applied. Hire was accepted into the class of 1981, the second year the Naval Academy included women.

Upon graduating with an engineering degree, Hire reported for flight training. Upon earning her wings in 1982, she was assigned to fly oceanographic research missions in the P-3 Orion aircraft. Over the course of three years, she participated in missions to twenty-five countries and became an airborne mission commander. She later became a master instructor and spent three years as the only Navy woman instructor teaching airborne navigation, communications, and avionics systems to hundreds of naval flight officers.[32]

In 1989 Hire transferred from full-time active duty to part-time Navy Reserve and went back to school for her master's degree. She began work as a space shuttle engineer with Lockheed Space Operations Company at the Kennedy Space Center (KSC). She first specialized in orbiter mechanical systems, then integration of all shuttle systems as a test project engineer and later rose to become engineering supervisor of orbiter mechanical systems and launch pad swing arms. For six years she participated in processing space shuttles from landing to launch countdown for more than forty missions.

During this time, in 1993 while in the Navy Reserve, Hire became the first woman in the US military to be assigned to a combat-designated position after the legal restriction barring women from combat was modified to allow women to serve in combat-missioned, land-based aircraft. She then flew in the P-3 Orion maritime patrol aircraft in Atlantic, European, and Caribbean operations.

Encouraged by NASA astronauts she worked with at KSC, Hire applied to become an astronaut. She was advised to finish her master's degree and reapply.[33] She earned an MS in space technology in 1991, then was selected by NASA in the next recruitment to begin astronaut training in 1995. Hire was the first astronaut selected from the KSC workforce. After completing initial training, she worked in mission control as a CAPCOM. Hire flew the first of her two missions three years after she joined the astronaut corps.

On STS-90 in 1998, the last Spacelab mission, Hire served as flight engineer for launch and entry and as a mission specialist for the Neurolab experiments. This sixteen-day mission was dedicated to studying the brain and central nervous system in microgravity. The crew operated twenty-six life science experiments and served as test subjects, along with rats, mice, crickets, and fish. After the flight, she supported ongoing space shuttle missions at Johnson Space Center and received Soyuz systems and survival training at Star City in Russia. She worked on flight software verification and in the payloads and crew equipment branches of the Astronaut Office, while also fulfilling assignments in the Navy Reserve.

History intervened, and the Navy recalled Hire to full-time active duty after the 9/11 attacks. From 2001 through 2003 she was assigned to the US Naval Forces Central Command staff.[34] During this time, she was promoted to the rank of captain. Upon returning to part-time Navy Reserve status and full-time NASA duty, she assumed new roles in both. At NASA, she provided astronaut support as a Cape Crusader for launch and landing operations at KSC. With the Navy she was the commanding officer of a

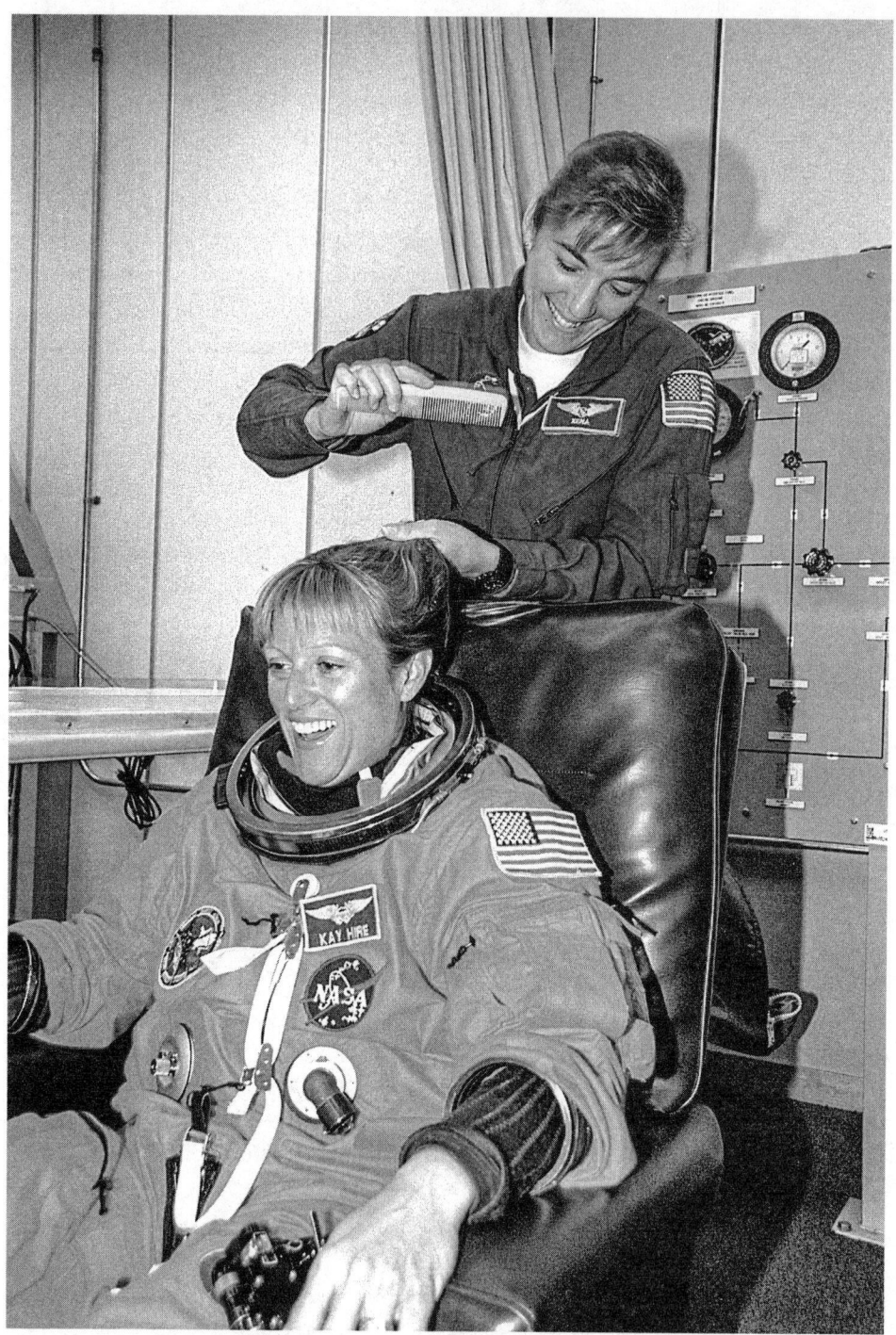

Heide Stefanyshyn-Piper braiding Kay Hire's hair before Hire heads to the launchpad, 1998. *NASA*

reserve unit for the Space and Naval Warfare Systems Command, then later commanding officer of an Office of Naval Research reserve unit.

Hire's second spaceflight took her to the International Space Station on STS-130 (2010), a two-week assembly mission. Hire operated the robotic arms to deliver and install two of the last major components on the station: the habitat Node 3 *Tranquility* along with the multi-windowed cupola. The STS-130 crew moved environmental control and life support systems from the US lab module into Node 3, thereby freeing space for more experiments in the lab. Hire also served as loadmaster for the transfer of two tons of equipment and supplies to the ISS.

After this mission, Hire resumed Cape Crusader duties for the remaining five shuttle flights and contributed her Node 3 and cupola installation experience to support ISS visiting cargo vehicle operations. In 2012 the Navy again recalled her into active duty at the Naval Research Lab. She later served as a space advisor in the Pentagon and then as an aerospace engineering instructor at the Naval Academy before retiring from the Navy in 2016.[35] Hire returned to NASA as a management astronaut and held positions at Wallops Flight Facility in Virginia and Goddard Space Flight Center in Maryland.[36] She fostered collaborative projects with NASA Goddard, Wallops, and the Department of Defense. Upon retirement from NASA in 2019, Hire founded an aerospace consulting firm, Astra Portolan Corporation, and she continues to consult, advise, and make public appearances to share her technical and management expertise.[37]

Hire remembers the early days of her career when the integration of women into male-dominant organizations was not smooth sailing. Her response to extra scrutiny was to work harder and be professional to demonstrate her value.[38] Even in retirement, Hire and other women astronauts realize that whatever they do still reflects on NASA, the astronaut corps, and women; it keeps them vigilant in their responsibility as role models.

Flights: **2** on space shuttle, **1** to the International Space Station • **26th** US woman astronaut in space • **34th** woman worldwide • Time in space: **29** days, **15** hours (**711** hours)

Janet Lynn Kavandi

From leading her high school class as valedictorian to serving as director of NASA's Glenn Research Center in Ohio to heading commercial spaceflight for the Sierra Space company, Janet Kavandi blazed a path to leadership at all stages of her career.

Kavandi remembers being interested in space throughout her childhood, especially looking at the night sky with her father and wondering what the view would be like from space. Early ventures into space, and especially the landing on the Moon, spurred her fascination with astronomy and spaceflight, but "of course at that time, there were no girls in space."[39]

Janet Kavandi training for EVA underwater in the Neutral Buoyancy Lab, 1997. *NASA*

Kavandi's interest in science led her to undergraduate and graduate degrees in chemistry during the 1980s. She worked in electrical power systems at Boeing for almost eleven years, earned two patents, and quickly moved into technical leadership roles. She credits the IMAX film *The Dream Is Alive* (1985) as part of her inspiration to become an astronaut. She submitted her application to NASA shortly after the *Challenger* tragedy and renewed it annually until the agency called her to interview and accepted her as a mission specialist in 1994.

Kavandi's early years at NASA coincided with preparations for space station assembly and operations. During her time on the ground as an astronaut, she held technical assignments in ISS payload integration, CAPCOM duties, and robotics training on both the shuttle and space station manipulator arms. Later, as International Space Station branch chief in the Astronaut Office, she was responsible for training, operations, safety, habitability, scientific payloads, and coordination between international partners for visiting vehicles and related operations.

Kavandi is a veteran of three space shuttle missions, with visits to two space stations. Her first mission was STS-91 (1998), the ninth and final Shuttle-*Mir* docking mission. She and Wendy Lawrence served on the crew together, carrying out science experiments (including the Alpha Magnetic Spectrometer particle physics experiment to collect data on antimatter and dark matter), helping to transfer tons of supplies to

Mir, and receiving experiments from *Mir* for return to Earth. This shuttle mission also brought home the last US astronaut from a long-duration stay on *Mir*.

On her second mission, STS-99 in 2000, Kavandi and the crew operated the huge Shuttle Radar Topography Mapping (SRTM) instruments to radar-image about eighty percent of Earth's land surface for highly accurate 3D topographical maps. Much of that area had been inaccessible for conventional mapping. The crew worked in two shifts continuously throughout the mission to manage the massive data collection and storage. Kavandi also served as flight engineer, assisting the commander and pilot during launch and entry.

Kavandi's last mission, STS-104 (2001), was an ISS assembly flight and the tenth mission to the International Space Station. Kavandi again served as flight engineer. The shuttle crew installed the joint airlock *Quest*, which could be used by ISS crewmembers wearing either American or Russian spacesuits. Kavandi operated the shuttle's robotic arm during three spacewalks, and she served as loadmaster for the transfer of supplies and equipment between the shuttle and ISS, working with the resident Expedition-2 crew, which included Susan Helms. Kavandi had trained for a contingency EVA but was not required to do one.

In 2003 Kavandi performed one of the most sorrowful duties imaginable. She was designated the lead casualty assistance calls officer (CACO) for the families of the last *Columbia* crew after their tragic deaths returning from space. Three of the seven astronauts had been in her 1995 class. She and a team of more than twenty colleagues helped care for the families, maintain their privacy, escort them through memorial ceremonies, schedule funerals, keep them informed about the investigation, and provide any other assistance needed.

After her last flight, Kavandi's leadership roles accelerated. After a couple of years as a branch chief, she became deputy chief of the Astronaut Office in 2005, then deputy director of Flight Crew Operations in 2008, and then Flight Crew Operations director in 2011, overseeing all activities of the astronaut corps and aircraft operations at Johnson Space Center. She chaired the Astronaut Selection Board for the 2013 class (the first and only class to date to include an equal number of women and men candidates), and she worked to address spacesuit size issues that affected the smallest women astronauts. In 2014 she moved into another part of the organization as deputy director of Human Health and Performance, working with the flight surgeons and research experiments involving humans. Kavandi was one of the first women to hold any of these positions beyond branch chief.

In 2015 Kavandi left JSC to become deputy director of NASA's Glenn Research Center in Ohio and a year later became its director, the first woman to hold the head position there and its first director to have flown in space. During that period the Glenn Center flourished while assuming major responsibilities for the Artemis program and Orion spacecraft testing, as well as advanced technology for commercial aviation.

During her twenty-five years of service, Kavandi received NASA's most prestigious awards. Upon retirement from NASA in 2019, Kavandi moved into leadership positions within the aerospace industry. Sierra Nevada Corporation initially selected her to be the senior vice president of its space systems business and later promoted her to executive vice president. In 2021, the space business side of Sierra Nevada became Sierra Space, focused on completing development of the Dream Chaser spaceplane to deliver supplies to the International Space Station. Kavandi served as president and chief science officer of Sierra Space and leader of its human spaceflight center and astronaut training academy, established in anticipation of operating a future commercial space station. She retired in 2023 but continues to consult in the aerospace industry and serve on nonprofit and educational boards.

Kavandi recalls that at NASA she "felt completely at ease being a female in a leadership position."[40] She is excited about prospects for women on the Moon. "I would have loved to be one of those candidates," she says.[41]

Flights: **3** on space shuttle • **27th** US woman astronaut in space • **35th** woman worldwide • Time in space: **33** days, **20** hours (**812** hours)

Pamela Ann Melroy

Pamela Melroy once remarked that the first thing people learned about her after her name was that she intended to become an astronaut. The more anyone tried to dissuade her about anything, the more determined she became.[42] And so, from flying her dolls like airplanes to watching the Apollo 11 lunar landing at age eight to decorating her dorm room with *Star Trek* posters in her college years until commanding a space shuttle mission at age forty-six, Pam Melroy plotted her trajectory and did exactly what was necessary to achieve her lifetime goal. She has said that since reading his book *Carrying the Fire* in college, Apollo 11 astronaut Michael Collins was her greatest inspiration.[43]

Melroy earned her undergraduate degree in physics and astronomy at Wellesley College and joined the Air Force ROTC program at MIT, where she went on to earn a master's degree in earth and planetary sciences. She completed pilot training in 1985 when the services were gradually admitting women pilots and seeing them excel; she was the only woman in her graduating class. She flew the KC-10, a large cargo transport and in-flight refueling tanker aircraft, for six years, advancing from copilot to aircraft commander to instructor pilot, again a rare woman flying that aircraft. During that time, she logged more than two hundred combat and combat support hours as a veteran of Operation Just Cause (Panama) and Operation Desert Shield/Desert Storm (Iraq). Along the way, she acquired the call sign and nickname "Pambo," echoing the *Rambo* films of the 1980s.

Melroy's next assignment was Air Force Test Pilot School in California, where she was only the third woman selected to attend the test pilot course. She graduated a year after Eileen Collins in 1992 and was assigned to the C-17 combined test force for this brand-new transport aircraft until NASA selected her as an astronaut candidate in 1994. She logged more than six thousand hours in more than fifty different aircraft by the time she retired from the Air Force as a colonel in 2007.

NASA selected Melroy and Susan Still together as pilot astronauts. Melroy's career at NASA, after initial training, began with astronaut support (Cape Crusader) duty for launch and landing at Kennedy Space Center. This involved monitoring vehicle and pay-load processing activities as a representative of the crew and the Astronaut Office. She was trained well by Marsha Ivins, then the leader of the Cape Crusader team, and soon became the team leader herself.[44] Melroy later worked on advanced projects, such as orbiter upgrades, after her first mission and after her second flight became a CAP-COM in Mission Control. She maintained her proficiency as a pilot by flying the T-38 training jets, the shuttle training aircraft (a modified Gulfstream II), and the shuttle cockpit simulators.

Only three women ever piloted or commanded space shuttle missions: first, Eileen Collins in 1995, 1997, 1999, and 2005; then, Susan Still in 1997; and Melroy in 2000, 2002, and 2007. All of them overlapped for about ten years in the astronaut corps, and no other women were selected as pilot astronauts after 1995. Melroy was the last woman to fly in the commander's seat of the space shuttle. Melroy, like Collins, thought her military experience flying a large aircraft was excellent preparation for flying the space shuttle, perhaps more so than flying small, nimble fighters with only two people on board.[45] Both were familiar with handling a large aircraft, but more important, they were accustomed to flying with a crew of specialists having different responsibilities during the flight, analogous to shuttle missions.

Melroy was the first of the women to make all of her shuttle flights in the 2000s, and all of them were space station assembly missions on which the crews delivered and installed major elements. Her first as pilot, STS-92 (2000) brought the Z-1 vertical truss segment to which the port and starboard trusses would later be attached as the foundation for solar arrays and radiators. They also brought a mating adapter port for future docking vehicles. This seven-person crew attached and configured the elements using the robotic arm and completed four spacewalks. Their work prepared the station for its first resident crew.

Melroy was back in space again as shuttle pilot on STS-112 (2002) for joint operations with the International Space Station Expedition 5 crew. They delivered and installed the first segment of the starboard truss, transferred cargo, and completed other tasks. Melroy served as the in-vehicle choreographer for three spacewalks as Sandra Magnus operated the shuttle's robotic arm.

When the *Columbia* tragedy occurred in 2003, Melroy was appointed to serve on the reconstruction team as crew module lead. She had the grim task of identifying debris items from the crew cabin and determining whether any of them belonged to a particular crewmember. After that work was completed, scraps of personal effects were returned to the deceased crew's families. Melroy then comanaged the *Columbia* crew survival investigation team, which assessed the debris, the launch-entry suit design, and the shuttle design to understand what the crew likely experienced during their final moments and recommend improvements to reduce the possibility of a repeated tragedy.

Melroy's final mission, this time as commander of STS-120 (2007), was her most dramatic and challenging. After successfully delivering the *Harmony* node to enable laboratory modules to be added to the space station, her crew relocated a solar array from the center to the end of the port side truss. As the folded array was being redeployed, a couple of the panels snagged and ripped. That damage prompted a contingency repair EVA never rehearsed on the ground. Melroy and the crew called it their "Apollo 13 moment." The shuttle and ISS crew, coached by support teams on the ground, had to handcraft some fasteners to close the tears and then send one of the EVA crew out to the electrically charged solar array at the farthest reach of the robotic arm to do the repair. Stephanie Wilson, nicknamed Madame Robotics Expert by the crew, handled the precision maneuvering. As crucial as the repair was, Melroy's greatest concern was the safety of their colleague riding on the arm, who was doing something never attempted in the precarious and hazardous space environment. She led the combined crews through a deliberate assessment of all the safety considerations they could think of to protect him from harm.[46] Fortunately, all turned out well, the array was repaired and restored to operation, and the EVA crewmember returned safe but

ISS commander Peggy Whitson and shuttle commander Pamela Melroy, the first time two women were in command of space missions simultaneously, 2007. *NASA*

exhausted. A less stressful dramatic moment was being greeted by ISS Expedition 16 commander Peggy Whitson, the first time two women were in command of spacecraft at the same time.

After fifteen years as an astronaut, Melroy left NASA in 2009 to work in private industry with Lockheed Martin in program management for space exploration initiatives. Two years later, she went back into government service at the Federal Aviation Administration as a manager for commercial space transportation. Two years after that, she moved to DARPA, the Defense Advanced Research Projects Agency, as deputy director of its tactical technology office, where she stayed until 2017. She and her husband formed a consulting company, and one of her projects took her to Australia to advise on the development of an Australian space agency.

After the 2020 presidential election, Melroy was tapped to serve on the transition team assessing NASA. That led to her nomination by President Joe Biden to be deputy administrator of the space agency and unanimous approval by the Senate. She was the first woman astronaut, but not the first woman, to hold that position. Through 2024 Melroy was a vocal advocate for NASA's programs and traveled extensively as the voice of NASA's leadership. She aimed to align the agency and its industry and international partners together on goals for the Moon-to-Mars effort and a follow-on to the International Space Station, both leading and learning more about effective strategy development while keeping her eye on the future.[47]

Melroy is satisfied that her career followed the trajectory she planned upon deciding to be a military aviator and astronaut. She says, though, "It took me a long time to figure out that there's not a women's leadership style and a men's leadership style. What it's really about is being an authentic human being. . . . I had to be my authentic self and bring my whole person."[48]

Flights: **3** on space shuttle to the International Space Station • **28th** US woman in space • **37th** woman worldwide • Time in space: **38** days, **20** hours (**932** hours)

Susan Leigh Still Kilrain

Susan Kilrain published a children's book, *An Unlikely Astronaut* (2023), a charming tale of the hurdles she faced in fulfilling her dream of flying, with a message that *unlikely* can become *likely* through hard work. Whether defying brothers who teased her or overcoming her perception that girls couldn't be pilots and astronauts, Still (Kilrain after marrying) persevered. She discovered that she was good at math, able to earn a pilot's license, accomplished enough to join the Navy and fly jets, and qualified to apply to become an astronaut. Selected into the astronaut corps as a pilot, Still became the second woman to pilot a space shuttle mission.

She earned her pilot's license and instrument rating by the time she was nineteen, did her undergraduate and graduate studies at Embry-Riddle Aeronautical University and Georgia Institute of Technology, and worked on wind tunnel projects for Lockheed,

Navy pilot and NASA pilot astronaut Susan Still Kilrain, now an author, 2023. *Susan Still Kilrain*

all in a largely male world.[49] Astronaut Dick Scobee, a friend of her Lockheed manager, encouraged her to enter military service to gain the flight experience credentials for becoming an astronaut, and she did. Still joined the Navy in 1985 and by 1987 was a naval aviator and flight instructor. She went on to fly EA-6A Intruders in a tactical electronic warfare squadron and, after completing naval test pilot school, flew F-14 Tomcat supersonic fighter jets. Along the way, she acquired her call sign "Shine" and earned Distinguished Graduate status in both aviation officer candidate school and test pilot school, as well as a number of Navy medals.[50]

After initial astronaut training, Kilrain worked in the vehicle systems and operations branch of the Astronaut Office and as a launch and entry CAPCOM for shuttle missions. Two years after arriving at NASA, she was the first in her astronaut class to fly a mission. She and the crew of the Spacelab Microgravity Science Laboratory mission (1997) have the unique distinction of flying the same mission twice within three months. Their initial mission, STS-83, was cut short on the third day when one of the shuttle's electricity-generating fuel cells malfunctioned and the crew was called home. The shuttle was loaded with experiments scheduled for sixteen days in orbit. Because this mission was meant to lay the foundation for microgravity research on the space station, NASA decided to leave everything on board, repair the fuel cell, and send the same crew back to carry out the full mission (STS-94). Kilrain and mission specialist Janice Voss were both in the seven-member crew. In addition to her pilot responsibilities, Kilrain handled Earth observation photography, in-flight maintenance, and acted as the medical officer, among other duties.

Soon after these missions, Kilrain became pregnant, bowed out of the flight rotation, and held other positions, including duty in the Office of Legislative Affairs at NASA headquarters in Washington, DC. She was considered for a third flight but left NASA in 2002 to keep her family together when her senior naval officer husband was transferred to Puerto Rico.[51] Kilrain retired from the Navy as a commander in 2005, having logged more than three thousand hours in more than thirty aircraft during her twenty years in service. She satisfied her dreams to be a pilot, astronaut, wife, and mother, bearing four children in seven years. She notes that, by setting priorities, "If you can't have it all at once, you can have it all in your life."[52]

Since leaving NASA and the Navy, Kilrain has been active as a motivational speaker and author. She makes public appearances worldwide and enjoys speaking to young people in places where they might not receive encouragement to follow their dreams.[53] She has given talks in various locations in the Middle East, where educational and professional opportunities for girls are limited. As a child who was both a tomboy and loved to play with Barbie dolls, she says, "You can be girlie and good in STEM, too."[54] Kilrain's next book project has a target audience of women and minorities who face obstacles and need encouragement to achieve their goals.[55] Kilrain's advice to all who are making career decisions is, "Pick your journey, forge your journey, and enjoy your journey, whatever it may be and wherever it may lead you."[56]

Flights: **2** on space shuttle • **24th** US woman in space • **32nd** woman worldwide •
Time in space: **19** days, **15** hours (**471** hours)

The Class of 1996

An identical recruitment announcement to the last one yielded the largest class ever: thirty-five US candidates joined by nine internationals for a total of forty-four.[57] Astronaut Group 16 was nicknamed the Sardines; their number stressed the training system and crowded the facilities shared with flight-assigned crews, so they were split into two staggered groups for more efficiency.

The 1996 class included a record number of American women, eight, and Canadian astronaut Julie Payette made nine. Among the Americans were three African American women: Yvonne Cagle, Joan Higginbotham, and Stephanie Wilson. (Mae Jemison, previously the single Black woman in the astronaut corps, had already left NASA.) Laurel Clark and Lisa Nowak, both from the Navy, and Sandra Magnus, Heidemarie Stefanyshyn-Piper, and Peggy Whitson rounded out the group. All of these

(*left to right*) Stephanie Wilson, Joan Higginbotham, Mae Jemison, Yvonne Cagle, and spacesuit team manager Sharon McDougle, 2004. *Sharon McDougle*

women were selected to be mission specialists. Two of them (Whitson and Magnus) first flew in 2002, and Clark flew on the ill-fated *Columbia* mission in 2003.

After that tragedy, the shuttle was grounded until mid-2005 and then again after one flight until mid-2006. Some members of this astronaut class had a long wait before their first mission, or they flew only once, in the effort to ensure that all active astronauts gained flight experience on the space shuttle.[58] The astronaut corps was approaching its maximum number of 150, and there were more astronauts than missions could accommodate. Shuttle missions to the International Space Station typically had six or seven crewmembers—commander, pilot, one or two exchange crewmembers to start or end an expedition, two EVA crewmembers, and one primary robotic arm operator. That left no, or perhaps one, available non-EVA mission specialist seat per mission. This math heightened competition and preyed on morale as astronauts waited hopefully for their assignments and worried about being passed over. In addition, the criteria for ISS expedition crews changed. To be eligible for long-duration stays, an astronaut had to be certified in both EVA and robotic operations to be able to multitask on the ISS. For anyone who had EVA suit fit issues, this requirement effectively reduced the number of flight opportunities and ruled out a stay on the ISS. Most of the women selected in 1996, 1998, and 2000 waited eight to ten years before their first flights, and five of them had only one flight before the space shuttle program came to an end in 2011. Despite the hard work and excitement of being an astronaut, waiting years to fly was, for some, an unsettling reality.

Yvonne Darlene Cagle

Yvonne Cagle first thought about going into space as a ten-year-old watching the Apollo 11 lunar landing on television. The primary path she chose to prepare for that possibility was medicine. Fascinated with images from her father's work as an X-ray technician, she studied biochemistry, went to medical school and aerospace medical school, and became certified as a family practice physician, flight surgeon, and senior aviation medical examiner.[59] She also followed her parents' lead into the US Air Force, where she gained flight experience on medical evacuations and other aeromedical missions. She served on active duty in England and at the Pentagon Flight Medicine/Special Mission Clinic, and she retired from the Air Force in 2009 with the rank of colonel.

In 1989 Cagle volunteered to be the Air Force medical liaison officer for the STS-30 space shuttle mission. She was assigned to the contingency landing site in Banjul, the capital of the West African nation of The Gambia, to help rescue and evacuate the shuttle crew if an emergency forced a landing there. After serving as a flight surgeon in the Air Force for six years, in 1994 Cagle joined NASA's Johnson Space Center occupational health clinic as a physician responsible for conducting job-related health screenings and exams, caring for work-related injuries and illness, and developing protocols for remote duty operations. She also applied to join the astronaut corps.

Like most mission specialists in the forty-four-member 1996 astronaut class, Cagle faced a long wait for a flight assignment, compounded by delays after the 2003 loss of *Columbia*. After her basic training, Cagle served in the operations planning branch supporting shuttle and ISS missions and traveled to Russia in a task group to work on international medical standards and procedures for spaceflight.

As yet without a flight assignment in 2005, Cagle accepted a special assignment as a management astronaut at NASA's Ames Research Center in California, where she held a variety of posts in medical research, program development, and relationship building. She became the agency's lead astronaut and science liaison to Google and other Silicon Valley entities with interests in human health and innovative biomedical technology. She also was a consulting professor to Stanford University's departments of cardiovascular medicine and electrical engineering and served on the faculties of the University of Texas Medical Branch in Galveston and the Volunteer Family Practice Clinic at the University of California in Davis. Cagle remains employed at Ames as of 2024.

Because she did not fly in space, Cagle is sometimes overlooked in lists or books about women astronauts. Her status as one of only six African American female astronauts thus far is noteworthy, as is her three-year term as a visiting professor at Fordham University, which conferred upon her an honorary doctorate for her service in establishing a unique research collaboration between the university, NASA, and other partners. Although she works remotely from the Astronaut Office and thus is rarely seen in the day-to-day activities of the astronaut corps, she consults with NASA on space telemedicine, is an advisor for NASA's flight opportunities program, does research related to astronaut health, and is active as a public speaker and in the partnership programs she has cultivated for NASA.

One of Cagle's most visible moments occurred in 2017: she escorted former NASA mathematician Katherine Johnson onstage at the Academy Awards program, when the film *Hidden Figures* was nominated for three awards. Johnson, Cagle, and the actors who portrayed the Black women "computers" of the early space age appeared together at the ceremony.[60] Cagle has been mindful of Black women pioneers in other disciplines who helped pave the way for her opportunities and success, and she especially credits Katherine Johnson in an autobiographical poem delivered as a TEDx talk, titled "Poetry of Space on Earth."[61] In talks with students, she shares the same encouragement offered by other astronauts: dream big dreams and give your best effort to achieve them. Her motto is "Space for All."[62]

Laurel Blair Salton Clark

Laurel Clark was a specialist in underwater medicine and a flight surgeon. Her career began in the sea and ended in 2003 high above Texas as the space shuttle orbiter *Columbia* disintegrated on its way to a planned landing in Florida. Clark was one of the seven crewmembers who died on *Columbia*, her first mission in space.

Clark first wanted to be a veterinarian and majored in zoology as an undergraduate. She loved animals, nature, and outdoor activities. Instead, she joined the Navy, went to medical school, and began to work in diving medicine. She had postgraduate training in pediatrics at the Naval Hospital in Bethesda, Maryland, but then decided to pursue undersea and diving medicine. Clark completed Navy training to earn certifications as an undersea medical officer, diving medical officer, radiation health officer, and naval submarine medical officer. Although the Navy did not yet allow women to serve aboard submarines, she was among the first women to earn the submarine medical officer title.

Clark then pursued aeromedical training and became a naval flight surgeon. She dove with Navy divers and Navy SEALs, performed many medical evacuations from submarines, flew with a Marine Corps Harrier squadron, served on numerous deployments, and flew on multiple aircraft. She also was certified as a basic life support instructor, an advanced cardiac life support provider, and an advanced trauma life support provider. Of all the medical doctors who became astronauts, Clark probably had the most varied credentials and unusual clinical experience, having practiced medicine in a variety of operational environments.

After ten years in the Navy medical corps, Clark joined NASA as an astronaut candidate in 1996 and completed training and evaluation as a mission specialist. For three years she fulfilled technical assignment in the payloads/habitability branch of the Astronaut Office. When her flight assignment came, it was for the STS-107 laboratory mission on *Columbia*, the first flight of 2003. She said before the flight, "I was thrilled to be assigned to any mission. But I was especially excited to be assigned to this one" because it had so much medical and life science research.[63] As a physician, she relished studying what happens to the body in space.

As NASA moved into the space station era at the turn of the century, STS-107 was the last space shuttle mission dedicated to scientific research. Thereafter, shuttle missions would focus on ISS assembly, maintenance, resupply, and crew exchanges, and most research would shift to the space station. For sixteen days, the *Columbia* crew worked on two shifts around the clock to carry out some eighty investigations in various fields of research. Most of their work took place in the Spacehab laboratory module mounted in the orbiter's payload bay. Clark served on the shift with commander Rick Husband, Kalpana Chawla on her second scientific research mission, and Israeli astronaut Ilan Ramon; payload commander Mike Anderson, pilot Willie McCool, and fellow Navy flight surgeon David Brown made up the other shift.[64]

Clark's assignments were primarily the biology and biomedicine experiments, and she was enthusiastic about the opportunity to learn more about the effects of microgravity on living things, including the crew. She found it magical that life thrived in the onboard experiments; roses bloomed, moths developed, and even worms adapted to spaceflight.[65] Astronauts cannot be required to participate as medical test subjects for invasive procedures such as blood draws, so all the crewmembers for this mission were volunteers, including Clark.[66]

During their return from space, Clark was seated on the flight deck with Chawla, behind the pilot and commander. She was using a small digital video camera to record their activity and conversation during descent. Improbably, that videotape with thirteen recorded minutes survived the breakup of the shuttle and was found among the ground debris.[67] It showed a happy crew excited to come home after a successful mission and unaware of the fate that would claim them minutes later. Clark's last act gave the crew's families a final glimpse of their loved ones. Her own last email message to her family and friends, sent hours earlier, was a poignant recap of the experience of being in space, her sense of Earth's beauty, and her gratitude to all who had supported her along the way.[68]

Captain Laurel Clark's gravesite is in Arlington National Cemetery near the monument to the STS-107 crew and the graves of two crewmates. She was buried with military honors on her forty-second birthday. Within twenty years, another Laurel Clark, her baby granddaughter, brought joy to the family.[69]

Flights: **1** on space shuttle • **31st** US woman astronaut in space • **40th** woman worldwide • Time in space: **15** days, **22** hours (**382** hours)

Joan Elizabeth Higginbotham

Joan Higginbotham did not dream of becoming an astronaut as a child. She was five years old when men walked on the Moon. Growing up, she was fascinated with math, science, and taking things apart to see how they worked; she felt she was born to be an engineer.[70] She graduated from high school in Chicago the year before Sally Ride first flew in space and was already committed to becoming an electrical engineer. NASA and spaceflight were not on her radar. However, NASA found Higginbotham among the outstanding engineering graduates of 1987 and offered her a position in space shuttle operations at Kennedy Space Center. She interviewed and decided to become part of the space adventure shortly after the 1986 *Challenger* tragedy. Thus began Higginbotham's unplanned, unexpected, life-changing journey to space.

Higginbotham's career thrived on taking advantage of opportunities that arose. She says that almost everything she planned for her career did *not* happen, but she was well poised to take unexpected chances.[71] Having interned at IBM while in college, Higginbotham planned to find her first job there. However, a temporary hiring freeze on engineers led to an offer to be a sales associate, not what she had in mind. Unbeknownst to her, the university had forwarded information about their outstanding electrical and mechanical engineering graduates to NASA, and Higginbotham received a call asking whether she would be interested in coming to Florida to work on the space shuttle. Curious but skeptical, she checked it out, found it exciting, and promptly decided to give it a try for at least five years.

During her time at Kennedy Space Center, Higginbotham began in a payload electrical engineer position where she managed the wiring configuration of the payload bay and performed electrical tests for all payloads flown on the shuttle, and she was the lead engineer for orbiter experiments on *Columbia*. She rapidly advanced to become backup orbiter project engineer for *Atlantis*, then lead orbiter project engineer for *Columbia*, then lead government engineer in the firing room during launches. Higginbotham participated in fifty-three space shuttle launches during her nine years at KSC. While working full-time, she also earned a master's degree in management.

In her seventh year, her manager suggested that Higginbotham had the right stuff to be an astronaut and encouraged her to apply. She did and was one of 122 persons interviewed from a pool of six thousand applicants for the 1994 class. After waiting six months, she learned that she was not one of the fifteen people selected to be an astronaut candidate. By then determined to become an astronaut, she sought feedback from an Astronaut Selection Board member and was told that her first master's degree was not technical enough. To better position herself for astronaut candidacy, she returned to school and earned a master's in space systems, and when she reapplied for the 1996 class, she was selected.[72]

Joan Higginbotham and the astronauts' signature aircraft, the T-38 jet, 2003. *NASA*

Because the 1996 class was so large, years passed before each member received a flight assignment. Higginbotham's wait was five years, during which time she held a variety of technical positions in payloads, avionics, operations support, CAPCOM, robotics, and evaluation of space station hardware. At last, she was assigned to an International Space Station assembly mission, first to STS-117 and then moved up to STS-116, which seemed a perfect match for her electrical engineering and robotics skills. However, the mission was delayed three years by the 2003 *Columbia* tragedy. Higginbotham never wavered in her determination to fly in space.

Flying with Suni Williams on STS-116 in 2006, Higginbotham was the primary operator of the station's robotic arm, assisted by Williams. They used it to transfer a truss segment from the shuttle and support its installation by the EVA crew. They supported three more spacewalks to rewire the station's power system to prepare for the arrival of the European and Japanese laboratory modules and to retract and stow a solar panel. Higginbotham also served as loadmaster for the transfer of tons of supplies and equipment between the shuttle and station.

After that successful twelve-day mission, Higginbotham was quickly assigned to a second flight, STS-126, planned for 2008. It was another ISS assembly mission, and she was to reprise her roles as robotic arm operator and loadmaster and serve as flight engineer assisting the pilots during launch and return. She began training but unexpectedly resigned and retired in late 2007 to accept an opportunity in the corporate world.[73]

Higginbotham moved on to a managerial role with Marathon Oil as senior technical consultant. After rotating through several departments, she opted to work in the corporate social responsibility organization and lead the malaria control project, a community health initiative in Equatorial Guinea, Africa.[74] Subsequently she worked in corporate social responsibility and community relations with Lowe's Home Improvement and United Technologies Aerospace Systems (UTAS). She also worked in supplier diversity and global sourcing. When UTAS became Collins Aerospace, she became the Director of Human Exploration.

After more than ten years in the corporate world, Higginbotham launched her own consulting firm in 2022 to offer her technical and business expertise to space industry clients. She is also a popular motivational speaker to professional groups about her career journey; managing stress in high-stakes environments; corporate social responsibility and social impact; and diversity, equity, and inclusion. She serves as a board member for several organizations and makes public appearances at schools, universities, and elsewhere.

Higginbotham spent nine years with NASA as an engineer and eleven years as an astronaut. Although it was not the career she initially planned and aimed for, she became the third African American woman to fly in space, following Mae Jemison and Stephanie Wilson. She is the recipient of numerous awards and accolades, including the National Space Medal. She also appeared in Alicia Keys's *Superwoman* video.

Although she bowed out of her second flight, Higginbotham has said, "I'd fly tomorrow if they called me up and asked me."[75]

Flights: **1**, space shuttle to the International Space Station • **36th** US woman astronaut in space • **46th** woman worldwide • Time in space: **12** days, **20** hours (**308** hours)

Sandra Hall Magnus

Although there were no women astronauts yet, Sandy Magnus decided in middle school to become one. With an affinity for math and science, she was curious about how and why things worked and was fascinated with space exploration. While she was in high school, NASA selected the first women astronauts, and Magnus saw a path and made her plan. She would take all the advanced science classes she could before graduating, major in physics in college, obtain master's and doctoral degrees in physics, and then apply to NASA.[76]

Magnus completed the first two steps of the plan before her path diverged. She discovered engineering, the practical application of physics. After completing her undergraduate studies in physics, she went to work at McDonnell Douglas Aircraft Company as an engineer in stealth technology and aircraft design and simultaneously earned her master's degree in electrical engineering. During those five years, her interests expanded to the science of materials, and she returned to graduate school full-time on a NASA fellowship to earn a PhD from the School of Materials Science and Engineering at Georgia Tech. Finally, she felt qualified to apply to NASA and was selected in the 1996 astronaut candidate class, arriving by a path different than the one she had plotted.

Like many of the women astronauts, Magnus was often the only female in her advanced classes and was the only one in her work unit at McDonnell Douglas. Unlike some, though, she was encouraged by her professors and supervisors and instilled with confidence and independence by her parents. As an intercollegiate soccer player, she had learned much about leadership, teamwork, and discipline, all helpful in the corporate world, where she handled an early leadership test—getting subordinates to accept direction from a woman—with skill and respect. Like many women pathbreakers, she felt that she had to do her best, or even better than that, in case women coming after her were judged based on her performance. She didn't want anyone to be handicapped by her example.[77]

Magnus spent her first years at NASA completing basic astronaut training and a variety of assignments in the payloads/habitability branch of the Astronaut Office. She worked with the European Space Agency, Japan's space agency, and Brazil on hardware for scientific research, and lived in Russia to support ISS hardware development and testing. She also worked with the Canadian Space Agency to prepare its two-armed teleoperated robot (Dextre) for installation on the International Space Station, where it is used for some routine maintenance tasks in lieu of EVAs. Magnus enjoyed the travel and international nature of her early career. After serving for a year as a CAPCOM for ISS expeditions, she received her first flight assignment in 2001.

As a mission specialist on shuttle mission STS-112 in 2002, an ISS assembly mission, Magnus operated the station's robotic arm for three spacewalks to install and activate the fourteen-ton starboard-1 truss structure. The crew delivered an equipment cart for use during EVAs and bioprocessing and crystal growth equipment for the labs, transferred cargo between the shuttle and station, and used the shuttle to boost the station's orbit. Magnus was loadmaster for the cargo transfers. Pilot Pamela Melroy also was on this flight, and Peggy Whitson of the ISS Expedition 5 crew greeted the shuttle crew and worked alongside Magnus during robotic arm operations.

Shortly after returning from her first mission, Magnus became the Astronaut Office lead representative in the return-to-flight effort after the 2003 loss of *Columbia* and crew. She was immersed in systems engineering efforts to understand the technical issues and determine when it was safe to fly again, communicating all information back to the Astronaut Office and carrying forward their questions and concerns. She deciphered and translated heavily technical work to keep everyone informed and considers this work her most important contribution as an astronaut.[78]

In 2005 Magnus was assigned to the ISS expedition corps and started training for a long-duration stay on the space station. Training included serving as commander of a week-long underwater NEEMO mission with a group of astronauts simulating a lunar expedition. She completed advanced neutral buoyancy training in case she needed to perform a spacewalk. Magnus joined ISS Expedition 18 in late 2008 and returned in spring 2009, completing a four-and-a-half-month stay as flight engineer and science officer. She traveled round-trip on the shuttle, arriving on the STS-126 flight that brought all the equipment needed to upgrade the station to support a crew of six. The Expedition 18 crew's primary task was to install a water regeneration system, a second toilet, two new crew sleep stations, and a new exercise device in preparation for the larger crew coming next. As the last three-person ISS crew, they completed two spacewalks and twice as much scientific research as expected.

Despite the busy schedule of required activities on the space station, Magnus found time to experiment with cooking to vary the crew's standard menu of packaged foods.[79] She arranged for a variety of unusual condiments to be sent up—seasoning pastes, mustards, olive oil, balsamic vinegar, and others—and used them to flavor bland items. She also tried combining some of the packaged foods into new entrees. Her journal relates some creative engineering without the aid of gravity, using plastic bags and duct tape to keep things organized, chop ingredients without scattering, and fashion a mixing bowl. She repacked her creations into foil or plastic bags for heating in the food warmers. Magnus demonstrated that it is possible to "cook" in space, albeit messier and more time-consuming in microgravity. Still, she said, "It's fun and certainly an adventure," and her crewmates liked the outcomes. Other astronauts find their own creative outlets to make life in space more normal by doing what they enjoy at home.

After completing her post-flight duties, Magnus served a six-month assignment in the Exploration Systems Mission Directorate at NASA headquarters in Washington, DC, where she was involved in planning and policymaking. NASA was preparing to retire the space shuttle and working on what-next and where-next options for human spaceflight, as well as encouraging the development of commercial spaceflight for crew and cargo service to the International Space Station.

As the shuttle program approached its end, Magnus was assigned to fly on the final shuttle mission, STS-135 (2011). The four-member *Atlantis* crew delivered tons of supplies, food, clothing, spare parts, and other equipment to the ISS, taking advantage of the

ISS chef Sandy Magnus displaying one of her creations, 2009. *NASA*

shuttle's cargo capacity to stock up the space station for at least a year. Until the United States had other spacecraft available, future cargo deliveries would be limited to the much smaller Progress vehicles flown by Russia. Magnus served as loadmaster for off-loading and stowing the cargo and then loading everything to be returned to Earth, a task she described as working "a giant three-dimensional puzzle" during the hectic thirteen-day mission.[80] She also was the primary robotics officer and supported a space-walk by the Expedition 28 crew. A highlight of this end-of-the-shuttle-era mission was a call from President Barack Obama to congratulate the combined crews and all of NASA on the many successes of thirty years of spaceflight since the shuttle's 1981 debut.

Magnus served briefly as deputy chief of the Astronaut Office in 2012 but left NASA after sixteen years to become executive director of the American Institute of Aeronautics and Astronautics (AIAA), the largest technical society in the world. She spent almost six years leading, representing, and advocating for the aerospace industry. She then formed a consulting firm but soon went to work in the Department of Defense as deputy director for engineering and chief engineer within the office of the Under Secretary of Defense for Research and Engineering. Magnus is a professor of practice in engineering at Georgia Tech, holding joint appointments in the School of Aerospace Engineering, the School of Materials Science and Engineering, and the School of International Affairs to provide advice, leadership, and mentorship. She is also working with the Office of Space Commerce to establish a civil space traffic coordination system.

Magnus has remained active as an advisory board member in the commercial space sector and as a speaker on aerospace topics and leadership. In 2019 she and astronaut colleagues Jan Davis and Susan Helms established AstraFemina, a nonprofit organization of professional women in science and technology who engage in outreach to inform and inspire girls and young women about careers in STEM fields. She is convinced that seeing and hearing successful, passionate women in STEM is a vital motivator.

Flights: **3**, including a long-duration stay on the International Space Station • **30th** US woman astronaut • **39th** woman worldwide • Total time in space: **157** days, **8** hours (**3,776** hours)

Lisa Marie Nowak

Lisa Nowak focused primarily on naval aviation and flight test before joining the 1996 astronaut class. A 1985 graduate of the US Naval Academy, where she competed on the track team and earned a bachelor's degree in aerospace engineering, she went on to earn a master's degree in aeronautical and astronautical engineering at the Naval Postgraduate School.[81] After earning her wings as a naval flight officer in 1987, she went to electronic warfare school and was assigned to an electronic warfare aggressor squadron in California, where she participated in reconnaissance exercises and qualified as an electronic warfare mission commander. After these assignments and graduate school, Nowak was sent to work in systems engineering tests at Naval Air Station Patuxent River in Maryland, and a year later was selected to attend Naval Test Pilot School there. After graduation, she stayed there to work as an aircraft systems project officer in the Air Combat Environment Test and Evaluation Facility and was assigned to perform flight tests in F/A-18 and EA-6B aircraft in the Strike Aircraft Test Squadron. After that, she moved to the Naval Air Systems Command.

Before she went to flight school after graduating from the academy, Nowak had a tour of duty at NASA's Johnson Space Center to provide engineering support to the shuttle training aircraft team. That is when she began to think seriously about becoming an astronaut, a career in the back of her mind since high school.[82] Ten years later, she was selected for astronaut training as a mission specialist. Her initial technical assignments were in operations planning, robotics, and CAPCOM duty. Like many in the large 1996 class, she had a long wait for a flight assignment. International Space Station assembly missions had few slots for astronauts who were not designated as EVA or expedition crew. Nowak specialized as a robotic arm operator. She received an initial flight assignment, but grounding of the shuttle after the 2003 *Columbia* tragedy caused subsequent flight delays and rearrangement of mission priorities. When a new flight was inserted into the sequence, she was moved up.

Nowak flew on the STS-121 (2006) space shuttle mission, the second post-*Columbia* return-to-flight test mission to the International Space Station. She was the prime robotic arm operator for three EVAs and for inspection of the orbiter's entire heat shield using the new extended arm and sensor system. She also was the flight engineer assisting the pilots on the flight deck during launch and return. The entire crew, which included classmate and first-time flyer Stephanie Wilson, participated in testing new safety equipment and procedures. Nowak and Wilson partnered in robotic arm maneuvers using both the shuttle and ISS arms. The crew transferred tons of supplies and equipment to the station, including an exercise cycle, a freezer for scientific samples, and an incubator for biological experiments, and they completed various maintenance tasks. A highlight was delivering a third crewmember to the ISS to restore full occupancy for the first time since 2002.

After this flight, Nowak returned to Navy duty in 2007 in the Naval Air Training Command and retired in 2011 with twenty-six years of service. She then started a new career as an engineer in the private sector.[83] Nowak is active with fellow astronauts in the volunteer AstraFemina organization of women who promote STEM education for careers in science and technology.

One of the hardest, and fortunately rarest, duties of an astronaut is to serve as a Casualty Assistance Calls Officer (CACO) to care for and assist the family of an injured or deceased crewmate. Nowak was a devoted member of the CACO team for the family of her friend, classmate, and fellow naval officer Laurel Clark, who perished on *Colum-bia*'s return home.[84]

Flights: **1**, space shuttle to the International Space Station • **33rd** US woman astronaut in space • **42nd** woman worldwide • Time in space: **12** days, **18** hours (**306** hours, **37** minutes)

Heidemarie Martha Stefanyshyn-Piper

Heide Stefanyshyn-Piper mastered ocean depths as a Navy diver and low Earth orbit as an astronaut. She never thought of becoming an astronaut until she was in her late twenties. She had already completed her education and advanced in her Navy career when she learned about NASA's plans to build a space station. She realized that as a ship maintenance engineer and diver, she had unique experience to offer.[85] She applied to NASA and was not selected on her first attempt but was encouraged to reapply and was selected the next time.

Heide earned her bachelor's and master's degrees in mechanical engineering from MIT and was commissioned into the Navy through the ROTC program. She trained as a diving and salvage officer, then served as an engineering duty officer for ship mainte-nance and repair at Pearl Harbor Naval Shipyard. Before applying to NASA she worked as an underwater ship husbandry operations officer, planning salvage projects and the repair of waterborne naval vessels. With almost ten years of experience in diving, marine maintenance and repair, and underwater salvage, she realized that her work had much in common with the assembly and maintenance of a space station. In fact, as she watched astronauts on spacewalks, she was struck with the similarity to diving activities.

After she completed basic astronaut training, Stefanyshyn-Piper received a favor-ite technical assignment: serving as a Cape Crusader for astronaut support at Kennedy Space Center. She also was the lead astronaut representative for ISS payloads and worked in the EVA branch of the Astronaut Office and later worked on EVA spacesuit techni-cal issues.

Her first mission was scheduled to launch in 2003 but was delayed by the *Columbia* tragedy until 2006, ten years after she reported to JSC. Stefanyshyn-Piper flew as an EVA mission specialist on STS-115, the mission that resumed assembly of the International Space Station. The crew delivered and installed a massive truss segment and two sets of solar arrays, during which she maneuvered the ISS robotic arm for the hardware trans-fers. Stefanyshyn-Piper performed two of the three spacewalks to install these elements and tend to a radiator, antenna, and external science payload, and she operated the ISS robotic arm for the other spacewalk. She also led the equipment transfer from the shut-tle middeck to the station and conducted science experiments inside the shuttle.

Stefanyshyn-Piper was in space again two years later on the STS-126 mission (2008). The shuttle crew outfitted the ISS living quarters to accommodate six residents, adding another galley, toilet, and sleep stations. The crew also had a busy EVA schedule to repair a damaged solar array rotary joint, lubricate another one, and install external cameras and an antenna. As EVA lead, Stefanyshyn-Piper conducted three of the four spacewalks. During one of her EVAs her toolkit escaped and drifted away to become a small satellite; it remained visible for several months before burning up in the atmosphere. STS-126 also delivered Sandra Magnus to the ISS for her long-duration stay on Expedition 18.

Between these two flights, Stefanyshyn-Piper led NASA's NEEMO-12 underwater mission in the Aquarius habitat near Key Largo, Florida. Four crewmembers—two physicians and two astronauts—supported by two engineers tested telerobotic surgery and lunar geology tools and techniques, and also built an undersea structure, during their twelve-day simulated space mission. These tasks were relevant to planning future long-duration missions on the Moon and Mars.

After Stefanyshyn-Piper hefted a suited astronaut by the parachute straps during a training exercise, witnesses impressed with her strength called her Xena, after the Warrior Princess fantasy series that aired on US television from 1995 to 2001. The nickname stuck.[86] The daughter of two post–World War II immigrants, her father from Ukraine and mother from Germany, she grew up in the Ukrainian émigré community in St. Paul, Minnesota, where she was taught that education and effort were the keys to success. She excelled in math, science, and sports, earned a Navy ROTC scholarship to her chosen university, was named the most valuable player of her MIT women's crew team, advanced in rank through the Navy to command leadership positions, and became an astronaut despite the formidable odds and requirements of this elite profession. It is often said of first-generation Americans who achieve notable success, sometimes despite humble origins, that reaching such heights so quickly would be improbable anywhere else.

It is a further accomplishment for a woman to ascend in such predominately male organizations as the military services and the Astronaut Office. Working in diving and salvage, Stefanyshyn-Piper was often the only woman, or one of very few women, but her rank as an officer elicited respect.[87] Beyond that, she relied on self-confidence in her abilities, did the jobs, and proved her value in the Navy and at NASA.

Stefanyshyn-Piper left NASA in 2009 to complete her Navy career, first as naval systems engineering chief technology officer for the Naval Sea Systems Command, then as commander of the Naval Surface Warfare Center, Carderock Division, in Maryland, and finally as commanding officer of the Navy's Southwest Regional Maintenance Center in San Diego, California. In 2015, she retired from active duty as a captain with thirty years of service. Since then, she has been active as a motivational speaker and as a STEM mentor through the AstraFemina organization. Among other honors received, she was inducted into the Women Divers Hall of Fame. As of 2024, Stefanyshyn-Piper is one of only four women astronauts to have logged five or more EVAs.

Flights: **2**, space shuttle to International Space Station • **34th** US woman astronaut in space • **43rd** woman worldwide • Time in space: **27** days, **15** hours (**664** hours) • EVA time: **33** hours, **42** minutes

Peggy Annette Whitson

Peggy Whitson has one of the longest and most impressive résumés among the astronauts, all of whom are impressive in their own right. However, during her thirty-year NASA career and her post-NASA work, she has held a remarkable variety of positions and set or broken more spaceflight records than anyone would have predicted. Kent Rominger, chief of the Astronaut Office, praised Whitson as being "gifted in everything—not only in space and life sciences, but also in operations, from robotics to spacewalking."[88] Before she retired, one of her admiring colleagues called her an "American space ninja" for being skilled, always prepared, and able to do anything well.[89] Whitson's humble take on this is simply that all of her experiences combined to make her a better astronaut than she would have been otherwise.

Whitson grew up in rural Iowa, helping her parents and siblings work on the family's farm. She was nine years old when she watched the Apollo 11 lunar landing but didn't think she could be an astronaut until years later. She was a high school senior in 1978, the same year NASA first admitted women as prospective astronauts. She has said that among those selected women, biochemist Shannon Lucid and physicist Sally Ride showed her that it was possible to be a woman, scientist, and astronaut.[90]

With new inspiration, Whitson attended Iowa Wesleyan College, Iowa's first coeducational liberal arts campus, where she majored in biology and chemistry and graduated first in her class. While there, two advisors tried to steer her to medical school, and an alumnus, space scientist James Van Allen, who was a critic of human spaceflight and advocate of automated spacecraft, asserted that being an astronaut was not a worthwhile profession; he predicted that NASA would soon have no need for astronauts.[91] Whitson ignored the advice and moved on with her plan.

When Whitson chose a graduate school for her doctoral and postdoctoral education, Rice University in Houston appealed to her for its proximity to NASA's Johnson Space Center. Whitson started working at JSC as a National Research Council resident research associate in 1986, and that opened the door to eventually becoming an astronaut a decade later. During her initial years at NASA, as a research chemist, Whitson gained increasing responsibilities in laboratory research and experiment development, and she joined the adjunct faculty at the University of Texas Medical Branch in Galveston. She was a member of the US-USSR Working Group in Space Medicine and Biology, a role that culminated in her becoming the project scientist for the Shuttle-*Mir* program. When she was selected for astronaut training in 1996, she was deputy chief of the Medical Sciences Division at JSC and as cochair of the US-Russian Mission Science Working Group she was leading international planning for scientific research on the space station. The next year she assumed an adjunct faculty position at Rice University.

In ten years, she advanced academically and managerially, increasing her qualifications to be an astronaut.

After completing her PhD, Whitson submitted her astronaut application every two years and received four polite rejection letters before succeeding on her fifth attempt. "I never let rejection hold me back," she said. "Instead, I used the time to explore opportunities to strengthen my skills and grow. . . . As much as I wanted to be an astronaut before I was 30, I was a much better astronaut getting in later."[92] She completed basic training in 1998 and several technical assignments in operations planning and crew support, including a year in Star City as a Russian Crusader before being assigned to her first space mission.

Whitson flew three times for NASA, each time as a long-duration resident on the International Space Station. She flew on the space shuttle only to arrive and depart from ISS Expedition 5 in 2002, when she was in charge of the US facilities and science on the station. She returned in 2007 as commander of Expedition 16, this time traveling via Soyuz, and again in 2016–2017 as commander of Expedition 51 while aboard for the entire Expedition 50/51/52 series, traveling again via Soyuz. Two of her stays lasted six months and the third extended to almost ten months; altogether, she logged a total of 665 days in space. In the course of her missions, the ISS expanded in size with new structures and facilities, improved in research capabilities with new scientific equipment and experiments, grew in resident crew size from three to six members, welcomed visiting space shuttle crews and their cargo, and became a traffic hub for a variety of automated resupply vehicles from the United States, Russia, Europe, and Japan. As commander responsible for overall operations on the ISS, Whitson stayed equally involved in carrying out research in hundreds of experiments. She also was a prime crewmember for robotic operations and EVAs. On her longest expedition, she said, "I love being up here. Living and working aboard the space station is where I feel like I make the greatest contribution."[93]

Between missions, Whitson held other influential positions, first as deputy chief of the Astronaut Office, then as commander of the two-week NEEMO-5 undersea mission, as branch chief for station operations, as a member and then chair of the Astronaut Selection Board, and finally as chief astronaut from 2009 to 2012. Whitson was the first mission specialist, first scientist, first civilian, and first woman to head the astronaut corps; the chief astronaut had always been a military pilot.

By the time Whitson had completed her third expedition on the International Space Station, she had incidentally achieved an impressive list of firsts and records, including first woman and first scientist commander of the ISS, first woman to serve as chief astronaut, most spacewalks by a woman (10), oldest woman in Earth orbit and oldest woman to spacewalk (57), most cumulative time in space by a US astronaut and by a woman (665 days) and crewmember on the first spaceflight with more women than men (Soyuz

Peggy Whitson near an ISS solar array on one of her ten EVAs. *NASA*

TMA-11, 2008). She and Suni Williams have alternated ranking as the woman who has spent the longest time in EVA.

Upon retiring from NASA in 2018, Whitson became a consultant in the commercial spaceflight industry and then took a position with Axiom Space as an astronaut and director of spaceflight operations. Axiom partners with SpaceX to fly short research missions to the ISS using the Crew Dragon spacecraft. Whitson commanded the Axiom-2 (2023) private mission to the ISS and is scheduled to command a similar Axiom-4 mission. She oversees Axiom's mission planning and astronaut training, applying her thirty years of NASA experience as an astronaut and manager. In 2023, while working at Axiom, Whitson became the first woman to command a private commercial space mission, increased her cumulative time in space to 675 days, and reset her own record as the oldest woman in Earth orbit at age sixty-three. These numbers will rise if she flies again.

When she retired from NASA, the administrator called Whitson "a testament to the American spirit," citing "her determination, strength of mind, character, and dedication."[94] Whitson once modestly wondered, "Who would believe that a girl from a hog and soybean farm in Iowa would become an astronaut?"[95]

Flights: **4**, to the International Space Station via space shuttle, Soyuz, and Axiom • **29th** US woman astronaut in space • **38th** woman worldwide • Time in space: **675** days (**16,200** hours) • EVA time: **60** hours, **21** minutes

Stephanie Diana Wilson

Stephanie Wilson was curious about astronomy and the universe as a teenager and also about how and why things work. Those interests led her to see engineering, her father's profession, as a pathway to space.[96] Both of her parents worked with aerospace firms, so she could see at close range some avenues to pursue her interests.

After earning an undergraduate engineering degree from Harvard University, Wilson began working at Martin Marietta Astronautics on the Titan IV heavy-lift launch vehicle. Two years later, she accepted a NASA research fellowship to attend graduate school at the University of Texas and earn a master's degree in engineering. That led her to a position at the Jet Propulsion Laboratory working on the attitude and pointing control systems for the *Galileo* probe launched to Jupiter in 1989. While handling various other assignments at JPL, Wilson decided to apply to the astronaut corps and was accepted in 1996 on her second attempt. Being an astronaut "was always something in the back of my mind," she said.[97] Ten years later, she became the second African American woman to fly in space. When asked about the challenges of being an astronaut, she cited learning to swim as the hardest and doing her part to ensure mission success the most unrelenting.[98]

After Wilson completed astronaut candidate training, she moved into several technical assignments in space station payloads, space shuttle propulsion system operations, and CAPCOM duty. She received her first flight assignment a year after the *Columbia* tragedy as a mission specialist on the second return-to-flight mission, STS-121 (2006). This mission had a dual purpose: to test the shuttle's new safety equipment and procedures and to deliver about seven tons of supplies and equipment to the International Space Station. Wilson was the robotic arm operator for orbiter inspection and maneuvering the *Leonardo* cargo module between the shuttle and station and was loadmaster for unpacking and repacking the cargo carrier during the twelve-day mission. She was flight engineer during ascent and participated in the shuttle's rendezvous and docking/undocking operations.

Wilson returned to space promptly the next year on STS-120, an ISS assembly mission commanded by Pamela Melroy. The crew's major task was to deliver and install the *Harmony* node, a connector unit where Europe's and Japan's laboratory modules would be attached. This node enabled expansion of the space station research capabilities and workspaces, a giant step toward achieving the full potential of the ISS. This time, Wilson was the flight engineer assisting the commander and pilot on the flight deck during launch and return. She was again the primary robotic arm operator for vehicle inspection and spacewalks during this fifteen-day mission. Wilson witnessed an historic moment when ISS commander Peggy Whitson welcomed shuttle commander Melroy aboard the space station; it was the first time that two women commanded space missions at the same time.

During STS-120, an unexpected situation occurred when a solar array snagged and tore as it unfolded. Repair was essential and urgent, so the combined ISS and shuttle crews and all the relevant support teams on the ground swung into action to devise a contingency (unplanned, unrehearsed) spacewalk. As in the Apollo 13 emergency, the crew had to improvise with only onboard equipment as tools and supplies. Wilson's primary role was to maneuver the astronaut riding on the robotic arm to the full extent of the arm's reach and precisely position him to do the repair without touching the electrified solar array. Working with only a few inches of margin, she earned the respectful nickname "Madame Robotics Expert" for achieving this complex, untried feat.[99] Her robotics training on the ground and level of concentration during the ordeal, and indeed the skills of the entire space crew and team on the ground, were crucial to the success of this dramatic EVA.

Wilson's third mission, STS-131 (2010), resupplied the ISS with almost fourteen tons of cargo. Before the shuttles were retired in 2011, the last missions aimed to stock the space station for every foreseeable need, because no other spacecraft had comparable cargo capacity. Again, Wilson was responsible for robotic arm operations for spacewalks and for maneuvering the *Leonardo* cargo module into place. A highlight of this fifteen-day mission was delivery and installation of the multi-windowed cupola observation post, used like an air-traffic-control station as other spacecraft come and go and also as a place for Earth observation photography. The cupola is a favorite place to relax and enjoy the view when time permits. The STS-131 crew included three women: Wilson, Dorothy Metcalf-Lindenburger, and Japan's Naoko Yamazaki. When they entered the space station, Expedition 23/24 resident Tracy Caldwell Dyson greeted them. It was the first time four women had orbited Earth in the same spacecraft.

Since last returning from space, Wilson held a variety of assignments: working on ISS, payloads, procedures, operations, crew support, *Orion*, and other matters. She served on astronaut selection boards and special teams, was a robotics mentor and instructor, and became a branch chief in the Astronaut Office. In 2013 she completed a nine-month management stint in the Spaceflight Systems Directorate at the Glenn Research Center in Ohio.

In 2020, NASA named Wilson to be one of eighteen astronauts on the Artemis team, eligible for missions to the Moon.[100] As of 2024 she is the most senior member of that team, having been an astronaut for almost thirty years and flown on three missions to log forty-two days in space. NASA also announced its intention to send the first woman and first person of color to the Moon.[101] Wilson is a match for both and is on track for the requisite training and experience. She was scheduled to begin a six-month ISS expedition in 2024 by traveling on the SpaceX Crew-9 flight, but plans changed, and she remains eligible for a different flight.

A record number of women on the ISS: (*clockwise from left*), Tracy Caldwell Dyson, Naoko Yamazaki, Dottie Metcalf-Lindenburger, Stephanie Wilson, 2010. *NASA*

Among her unusual distinctions, Wilson flew all three of her missions on the orbiter *Discovery*, the only astronaut with that record. She experienced the two coincidental, unplanned women's "firsts" in space on her second and third missions, mentioned above. She also was the lead CAPCOM during the ill-fated last *Columbia* mission (STS-107) and again for the first all-woman spacewalk by Christina Koch and Jessica Meir (2019).

Although the next couple of years may be suspenseful for Wilson as she prepares to fly again, she is a person of faith who knows her purpose in life and believes she is where she is meant to be.[102] She is confident that things work out with hard work, patience, and perseverance. Besides tending to her NASA career, Wilson makes time to volunteer as a tutor and mentor to minority students in math and science, helping prepare them for their futures. "Understanding that the number of minority astronauts is small . . . and doing the best job for NASA that I can, we want to see more young people of color becoming more interested and showing strengths in aviation, math, science, and engineering," she said years ago.[103] She hopes that will eventually increase the pool of eligible minority astronauts. As one of the first and the few, she aims for a future where a person of color will be as common as a woman in space.

Flights: **3** on space shuttle to International Space Station • **32nd** US woman astronaut in space • **41st** woman worldwide • Time in space: **42** days, **23** hours (**1,031** hours)

The Class of 1998

The 1998 recruitment brought in another large class of candidates: twenty-five US citizens (eight pilots, seventeen mission specialists), joined by seven international members of partner space agencies for a class of thirty-two. The call for applications announced, "We are looking for individuals who not only are outstanding in their chosen field of work, but also possess the ability to get along with others and work in a diverse, multicultural environment. We are in a very challenging and dynamic time in human space exploration, and the people we choose will be an integral part of this nation's reach for the stars."[104] This year saw the first Russian module of the International Space Station placed in orbit and the first shuttle mission to deliver and install a major US component, the *Unity* node. The ISS would be the focus of most astronaut activities for the foreseeable future. Group 17 included four women, all mission specialists: Tracy Caldwell (later Dyson); Barbara Morgan; Patricia Hilliard (Robertson), who died before being assigned to a mission; and Sunita Williams.

Tracy Ellen Caldwell Dyson

Tracy Caldwell made a list in high school of what she enjoyed doing as she tried to envision a career. She listed being athletic, working with tools, doing science, learning languages, and learning about other cultures.[105] About the same time, she noticed that NASA was sending teacher Christa McAuliffe into space and learned that astronauts could be scientists and engineers as well as pilots. McAuliffe's selection and a new version of "the right stuff" inspired her to settle on astronaut as a good match to her interests.[106]

The first in her family to go to college, Caldwell was fortunate to find supportive mentors as an undergraduate chemistry major and lab research assistant.[107] She had rigorous coaching and practiced teamwork through intercollegiate track and field competition. She found comparable faculty encouragement as a graduate student, with advisors guiding her original research in physical chemistry and giving her opportunities to grow through academic publishing and presentations. Upon completing her PhD, she immediately applied to NASA and took a postdoctoral research position as she waited almost a year for the call to interview.

Upon selection, Caldwell fulfilled assignments in shuttle flight software verification and in the International Space Station operations branch of the Astronaut Office.

That included duty in Star City as a Russian Crusader to oversee hardware and software testing and integration for the space station. She also was a Cape Crusader supporting space shuttle launch and landing at the Kennedy Space Center. She served as prime crew support astronaut and representative for the ISS Expedition 5 crew and as CAPCOM for both shuttle and station operations.

Caldwell's first flight assignment was the STS-118 space shuttle mission (2007), an ISS assembly mission. The crew, which included classmate Barbara Morgan, delivered a large truss for the station's backbone that supports the solar arrays and radiators and also a new gyroscope and stowage platform. As flight engineer, she assisted the pilots during ascent and rendezvous/docking with the ISS. Her other main roles were to operate the shuttle's robotic arm to transfer the truss segment to the space station and to be the choreographer assisting the EVA crew before, during, and after four spacewalks.

Two and a half years later, newlywed Caldwell, now Dyson, was in space again, this time as a flight engineer on ISS Expedition 23/24 (2010). She traveled to and from the space station via Soyuz. During her six-month stay, she worked on science and technology experiments and the various routine housekeeping and public engagement activities that all the crewmembers share. However, when a pump module in the station's cooling system failed, Dyson performed three urgent contingency spacewalks. She and her EVA partner were generally familiar with the pump module but had not practiced this task before flight. The support team on the ground coached them through procedures to remove and replace the failed item successfully.

While she was on the space station, *Discovery* arrived on its STS-131 (2010) mission with three women among the crew: fellow astronauts Stephanie Wilson and Dorothy Metcalf-Lindenburger, and Japanese astronaut Naoko Yamazaki. It was the first and so far only time that four women have been in space together. About such occurrences deemed remarkable for women, Dyson says, "I look forward to the day when it's not so noteworthy."[108]

Part of Dyson's life story is quite different from her peers' histories. She learned to handle tools and develop mechanical skills from her master electrician father, whom she accompanied to his office and jobsites from a young age.[109] By the time she was seven years old, she had her own tool belt and set of tools for basic wiring work. Her father taught her skills and problem solving as she worked with him on weekends and school breaks until her mid-twenties. She brought all that construction experience to bear on her job as astronaut and thought about her father as she used her EVA tools.

Dyson has another skill unique among astronauts: she is fluent in American Sign Language (ASL). During her stay on the ISS, she created a video using ASL to give hearing-impaired viewers more direct insight into life in space. She also conducted interviews and answered student questions in ASL. Among her other talents, Dyson is a private pilot and one of the lead vocalists for the all-astronaut rock band Max-Q. She has hosted a NASA television program, *StationLife*, about life aboard the ISS, and she was a consultant to actor Jessica Chastain, interplanetary mission commander in the 2015 film *The Martian*. She enjoys doing home and auto repair.

As of 2024, Dyson had been an astronaut for twenty-six years and was still on active duty. After her return from space in 2010, she led a multidisciplinary team to improve ISS stowage and cargo transfer operations, which are crucial for efficiency and knowing where everything is located. She also worked with newer astronauts to prepare them for missions. She returned to flight training as backup to Loral O'Hara, a member of the ISS Expedition 69/70 crew. Dyson returned to the space station on Expedition 70/71 (2024) for another six-month stay. "I feel like I've grown up with the space station," she remarked.[110]

> Flights: as of **2024**, **3**, on space shuttle and Soyuz, including two long-duration stays on the International Space Station • **37th** US woman astronaut in space • **47th** woman worldwide • Time in space as of **2024**: **372** days, **18** hours (**8,946** hours) • EVA time: **4** for a total of **23** hours, **20** minutes

Barbara Radding Morgan

Barbara Morgan came into the astronaut corps by a unique path. She was an elementary school teacher who participated in the competitive Teacher-in-Space program announced by President Ronald Reagan to fly a teacher as the first private citizen, or spaceflight participant, on the space shuttle. Morgan was selected to be the backup for the chosen teacher, Christa McAuliffe. The two women went through the same five-month training together at NASA in 1985, and Morgan was prepared to make the flight if McAuliffe could not go. McAuliffe flew on space shuttle mission STS 51-L in 1986 as planned.

After McAuliffe and the *Challenger* crew died when the shuttle failed catastrophically, Morgan returned to teaching but continued working with NASA part-time for more than twenty years. Named the Teacher-in-Space Designee, she was a speaker, education advisor, and ambassador for space exploration and STEM education. Over time, NASA recast its plan to fly a teacher in space and selected Morgan as a member of the 1998 astronaut class.[111] As a NASA employee like others in the class, she would train as a mission specialist and enter the astronaut corps eligible for flight, having regular astronaut duties as well as the role of educator. Morgan became the first astronaut who was an educator by profession. She vowed to go into space "with the eyes, ears, heart, and mind of a teacher."[112]

Born in 1951 like Sally Ride and Kathy Sullivan, she might have fit more naturally into the group of women selected during the 1980s, but she was a welcome presence among the younger women selected from the mid-1990s into the early 2000s. The four women in the 1998 class were nicknamed the Spice Girls, after a then-popular pop music group by that name. Chemist Tracy Caldwell was Helium Spice, Patty Robertson was Peppermint Patty Spice, Suni Williams was Sunny Spice, and Morgan's "stage name"

Teachers Barbara Morgan (*l*) and Christa McAuliffe, 1985. *NASA*

was Old Spice.[113] Morgan finally flew in space at age fifty-five in 2007, after serving two decades as a dedicated and patient advocate for space exploration.

Morgan's credentials were already impressive in 1985, although not the ones typically found on astronaut candidate resumes. She held a bachelor's degree in human biology from Stanford University, where she graduated with Phi Beta Kappa honors, and had been teaching young children for ten years when she applied for the Teacher-in-Space program. Morgan had taught in an inner-city Bay Area school in California; on the Flathead Indian Reservation in Montana, at the Colegio Americano in Quito, Ecuador; and in the elementary school grades in McCall, Idaho. Morgan was known for innovative teaching methods meant to bring the world into her isolated, small-town classrooms.[114]

Always seeking to become a better teacher, Morgan continued to take university courses and attend professional development workshops and also served on district and state education committees and textbook adoption committees. She advised her school's student council; organized its newspaper; taught enrichment classes in folk dance, music, and computers; and held evening astronomy programs for the school and community. Beyond her teaching work, she found time to volunteer as the musical director of the McCall Chamber Orchestra, Chorus, and Mandolin Orchestra and also worked with the McCall Folklore Society and Summer Folk Festivals, ran the Red Cross summer swimming program, and organized McCall's participation in Idaho's Special Olympics. Morgan's varied interests and organizational abilities, as well as her ability to communicate her philosophy of life and learning, propelled her to the top among 11,000 Teacher-in-Space applicants and 114 semifinalists.

By the time Morgan flew on the STS-118 mission (2007) to help construct the International Space Station, she had been teaching for twenty-four years, completed two years of astronaut candidate training, and moved into technical assignments in space station operations, robotics, and CAPCOM duty. As she was preparing for a planned 2004 mission on *Columbia*, the second shuttle tragedy occurred, and her flight was delayed in the aftermath. The STS-118 mission crew, which included her classmate Tracy Caldwell Dyson, delivered and installed a truss segment, gyroscope, and external stowage platform as well as supplies and equipment, and also carried out four spacewalks. As primary robotics operator, Morgan used the shuttle and station manipulator arms to support the spacewalks, and she used the extended shuttle arm to inspect the bottom of the orbiter to ensure there was no damage from launch. She also served as loadmaster for the transfer of cargo between the shuttle and station, assisted the pilots during entry and landing, and held radio and video lessons with students during the mission. Morgan found it especially meaningful to fly on *Endeavour*, the replacement for *Challenger*.[115] *Endeavour* had been named through a nationwide student competition, and children from the elementary school where she taught were among the many who submitted that name.

Despite her proximity to both shuttle tragedies and the deaths of fourteen crew-members who were dear friends, Morgan did not waver in her desire to fly. She said that being an astronaut was in many ways like being a teacher: "We explore, we discover, and we share."[116]

Morgan retired from NASA in 2008 and returned to Idaho as the distinguished educator in residence, now emerita, at Boise State University, which granted her an honorary doctorate. She continues to work with national and international education organizations, nonprofits, commercial entities, and NASA as an advisor and advocate. Her legacy includes curriculum materials, education policy, and program development, particularly focused on STEM. Morgan has received many awards for her work as an educator and NASA astronaut. Among the most significant, a new elementary school in McCall, Idaho, is named for her.

> Flights: **1**, space shuttle to the International Space Station • **38th** US woman astronaut in space • **48th** woman worldwide • Time in space: **12** days, **17** hours (**305** hours)

Patricia C. Hilliard Robertson

Patty Robertson died in 2001 at age thirty-eight, after completing astronaut training but before being assigned to a mission. She suffered fatal burn injuries in the crash of a small private airplane while serving as flight instructor for the aircraft's owner. She was a multiengine rated flight instructor, owned an airplane, was accomplished in aerobatics, and loved to fly. The probable cause of the accident was determined to be loss of control, with no details. At the time of her death, Robertson was serving as a crew support astronaut for ISS Expedition-2.

Robertson was selected to be an astronaut candidate nine years after receiving her medical degree in 1989. She completed a residency in family medicine and was board certified in family practice. She soon joined the staff of a hospital as clinical coordinator for training medical students and resident physicians. In 1995, she was selected to study aerospace medicine in a joint program with NASA's Johnson Space Center and the University of Texas Medical Branch in Galveston. Upon completing that program, Robertson joined the Flight Medicine Clinic at JSC as a flight surgeon.

Because she did not fly in space, Robertson is often overlooked as a NASA astronaut, but during her brief time in the corps, she was influential. She participated in KC-135 flights to test equipment and techniques for life science experiments and crew exercise in weightlessness, helping scientists and engineers refine their hardware and procedures for flight experiments. Coworkers appreciated her work in validating procedures to ensure clarity.[117] Upon completing astronaut training in 2000, Robertson received the Distinguished Alumni award from her alma mater, Indiana University of Pennsylvania, and after her death the Indiana Regional Medical Center named its Center for Aviation Medicine for her. In January 2024 Northrop Grumman launched one of its Cygnus ISS-resupply vehicles named in Robertson's honor.

Sunita Lyn Pandya Williams

Suni Williams has set a few records as an astronaut, and she may well set others. In 2007, on her first mission, Suni became the woman who had spent the longest time in space on a single mission, a record that stood for eight years, and also the woman who had completed the most spacewalks (four). She and Peggy Whitson twice took turns holding the women's spacewalking record, with Williams's increase to seven surpassed after five years by Whitson's ten. Williams says, "Records are for breaking. . . . I don't think too much about them."[118] Williams was in space again in 2024–2025 for almost nine months, rising higher on the list of astronauts ranked by time spent in space. During that stay on the space station, she set a new record as the woman with the most cumulative EVA time. Two records—as the first person to run the Boston Marathon in space and the first person to complete a triathlon in space—are hers forever.

Williams first established herself in military aviation. A 1987 engineering science graduate of the US Naval Academy, she was commissioned into the Navy and trained as a basic diving officer. She then went to flight school and earned her wings as a naval aviator. She hoped to fly jets, but at the time there were limitations on which aircraft women were allowed to fly, so she opted for more flight opportunities in helicopters. Williams trained for helicopter combat support and was assigned to a squadron that deployed overseas to the Mediterranean, Red Sea, and Persian Gulf. She flew both combat support and humanitarian aid missions until 1993, when she was selected for Naval Test Pilot School. Upon graduation, she was assigned to be a project officer, chase pilot, and test pilot in a variety of rotary wing aircraft. She also served as a rotary wing instructor at the Test Pilot School and as the school's safety officer. Her last naval assignment before becoming an astronaut was deployment on the USS *Saipan*, an amphibious assault ship, where she was the aircraft handler and assistant air boss. In her Navy career, Captain Williams logged more than three thousand flight hours in more than thirty different aircraft. She also earned a master's degree in engineering management.

Williams' ambition turned toward NASA during a visit to the Johnson Space Center, when she met legendary astronaut John Young.[119] As he described the Apollo lunar landing vehicle, she realized that helicopter skills and experience might be useful in spaceflight. She decided to apply, didn't make it on her first attempt, but applied again, and was selected into the 1998 astronaut candidate class as a mission specialist. After completing the initial training, she volunteered for a technical assignment as a Russian Crusader to support the astronauts training for the first assembly mission to the International Space Station. She and several other new astronauts took turns going to Moscow and Star City to work with Russian and NASA personnel as everyone was learning how to collaborate and help the Expedition 1 crew prepare for their mission. One task she remembers was to verify that procedures for operating Russian hardware—composed in Russian and translated into English—were accurate and made sense.[120]

After ISS Expedition 1 (2000/2001), Williams focused on robotics, working on development of the station's robotic arm and its smaller dexterous manipulator (Dextre). She was a crewmember on the nine-day NEEMO-2 undersea mission, simulating spacewalks and other spaceflight activities. When shuttle flights resumed after the *Columbia* tragedy, Williams was trained and ready to catch a shuttle ride to her first mission on the space station as a member of the Expedition 14/15 crew. She spent six months on the station, arriving in December 2006 on STS-116 and returning in June 2007 on STS-117.

Designated as an ISS flight engineer, Williams was an orbital construction worker. To this end, she completed four spacewalks to install, relocate, service, or repair various elements of the station. There was also interior work to do—rewiring, installing equipment, reconfiguring hardware—to improve the utility of the space station. The ISS crew also managed arrivals and departures of Russian Progress resupply ships bringing food, water, clothing, and other necessities, and worked in concert with the space shuttle crews who brought not only supplies but also the large trusses and solar arrays to power the station.

After her return, Williams served as deputy chief of the Astronaut Office and then trained for her next mission to the ISS on Expedition 32/33 in 2012. She was flight engineer for the first two-month segment (Expedition 32) and commander of the two-month Expedition 33 segment. During this time she completed three more spacewalks and spent more time than before on experiment operations, as the ISS was then outfitted for research. For this mission she traveled to and from the space station on a Russian Soyuz craft. When she landed, she had a spent almost 322 days in space on the two missions.

Physical exercise had always been part of Williams's daily routine. She regularly biked, ran, swam, and played various team sports. She had run marathons and qualified for the Boston Marathon shortly before her first mission. Not wanting to miss it,

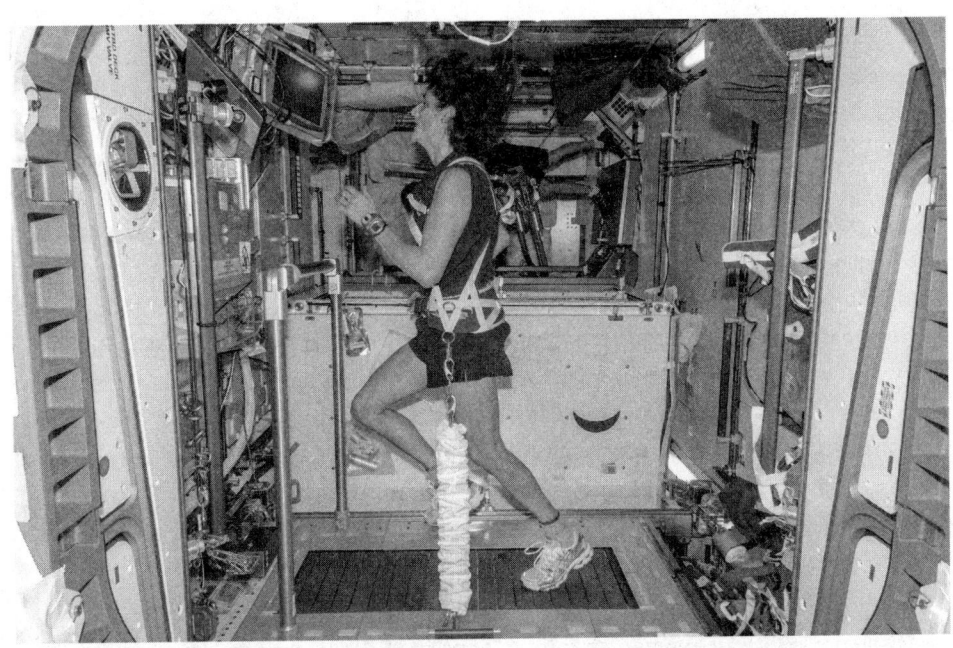

Suni Williams running on the ISS treadmill, part of the crews' exercise regimen, 2007. *NASA*

she hatched the idea of running the same distance on the ISS treadmill at the same time as her sister and astronaut Karen Nyberg ran the actual race together in Boston.[121] Williams completed her run in a respectable four hours, twenty-three minutes, cheered on by her crewmates. She thought that the publicity surrounding her space marathon might increase awareness of exercise as a lifestyle choice for good health. On her second mission and for the same reason, she completed a California triathlon using the space station's stationary bike, treadmill, and resistance device in lieu of swimming.

Twelve years passed before Williams was in space again. During that time NASA was working on its own next-generation space transportation system to replace the shuttle and also working to stimulate commercial space transportation options in private industry. In 2015, NASA named Williams and three other astronauts to work with SpaceX and Boeing as they brought their Crew Dragon and Starliner spacecraft, respectively, from development to test flights to operational service.[122] The Commercial Crew Team astronauts shared their expertise with the commercial partners to ensure that safety and user standards were met as they trained for the first flights on the new spacecraft.

Three years later, NASA announced expansion of this team to nine astronauts and assigned pairs to each company for their test flights and first operational missions.[123] Williams was assigned to fly on Boeing's first Starliner mission but was later reassigned to the Starliner crew flight test, which unfortunately was delayed by technical issues. Starliner launched in June 2024 but had some thruster problems en route to the space station. When NASA decided not to send Williams and the commander back home in the Starliner but on the next returning Crew Dragon craft, their planned ten-day stay turned into a nine-month sojourn, with an expected return in March 2025. In the meantime, they were integrated into the Expedition 71/72 research and maintenance activities on the station, and Williams was named commander for Expedition 72.

With her family steeped in two cultures—her father born in India and mother of Slovenian descent—Williams especially appreciates the internationalism of the space station program.[124] Through training and working with spaceflight professionals and scientists in other countries, learning their traditions, studying Russian language, and living for months with crewmates of other cultures, she has been most impressed with the international partnering that has made the space station successful. For her, the ISS dinner table is the symbol. There, cooperation is most evident in the sharing of food and conversation, planning and discussion of each day's activities, problem solving, professional and social camaraderie, laughter and genuine caring for one another, and feeling the privilege of being there. In her view, the most significant legacy of the International Space Station is the trust it has fostered, person to person and agency to agency. The ISS partnership is itself a remarkable record.

Flights: **3** to the International Space Station on the space shuttle, Soyuz, and Starliner • **35th** US woman astronaut in space • **45th** woman worldwide • Time in space: **608** days (**14,592** hours) • EVA time: **62** hours, **6** minutes

The Class of 2000

The 1999 astronaut recruitment notice was the first to mention the International Space Station.[125] It stated that NASA was seeking candidates "to join the Agency as it enters the era of International Space Station and continues the exploration of space," with selections to be made in 2000. Astronaut and Director of Flight Crew Operations James Wetherbee encouraged "those who are leaders in their fields and work well on diverse, multi-cultural teams to join us as we move into the new millennium." He added, "This is a very exciting time to be in space." More than three thousand applications arrived, from which NASA chose a group of seventeen US candidates, three of them women: Megan McArthur, Karen Nyberg, and Nicole Stott, all mission specialists. This class was nicknamed the Bugs, in recognition of the so-called Millennium Bug (a.k.a. Y2K) that might affect computers as 1999 turned to 2000 on New Year's Eve. Within a decade, two astronaut couples from this group married: Megan McArthur and Robert Behnken, and Karen Nyberg and Douglas Hurley. Each eventually added a son to their family, as did Nicole Stott.

This was the first of smaller selection groups for the next twenty-plus years; they ranged in number from eight to fourteen candidates, even though the number of applications kept increasing. After reaching a peak of 150 active astronauts in 2000, NASA had more than enough for projected space shuttle flights, space station assembly, and space station expedition crews. By the time the space shuttle program ended in 2011, NASA needed to right-size the astronaut corps through attrition and reduced recruitment.

K. Megan McArthur

Katherine Megan McArthur holds an enviable title: last person to touch the Hubble Space Telescope. She earned that as the remote manipulator system (RMS) arm operator who released the telescope into orbit after the final servicing mission, STS-125, in 2009. McArthur used the robotic arm to capture the telescope, settle it into the payload bay of the space shuttle, move the astronauts around it during five spacewalks, and then gently lift it out of the payload bay and place it back into space. Although she didn't touch it with her own hands, through the mechanical arm she was indirectly its last human contact. McArthur also was one of only four women among the forty crewmembers serving on the Hubble Space Telescope deployment and servicing missions, six of the most demanding flights accomplished by shuttle-era astronauts.[126]

The daughter of a Navy pilot, McArthur was exposed to aviation and life on military bases from a young age, and like many children of pilots, she vaguely thought about becoming one, too. However, during her teenage years when her family was living at the Moffett Field Naval Air Station, next door to the NASA Ames Research Center, she noticed the astronauts flying in and out for training sessions and began to think that looked like an even more interesting career.[127] Becoming an astronaut stayed in the back of her mind as she went on to earn an undergraduate degree in aerospace engineering until she was well into her graduate studies in oceanography and applied ocean sciences at the Scripps Institute of Oceanography. She had almost completed her PhD when she applied to NASA and was selected into the 2000 class. By that time she had extensive experience in diving and seafloor operations, underwater engineering research, and scientific data collection; she had served as chief scientist on oceanographic projects and had planned and led at-sea missions. She also had conducted educational demonstrations for the public from inside the exhibit tank at the aquarium located at Scripps.

After initial training at NASA, McArthur had engineering-based technical assignments in shuttle systems and avionics. She was the crew support astronaut for the ISS Expedition 9 crew and spent six months in Russia in that role, and she then worked as a CAPCOM for various shuttle and space station missions. She also finished her PhD and supported the work of commercial companies seeking to deliver NASA cargo to the ISS. Later, between her missions, she was deputy chief of the Astronaut Office for two years.

McArthur and her husband, Robert L. Behnken, are one of nine couples who met and married while in the astronaut corps. Both went into space (separately) during their first year of marriage, he on a shuttle mission to ISS in 2008 and she on a shuttle mission to the Hubble Space Telescope in 2009, when she was flight engineer and prime robotics operator. More than ten years later, they flew back-to-back on the earliest SpaceX Crew Dragon flights, he on the demonstration flight to the ISS in 2020 and she as pilot of the Crew-2 flights for ISS Expedition 65/66 in 2021. He spent three weeks on the ISS, and she was there for six months engaged in research and operating the robotic arm. By that time, they had a young child and faced the same challenge as other astronaut families: preparing for a long parental absence. For all astronaut families, email and weekly video visits from space help bridge the distance and maintain normal relationships. One of McArthur's strategies was to fill a jar with chocolate kisses, one for every day she would be away, so their son could have a kiss from Mommy each day.[128] He said he wants to go to space someday, too, but still has years to find his true ambitions.

After her second mission, McArthur entered into a unique arrangement between NASA and Space Center Houston (SCH). She remained an active-duty astronaut eligible for flight assignment, and she also became the first chief science officer at SCH. Her role there is to ensure authenticity in the science and space exploration visitor learning experiences. She is involved in strategic planning and development of educational content in programs and exhibits. She makes public appearances for both NASA and SCH and continues to have responsibilities in the Astronaut Office.

The online Internet Movie Database (IMDb) lists astronaut McArthur as an actress, because she lent her voice to an animated character of herself in a televised episode of

Blaze and the Monster Machines in 2014.[129] She also appeared in the IMAX documentary film *Hubble 3D*, released in 2010 after the STS-125 Hubble servicing mission.

Flights: **2**, **1** on space shuttle and **1** on SpaceX to the International Space Station • **40th** US woman astronaut in space • **51st** woman worldwide • Time in space: **212** days, **15** hours (**5,103** hours)

Karen Lujean Nyberg

A mechanical engineer, Karen Lujean Nyberg became the fiftieth woman to fly in space on her first NASA mission in 2008. She was a mission specialist and primary robotic arm operator on the STS-124 space shuttle mission that delivered, installed, and activated the Japanese portion of the International Space Station: the laboratory module Kibō, its pressurized experiment storage module, and its remote manipulator arm. As the lead arm operator, she supported three spacewalks, maneuvered the Japanese components into place, and drove the shuttle, ISS, and Japanese robotic arms—the first person to have that opportunity.[130]

Nyberg returned to space five years later in 2013, this time on a Soyuz, for a long-duration stay on ISS as an Expedition 36/37 crewmember. Besides working on the range of onboard experiments, supporting spacewalks and docking operations, and handling day-to-day routines, she also had a role in saving astronaut Luca Parmitano, whose helmet began to fill with water from a leak in his spacesuit's cooling system while he was outside on an EVA.[131] He couldn't see, his communications system failed so he couldn't hear or talk, and he was in danger of drowning inside the helmet. He and his EVA partner made it back to the airlock, where Nyberg was ready to repressurize it. She removed his helmet as soon as it was safe to do so. It was a close call.

Nyberg began her almost thirty-year career at NASA in 1991 when she was a university student; she participated in the cooperative education program, alternating a semester on campus with a semester at the Johnson Space Center while she earned her degrees. One of her interests was human thermoregulation related to temperature control in space suits. Upon completing her doctorate in 1998, she took a full-time position as an environmental control systems engineer in the Crew and Thermal Systems Division at JSC. Two years later NASA selected her as an astronaut candidate, and after basic training she became a crew support astronaut for the ISS Expedition 6 crew. Over time she also served in the space shuttle and exploration branches and as chief of the robotics branch of the Astronaut Office, and she participated in the week-long NEEMO-10 underwater mission. Nyberg retired from NASA in 2020 after twenty years as an astronaut.

Searching for information about Nyberg online yields a long list of articles and interviews about something unusual that she did during her stay on the ISS. Many astronauts bring something personal—books, recorded music, a musical instrument—for relaxation when they are off duty. A crafter and seamstress since childhood, Nyberg brought a small sewing kit on board in case she had a chance to indulge in her hobby during her limited free time.[132] She fashioned three items while in space—a small stuffed dinosaur for her young son, a star-pattern quilt block, and a small Texas flag. She found that it was "tricky" to sew in space without everything floating around. She had to anchor herself in her sleep station and use a small Velcro covered panel to secure her needles, thread, fabric, and scissors drawn from a Ziploc bag.

Nyberg is the first person known to have sewn and quilted while in space, a novel activity in orbit. Unexpectedly, though, her space hobby ignited worldwide interest among quilters, mostly women who had no particular connection to spaceflight. In cooperation with the International Quilt Festival, Nyberg invited quilters around the world to submit their own star-pattern quilt block to be joined together into a quilt for display at the 2014 festival in Houston.[133] So many fabric blocks were submitted—about 2,400 from thirty countries—that twenty-eight large quilts were assembled and displayed. This outpouring of enthusiasm resulted in a global community quilt project and also a new relationship between quilters and spaceflight and for space enthusiasts with quilting. Not many people realize what a critical skill sewing is to space exploration in the fabrication of spacecraft insulation blankets, spacesuits, storage containers, and other items.

In retirement, Nyberg has turned her passion for textile arts and quilting into new ventures. She collaborated with a fabric company to release organic cotton fabrics in designs derived from Earthscape photos she took through the ISS windows.[134] One of her favorites captures the hexagonal pattern of the cupola windows, with clouds, ocean, or landforms in each window panel. She has also created quilted-fabric astronaut portraits of Valentina Tereshkova, Sally Ride, and her astronaut-husband, Douglas Hurley.[135] Nyberg has also collaborated with an apparel line to develop space-themed, gender-neutral fabric designs for children's clothing, including dinos in space. She, her fabrics, and her patterns have become popular in the community of quilters and fabric artists, and she now considers herself an artist as well as an astronaut and engineer. She continues to enjoy drawing and painting, too.

As one of several astronaut couples who have children, Nyberg has spoken candidly about the challenges of pursuing one's dream of spaceflight while attending to the responsibilities of parenting and marriage. She was not able to fly her originally assigned second mission in 2010, when she became pregnant. Their child was barely verbal when she was away for six months on the ISS. To stay connected, Nyberg made short daily videos demonstrating life on the space station or reading bedtime stories. She placed a little painting made "For Mommy" in her sleep station, and she prepared herself for her child's initial shyness upon her return. She, her husband, and son later were interviewed for a Netflix documentary titled *Like a Mother* about moms in unusual, highly demanding careers, and she also briefed actor Hilary Swank, who played a Mars mission commander and mother in the Netflix TV series *Away*.[136]

On a Mars mission, the time away and the distance will be much greater, and the communications delay will amplify remoteness. "If we were in a position where we were actually doing Mars missions right now, I don't know how I would feel [about going]," Nyberg muses.[137]

Flights: **2**, **1** on space shuttle, **1** International Space Station expedition via Soyuz • **39th** US woman astronaut in space • **50th** woman worldwide • Time in space: **180** days (**4,320** hours)

Nicole Marie Passonno Stott

Nicole Stott enjoyed an almost twenty-eight-year career with NASA. During that time, she participated in the space shuttle and International Space Station programs, working for ten years as an engineer at NASA's Kennedy Space Center and two years at Johnson Space Center before being selected as an astronaut candidate. She then spent the second half of her career, another fifteen years, as an astronaut. In her post-NASA life, Stott is an artist, public speaker, consultant, and cofounder of the Space for Art Foundation, traveling worldwide for appearances in events that link art and space exploration.

Stott was inspired by her parents, a mom who shared creativity with her and a dad who built and flew small planes as a hobby. Both welcomed her participation in the activities they loved.[138] As a family, they spent a lot of time at the local airport where she developed her love of flying and desire to know how things fly.[139] She was thrilled to discover that her high school offered an introduction to aviation course, and a program at the local junior college enabled her to earn her private pilot's license and instrument rating. These steps led her to pursue a bachelor's degree in aeronautical engineering from Embry-Riddle Aeronautical University and a master's in engineering management from the University of Central Florida.

After joining NASA as an engineer in 1988 during the return-to-flight period after the *Challenger* tragedy, Stott worked in several roles and facilities at KSC. During the busy 1990s with six to eight flights a year, she was an orbiter vehicle engineer, orbiter project manager, shuttle processing flow director, and launchpad convoy commander. She spent two years in California supporting International Space Station development as the NASA project lead for the Boeing-built ISS truss elements—the structural spine of the ISS. With a good understanding of the astronaut job and encouragement by NASA coworkers, Stott applied to become an astronaut in 1997. She interviewed and was surprised to be offered a different position at JSC as a flight simulation engineer on the shuttle training aircraft, helping train astronaut pilots to land the space shuttle. Having gained the operational experience that NASA sought, Stott reapplied in 1999 and was chosen in the 2000 class. She entered astronaut training with a solid understanding of the space shuttle and space station and the subtleties of working with a crew in a complex operational environment.

After completing basic astronaut training, Stott had technical assignments evaluating payloads for the space station operations branch and serving as a crew support astronaut and CAPCOM for space station missions. In preparation for long-duration spaceflight, she successfully completed ISS systems training at each of the international partner training sites in Star City, Russia; Tsukuba, Japan; Cologne, Germany; and Montreal, Canada; a six-week Russian language immersion class in Russia; intensive spacewalk, robotics, and payload training; and a variety of expeditionary survival training experiences.

Before heading into space, in 2006 Stott earned the title of aquanaut as a crewmember on the record-setting eighteen-day NEEMO-9 saturation dive mission at the Aquarius undersea research station. The crew carried out a lunar exploration analog mission, testing new robotic devices, spacesuit design concepts, and techniques for construction, communication, and telemedicine. Like other astronaut-aquanauts, she found the experience to be excellent preparation for living and working in space.

Stott's first spaceflight was to the International Space Station in 2009 as a member of the Expedition 20/21 crew for three months. She was the last person to both arrive and depart on space shuttles (STS-128 and STS-129) before the vehicles were retired in 2011. She participated in a six-and-a-half-hour EVA at the space station to perform both maintenance and assembly tasks. During her stay on the station, she was engaged in research and operated the station's robotic arm to capture the automated Japanese HTV cargo vehicle on its first mission to resupply the station. She gained notice as the first person to create a watercolor painting in space—an orbital view of a chain of islands off the northern coast of South America—using a small paint kit she packed in case she had free time. She didn't realize at the time how much this experience would inspire her future.

Stott was assigned to her second spaceflight while still on her first mission. It came quickly in 2011 on STS-133, the final flight of *Discovery.* This crew delivered and installed the last elements of the US portion of the space station, as well as critical spare parts and supplies. Stott operated the station's robotic arm and directed two spacewalks during the assembly process.

After this flight, Stott spent the next year on assignment at KSC as the Astronaut Office representative to the Commercial Crew Program. She returned to JSC in 2012 as branch chief for space station integration and then as chief of the vehicle integration test office and lead astronaut on the Orion landing and recovery team.

Since retiring from NASA in 2015, Stott published a book, *Back to Earth: What Life in Space Taught Me about Our Home Planet and Our Mission to Protect It* (2021), and she resumed painting for pleasure, working in watercolor, oil, acrylics, and mixed media. Her intention with both is to share the spaceflight experience with as many people as possible. One of the signature activities of the Space for Art Foundation is the Spacesuit Art Project, which involves children in hospitals and refugee centers around the world with the wonder of space exploration and the healing power of art. She showcases exhibits of children's art inspired by science and space and leads other space-themed art initiatives. An in-demand motivational speaker and advocate for STEAM education, Stott communicates with passion about humanity's shared role as crewmates here on Spaceship Earth. Her public persona in her new career is "Nicole Stott: Artist, Astronaut, Earthling."[140]

Flights: **2**, to the International Space Station on space shuttle and SpaceX • **41st** US woman astronaut in space • **52nd** woman worldwide • Time in space: **103** days, **5** hours (**2,477** hours) • EVA time: **6** hours, **39** minutes

Women in Space in the 1990s

Well before the end of the century, women were firmly established in the astronaut corps and achieving impressive flight records. However, a shift in generations was becoming evident. Many of the women selected in the 1980s, and one from the 1990s, had already vanished from the NASA Astronaut Office. More than half of the twenty women in the trailblazing generation were no longer flying. Mary Cleave, Jan Davis, Mae Jemison, Sally Ride, Susan Still, Kathy Sullivan, and Kathy Thornton had left the astronaut corps for other pursuits. Judy Resnik had died, and Rhea Seddon had retired. Bonnie Dunbar, Tammy Jernigan, and Shannon Lucid had flown the last of their five missions each, and Ellen Baker had completed her three missions. Anna Fisher was on leave during the first half of the decade.

During this decade, six women astronauts from Russian and international partner space agencies made their first journeys into space, several of them as payload specialists on shuttle missions. Helen Sharman of the United Kingdom spent seven days as a guest cosmonaut on *Mir* in 1991. Canadians Roberta Bondar and Julie Payette flew on shuttle missions in 1992 and 1999 respectively. Chiaki Mukai from Japan flew on two shuttle missions, in 1994 and 1998. Russian cosmonaut Yelena Kondakova spent five months on *Mir* in 1994 and then flew on a 1998 shuttle mission. Claudie André-Deshays from France spent sixteen days on *Mir* in 1996. All but Payette and Kondakova were the first women astronauts from their home countries.

Meanwhile, seven of the twenty-three women of the inspired generation celebrated their first, second, or third flights during the 1990s, while everyone from the 1996 and 1998 classes was either training or waiting for flight, and the 2000 class was just arriving. Their time to fly would come in the next decade with assembly and occupancy of the International Space Station, and the ranks of women astronauts gradually would be replenished by another generation.

The Empowered Generation: Astronauts for the Space Station and Beyond, 2004–2021

The turn of the twenty-first century brought changes to NASA and the astronaut corps. The International Space Station became a reality when the first two habitable modules were launched in 1998, the first shuttle crew assembled them in 1999, and the first two occupants began residence there in 2000. Thirty-seven shuttle missions were dedicated to ongoing assembly, resupply, and crew exchanges to support the first decade of continuous ISS expeditions. However, with the space station expected to be complete around 2010, the space shuttle had no new role, and talk of its retirement escalated. Such a large vehicle would no longer be needed; a smaller craft would suffice to support the station. Uncertainties about what would follow the space shuttle vehicle colored the next ten to fifteen years, with each presidential administration introducing its own ideas and priorities.

At the same time, NASA needed to field and train an astronaut corps to execute various proposed missions. For the near term before the shuttles were retired, enough astronauts were needed to crew the shuttle's assembly, resupply, and crew exchange missions (typically seven astronauts per mission) and enough qualified astronauts to be flight engineers or commanders of the ISS (at first, one or two Americans per ISS expedition lasting three, four, or six months). Once the shuttle was retired, there was neither a need nor a vehicle for large crews, which meant that the astronaut corps would have to shrink from its turn-of-the-century high of about 150 members to 40 to 50 members. Until the future was clearer, it was difficult to predict precisely what training and how many astronauts were needed.

The story of changing NASA programs and astronaut recruitment strategies is the background for the third generation of women astronauts. Women continued to apply and be selected from 2004 through the most recent recruitment in 2021, with each group from the nineteenth through the twenty-third class seemingly more highly qualified. But they differed in some ways from the two previous generations. Almost all of the eighteen women selectees since 2004 were born between 1975 and 1989, in population cohorts called Generation X and Millennials. They came of age in the 1990s and early 2000s, in a much different era from that of their predecessors, and most of them had a much different experience in plotting their futures.

Young women of their generation who aspired to be scientists, engineers, doctors, military pilots, or astronauts were confident they could pursue their chosen career. Women had become so visible in those professions, and so many of the obstacles and attitudes that previously limited women's success were gone, that young women were empowered. They saw the paths to their goals and marched ahead with support and encouragement. It wasn't a perfect world yet; there were still inequities in women's pay, advancement, and recognition, but the tide had turned since the 1960s, and they were prepared to dive in.

The United States set landmarks in the 1990s and early 2000s that reflected the increasing empowerment and success of women. One was in the political arena, when 1992 was hailed as the "Year of the Woman" because record numbers of women were elected at all levels of government from local to national.[1] The number of women in the House of Representatives increased from twenty-nine to forty-seven, a low percentage in that large body but the beginning of a steady influx that totaled 127 women in 2024. Four women won new Senate seats, and one was easily reelected. President Ronald Reagan had appointed Sandra Day O'Connor as the first woman justice on the Supreme Court in 1981, and in the next decade President Bill Clinton appointed the second, Ruth Bader Ginsburg, a champion of women's rights and equality, in 1993. As of 2024 there were four women on the Supreme Court, including the first Latina and African American women justices, and women held judgeships throughout the federal court system.

Women became visible in cabinet level posts during the 1990s and early 2000s: Elizabeth Dole as secretary of transportation in the 1980s, then as secretary of labor, and then as a senator; Janet Reno as attorney general and head of the Department of Justice; and Madeleine Albright. the first woman to serve as secretary of state. Hillary Rodham Clinton, an activist first lady like Eleanor Roosevelt, prominently advocated for women's rights, health care, and issues affecting children and families almost, some thought, as a co-president.

In other arenas than politics and public service, such as sports and entertainment, women were becoming noticeably powerful. Athletes and entertainers are widely admired by young people, and their popularity is a window into cultural trends and values. The 1990s and early 2000s saw a surge in women's professional sports with the founding of the Women's National Basketball Association, the Women's National Hockey League, and the National Women's Soccer League. Women figure skaters, gymnasts, golfers, runners, and other athletes regularly made headlines and won trophies. In the entertainment world, savvy female vocalists took charge of their careers to became superstars, earning fortunes. Talented film stars won accolades for portrayals of complex women characters. Women in journalism rose to prominence as network television news anchors. All these figures embodied independence, assertiveness, and accomplishment in the cultural milieu that surrounded the empowered generation of women as they were growing up.

The women who became astronauts after 2000 entered at a slightly older average age, thirty-three to thirty-four years in a range from twenty-nine to thirty-nine. Unlike the first generation, who were on average thirty-one years old in a range from twenty-six to thirty-five, the empowered generation did not come to NASA directly from graduate school or a briefly held first professional job. Instead, most of them had ten years of professional experience in military service or as working scientists, engineers, or physicians. Some had experience on remote military deployments or research expeditions; some had risen to positions of leadership; and some had accrued additional credentials to better prepare for astronaut duties. They had watched women astronauts in action and studied the results of the selection process for clues about how best to prepare themselves. There were no guarantees, of course, because in any given selection cycle NASA might seek more candidates in one discipline than another to meet its staffing needs, but at least applicants could see with certainty that women had a place in space. In fact, women had many places in space, and the opportunities were about to expand in new directions in programs initiated after the space shuttle era.

New Directions for Human Spaceflight

The first two decades of the twenty-first century proved tumultuous as NASA, four presidents, Congress, and the broader space policy community struggled to reach consensus on the future direction of human spaceflight. The space shuttle and space station had dominated the past quarter-century of humans in space, and both were approaching full realization. It was time to look farther ahead. In 1999, NASA Administrator Daniel Goldin tasked an internal planning group to outline NASA's next grand human spaceflight goal. He posed the basic questions: where do we want to be in twenty to thirty years and how can we get there? He insisted that the roadmap be based on a compelling rationale to fuel enthusiasm, both public and political.[2]

In 2003 the loss of *Columbia* called into question again the assumptions that human spaceflight was an essential, valuable, inspirational, national priority, albeit a costly and risky one. The space shuttle and space station were mutually dependent in the goal to establish a permanent human presence in space near Earth. Some critics proposed shutting down the shuttle program and abandoning the space station, ending US human spaceflight. The ISS program had to be completed to live up to international partnership agreements, but it was uncertain what further space effort NASA and the nation might undertake. The *Columbia* tragedy spurred a new strategy to address what would come next.

Shortly before the first anniversary of the tragedy, President George W. Bush announced his answer, derived from the forward-thinking NASA had done and recommendations from a team of agency representatives within his administration. In January 2004, at the beginning of his reelection year, President Bush articulated a new course for the nation's space program, "The Vision for Space Exploration," that would reorient NASA's human spaceflight program from activity in low Earth orbit to deep space, where robotic explorers had long operated.[3] The key points were to complete the space station, retire the space shuttles, develop a new crew exploration vehicle and launcher, return to the Moon with robotic and human explorers, and prepare for human exploration of Mars and beyond. To support the goal of human exploration far from Earth, US research on the ISS would focus on life sciences and the long-term effects of spaceflight on the human body. Bush set target years for most of these actions to be accomplished, offered NASA some startup funding, and told the agency to trim its costs elsewhere to fund the new developments. Corollaries to this vision were that private companies would increasingly handle transportation in low Earth orbit and international partners would be invited to participate in the exploration venture. Although the announcement soared with historical and inspirational allusions, this uplifting vision had a practical element. It called for an affordable, sustainable program, with affordable being the key, to extend human presence in the solar system. The "Vision for Space Exploration" implied realignments in future astronaut recruitment and training to match the new Moon and Mars focus.

Congress initially rebuffed the Bush vision, then endorsed it, but the space agency embraced it. NASA began to realign its organization and mission to these goals, naming the new program Constellation.[4] As planning progressed, NASA selected Lockheed Martin to design and develop the new Orion crew vehicle, an Apollo-type capsule, but larger and modernized for travel to the space station, the Moon, and beyond. Two new rockets were envisioned, Ares I for crew transport into Earth orbit and Ares V, an enormous heavy lifter for crew and cargo transport to the Moon and beyond. The Constellation program also included a large lunar landing craft. The point of returning to the Moon was to develop the technologies and techniques for living and working away from Earth, thus enabling discovery for the sake of science and possibly for such economic activity as using the Moon's resources. Work proceeded on all fronts from 2004 through 2011. Meanwhile, the space shuttle made its last flights in 2011, when the International Space Station assembly was declared complete.

However, Constellation ran into trouble keeping pace with the Vision's timeline due to underfunding and taking a year longer to complete the space station and retire the shuttles than predicted. NASA was juggling three very expensive programs—shuttle, ISS, and Constellation—cutting back on others, encountering technical problems, incurring unplanned costs, and falling behind. When President Bush left office and President Barack Obama arrived in 2009, Constellation was ripe for reassessment.

One of Obama's first space policy actions was to commission a study of NASA's Constellation program to make recommendations. This Review of US Human Spaceflight Plans Committee, led by aerospace executive Norman Augustine, concluded that the Constellation goals could not be met in the proposed time frame with then-foreseeable funding; it would require either a massive infusion of funds contrary to the administration's aim to reduce the national deficit, or a stretched-out schedule, or cancellation.[5] Obama considered Constellation over budget, behind schedule, and without innovation. He was less interested in a return to the Moon ("We've been there before") than efficient technology and a sustainable exploration plan to places no one has ever gone.[6] He notified NASA that he would not include funding for Constellation in the upcoming budget and would provide new guidance for a more affordable and achievable human space exploration program. The president and Congress formally canceled the Constellation program in 2011.[7]

When new guidance came, all that remained of Constellation was the crew exploration vehicle, Orion, repurposed as an ISS crew rescue vehicle. Gone were the two Ares launch vehicles and any lunar landing capability. Gone, too, was a return to the Moon. The review committee had laid out three optional strategies: Moon first, Mars first, and a "flexible path" that included other destinations, with Orion as the crew vehicle. Obama favored the flexible path strategy and a mission to a near-Earth asteroid, the Asteroid Redirect Mission, around 2025. That would allow an opportunity to develop and demonstrate innovative technologies for deep space exploration, investigate potential mineral resources in asteroids, and test a robotic technique to deflect an asteroid on a collision path toward Earth. Obama mentioned a possible Mars mission in the 2030s but bypassed the Moon. Orion, revived as the crew vehicle for any missions beyond Earth orbit, managed to survive further changes in the direction of human spaceflight under the next two presidents. Orion also became the reason to develop a powerful new heavy-lift Space Launch System (SLS) after Ares I and Ares V were canceled, because the United States did not have a rocket able to launch a human crew to deep space destinations.

The 2011–2016 Obama spaceflight program did not receive a formal name, but it retained some features of the Bush vision. Both Bush and Obama endorsed a new approach to spaceflight based on a partnership with US aerospace companies to provide access to low Earth orbit, specifically to deliver cargo and crews to the International Space Station. The Obama administration prioritized commercial activity in Earth orbit and spurred NASA to stimulate and purchase orbital transportation services for cargo resupply and crew transportation, formerly done by the space shuttle. Companies would develop, own, and operate new spacecraft, with NASA as the customer paying for their services. This was a seismic shift from NASA's history as the developer, owner, and operator of spacecraft. More vehicles meant more resilience in spaceflight, and buying access to low Earth orbit would free NASA to concentrate on deep space missions. NASA also opened the door for commercial enterprises to become involved in deep space exploration by issuing contracts for development of necessary elements, such as spacesuits and landers. Partnering with commercial enterprise allegedly would help contain costs and maintain schedule for a new era of human space exploration to begin in the 2020s. The vehicles, as well as NASA's Orion, would become the next generation of spacecraft for a new generation of explorers.

There were many dimensions to this commercialization effort, and the Commercial Crew Program (CCP) that began in 2010–2011 was the one most directly involved in human spaceflight.[8] It fostered growth of relatively new companies while continuing to work with the established aerospace giants. For example, in 2014, NASA selected SpaceX and Boeing as providers for ISS crew transportation services under the CCP. SpaceX developed its Crew Dragon capsule to transport up to four people to and from the ISS. SpaceX also offered an automated cargo version of the ship to resupply the station. Boeing's development of its crew capsule, called Starliner, suffered delays and fell behind SpaceX. As of 2024, SpaceX was operating crew rotation missions and Boeing was still in the test flight stage. Other new companies—Orbital Sciences/Orbital ATK and Sierra Space—developed vehicles for automated cargo delivery, Cygnus and Dream Chaser, respectively. They, like SpaceX and other firms, also were angling to provide spacesuits, landers, and other lunar mission elements.

By the end of Obama's presidency in 2016, none of the approved new spacecraft or NASA's Space Launch System had yet been tested, much less become operational. Since 2011 the United States had depended on the Russian space program for access to the International Space Station, paying for a seat on their three-seater Soyuz spacecraft that commuted to and from ISS up to four times a year. The fare was expensive, and sometimes technical problems caused delays or cancellations, which disrupted NASA's crew rotation schedule. It was a workable but not ideal situation. NASA was eager to launch American astronauts from US soil again by using the US-made Crew Dragon and Starliner spacecraft, like it was already using automated commercial spacecraft for cargo shipments to the ISS.

Upon taking office in 2017, President Donald Trump canceled the Asteroid Redirect Mission, already five years into planning, restored a return to the Moon as NASA's priority, and accelerated the pace, calling for a human landing on the Moon in 2024, the last year of a presumed second term of his presidency.[9] The Trump national space strategy echoed much of the "Vision for Space Exploration" in its call for "an innovative and sustainable program of exploration with commercial and international partners to enable human expansion across the solar system."[10] This administration continued to support a thriving commercial space transportation industry for Earth orbital flight and urged its potential to create a global and domestic market for American space goods and services. The administration also supported NASA's Space Launch System for deep space exploration. The goals of the "Vision for Space Exploration" and the Constellation program gained momentum again.

This return-to-the-Moon effort coalesced as the Artemis Program under NASA Administrator Jim Bridenstine, the first head of NASA born after the Apollo missions and a champion of the new "Artemis Generation."[11] Named for Apollo's twin sister, Artemis both harkened back to the first era of lunar exploration and signaled that women would go to the Moon this time. NASA committed to send "the first woman and next man" to the Moon. Artemis missions would use the Moon as a "proving ground," "test bed," and "stepping-stone" for long-term exploration and discovery in the solar system, including eventual missions to Mars. Commercial crew transportation providers, NASA's Orion, and the Space Launch System all progressed during Trump's term in office.

President Joe Biden and his incoming administration in 2021 endorsed the Artemis program, with one tweak: it would send the "first woman and first person of color" to land on the Moon, to be inclusive of America's diversity represented in the astronaut corps.[12] NASA and Congress welcomed the continuity. However, the unfinished Space Launch System prevented a lunar landing from happening in 2024. Meanwhile, commercial crew transportation efforts proceeded. Beginning in 2020, NASA contracted with SpaceX as a commercial partner for crew rotation missions to the ISS, while Boeing continued working toward its Starliner test flights and certification. NASA had already assigned several astronauts to work closely with SpaceX and Boeing on details of the new spacecraft designs and operations and to participate in their testing. Two women were on the team supporting Starliner; Nicole Mann was tapped for a test flight and Sunita Williams was chosen for the first operational mission.[13]

SpaceX Crew Dragon, with two NASA astronauts on board, passed its crewed flight test, a demonstration mission to the International Space Station in mid-2020, and the first operational mission to ISS, SpaceX Crew-1, launched in November 2020 with Shannon Walker and three male astronauts aboard. By the end of 2024, SpaceX had ferried astronaut crews, including seven NASA women, to the ISS eight times and had at least six more crew rotation missions under contract through 2030. Starliner had not yet completed test flights; its crew flight test to the ISS in 2024 experienced technical problems. Concerned for their safety, NASA and Boeing decided to send the spacecraft back in automated mode without its crew, keeping them on the space station to await the next scheduled SpaceX visit for their return flight. Starliner flight testing was expected to resume in 2025, with certification for crew transport to follow.

The uncrewed Artemis I mission in November 2022 successfully tested the powerful Space Launch System, the Orion spacecraft, and navigation and communications systems in a twenty-five and a half-day lunar flyby that looped far past the Moon before

Artemis II crew named in 2023: (*clockwise from left*) Christina Koch, Victor Glover, Jeremy Hansen, Reid Wiseman. *NASA*

the vehicle returned to Earth. The first crewed mission, Artemis II, was planned for late 2023 but postponed until late 2025 and then to late 2026 pending resolution of technical issues revealed by the test flight.[14] It is planned to be an elliptical figure-eight flyby, not a landing, that will further exercise all systems and maneuvers from low- to high-Earth orbit and then the translunar trajectory. The crew will practice rendezvous, docking, emergency, and other maneuvers during the ten-day mission and will check out the life support, navigation, and communications systems. Artemis III, planned for late 2027, includes a landing in the south polar region of the Moon, with the primary scientific task of exploring the lunar features of this unvisited area and investigating the presence of water. This mission is expected to demonstrate new spacesuits and other technologies for enabling a sustained human presence on the Moon. Increasingly complex Artemis IV and Artemis V missions would follow in the 2028–2030 timeframe.[15]

NASA announced the Artemis II crew in 2023, and as promised, it included a woman, Christina Hammock Koch, and a person of color, Victor Glover, as well as commander Reid Wiseman and Canadian astronaut Jeremy Hansen.[16] They would be the first humans on a NASA mission to the vicinity of the Moon since the Apollo missions ended in 1972.[17] The four Artemis II crew members immediately began training and making public appearances to promote the planned lunar missions. They were the only Artemis crew named as of 2024. The Artemis III landing mission is expected to include the first woman and first person of color to walk on the Moon.

After a twenty-year period of on-and-off progress, by 2024 the nation's human spaceflight program had settled into a period of certainty about its goals and plans. The flight schedule was no longer years in the future; the first Artemis flight had already occurred, and the next ones were queued up annually to 2030. Orion and the Space Launch System were on the verge of readiness. SpaceX was routinely shuttling crews to and from the space station and, with other providers, resupplying the ISS with everything needed. Political scientist Roger Handberg has described policymaking as "a dance between operational reality and political reality" with inevitable changes until the technical capabilities and political choices align.[18] It took about two decades for the dance partners to move to the same tune and rhythm, but they finally synchronized enough to execute the next decade in human spaceflight.

Recruiting Astronauts for a New Era in Spaceflight

In view of the changing priorities for human spaceflight, a third wave of recruitments from 2004 onward looked to future programs. NASA sought candidates to prepare for even longer-duration missions to the Moon and Mars or to work with commercial ventures in an expanding realm of space transportation. Just as astronauts selected in the 1990s stepped up from shuttle missions to ISS expeditions, the astronauts of the early 2000s would gain their defining spaceflight experience at the space station in preparation for more remote missions. Eighteen more women joined the astronaut corps in this phase. Their expertise included geology and geobiology for lunar and Mars missions, research missions in Moon- or Mars-like places on Earth, or leading military units or research teams. These women were far enough into their careers to have impressive resumes and managerial skills suited to the exploratory missions ahead.

After the year 2000, astronaut recruitments and selections occurred at regular four-year intervals to balance continuity of experience and leadership with diverse new talent as astronauts retired or left the corps for other opportunities. The application process was streamlined, and the application itself evolved to better solicit the kinds of information and qualities sought for the next era of exploration. While she was working in flight crew operations during 2003–2007, astronaut Ellen Ochoa led the effort to revise astronaut selection criteria, application forms, and reference questions to focus on competencies and traits necessary for long-duration expeditions in space.[19] She consulted with human relations, Astronaut Office, human behavior staff, and other resources to update the entire process from submission to final selection. The goal was to gain clear evidence of applicants' technical and operational expertise, professionalism, discipline, teamwork, and abilities as a leader and follower. The revised applicant package would help the rating panels and selection board better understand how an applicant's prior experience compared to the jobs that astronauts actually perform on the ground and in space and how prepared they might already be for the challenges of long-duration spaceflight. Ochoa oversaw initial use of this updated system in 2007 for selection of the Group 20 astronaut class in 2009.

The selection emphasis soon began to shift to operational experience and "expeditionary skills" such as teamwork, leadership and followership, self-care and team-care, communications, and conflict resolution.[20] NASA sought people with experience analogous to long duration stays in space, such as research expeditions in Antarctica and other remote locations. ISS expeditions lasting three to six months or longer, living and working with crewmembers from different nations and cultures, called for more refined interpersonal skills and personal traits than the relatively short space shuttle missions. It would be essential for the residents' mental health and productivity to communicate well, avoid or resolve interpersonal tensions, and serve alternately as leaders or followers depending on the task. The same would be true for monthslong to yearslong missions on the Moon or Mars. NASA also put increased emphasis on scientific competencies in geology and planetary sciences, in contrast to the disciplines previously desired for materials science and life science research on the shuttle and space station. Master's degrees, or two years in a PhD program, were required. Astronauts were no longer categorized as mission specialists and pilots. All served as flight engineers on board the ISS, and they were individually trained for any specialized roles on commercial spacecraft. As the astronaut corps decreased in size from a peak of 150 active astronauts in 2000 to about a third of that number in 2013, it was essential for each person to have diverse competencies and a broad spectrum of experiences.[21]

The application process itself needed to be modernized from paper to electronics.[22] In the 1990s the sorting, filing, and reviewing of thousands of multipage applications, transcripts, letters of recommendation, and health forms flooding in every two years was unwieldy at best. In 1996, the federal government moved its job application process to an online website, www.USAJobs.gov, and leading up to 2003 NASA tried its own version in a similar website, www.nasajobs.gov, first used in that year, and another, www.nasa.gov/astronauts/recruit.html, in 2007. Thereafter, NASA worked with the Office of Personnel Management to use the USAJobs portal for astronaut applications. Starting in 2011, applicants could build a standardized-format resume online and upload their academic, military, and health records, publications, and other required documents. Reference letters could be submitted privately online. Gone were the stacks of envelopes and paper files that had deluged JSC's astronaut selection office for three decades. The electronic system was a boon for applicants and reviewers alike. NASA's adoption of social media to advertise astronaut recruitment also reached far more people and more diverse populations than the former reliance on news releases and media announcements.

The new selection criteria and questions were implemented in the 2007 cycle that yielded the 2009 astronaut class. The first opportunity for applicants to use the USA-Jobs process was the 2011 recruitment for the 2013 astronaut class.[23] Recruitment announcements emphasized the opportunity to be an explorer in a different sense than in the 1980s and 1990s, now with the tantalizing prospect of leaving low Earth orbit. The 2009 and 2013 classes were the first selected specifically for long-duration missions to the Moon, Mars, or asteroids. The 2017 and 2021 classes became the core of the Artemis generation for the program dedicated to lunar missions. The recruitment and selection process thus paralleled and reflected changing priorities from the Vision for Space Exploration and Constellation through the Artemis programs.

NASA also instituted a second round of interviews, another change that helped ensure selection of the best candidates. Previously, 100–120 candidates were invited to JSC for a week of interviews, health assessments, briefings, and get-acquainted activities, and up to thirty-five candidates were selected from that number. Starting with the 2011 recruitment, NASA invited about fifty of those applicants back for second-round interviews and assessments before selecting ten to fifteen prospective astronauts for training.[24] The health and medical screenings became even more stringent than the NASA flight physical. All finalists had to undergo ultrasound and MRI (magnetic resonance imaging) examinations, in what was now called the "long-duration physical" focused on identifying any chronic or hidden issues, whether physical or mental, that might emerge as a problem in long-duration spaceflight. Examples are conditions that require ongoing medication (high blood pressure, thyroid disorders, depression, or bipolar disorder) or a congenital or incipient defect (such as an arteriovenous malformation or aneurysm) or tumors, kidney stones, or gallstones.

From 2011 onward, and especially as NASA moved into the Artemis program with the goal of returning to the Moon to set the stage for future journeys to Mars, the desirable criteria for astronauts evolved to match the expected training and flight experiences.

Recruitment and selection announcements outlined the five essential masteries now required of *every* new astronaut: operating and maintaining the International Space Station, spacewalks, complex robotics, operating a T-38 jet, and Russian language.[25] It was no longer optional, as it had been with a larger astronaut corps, to be certified in only some of those competencies. In addition, the new astronauts would need to become proficient in operating new spacecraft such as Orion, SpaceX, and Starliner, and training would include planetary science and geology fieldwork to inform surface activities on deep space destinations.

The 2015 recruitment announcement that led to the 2017 astronaut class made clear the connection between NASA's changing mission and the caliber of astronauts: "NASA's mission, and what we need from the astronauts helping to carry it out, has evolved over the years. We want and need a diverse mix of individuals to ensure we have the best astronaut corps possible."[26] Although some attributes had always been important—teamwork, for example—its meaning and practice changed somewhat from shuttle to space station to Artemis as missions became longer and more complex and, for Artemis, more distant. Crews would need to become even more skilled and autonomous the farther they ventured from mission control.

The astronaut selection process is so important to NASA's human spaceflight mission, and the astronauts are so important as the public face of NASA, that a tremendous amount of energy goes into finding, training, and cultivating talented individuals. The ideal astronaut is not a static concept; it evolves as the human spaceflight program and the operational culture of spacefaring evolve. This truism became evident in the astronaut recruitment and selection cycles from 2004 through 2021.

The Class of 2004

This recruitment cycle continued the practice of soliciting experienced jet pilots and mission specialist scientists and engineers. For the first time, application forms were available electronically online through the NASA Astronaut Selection Office website.[27] Also for the first time, teachers were invited to apply to join the regular astronaut corps as a new category, educator astronauts, eligible to be assigned to missions. They would share responsibilities with mission specialists, but their additional role was to conduct educational programming to connect with teachers and encourage students to study science and math. NASA Administrator Sean O'Keefe remarked that "the educator astronaut program will help us fulfill our mandate to inspire [the] next generation of explorers."[28] More than 1,600 teachers applied, from whom NASA selected three, one of them a woman, Dorothy Metcalf-Lindenburger. NASA also selected two pilots and six mission specialists, only one of whom, Shannon Walker, was a woman. The fourteen-member class of 2004 also included three astronauts from Japan's space agency, JAXA.

Introduced by NASA at the Smithsonian's National Air and Space Museum rather than the Johnson Space Center as usual, the 2004 astronaut class was presented as "the next generation of explorers" who would help develop the next generation vehicle and "lead us through the next steps in the new exploration vision."[29] This astronaut class was the last to train for space shuttle missions.[30]

Dorothy Marie Metcalf-Lindenburger

Dorothy Metcalf-Lindenburger was one of three teachers selected in 2004 to train as regular mission specialists within the astronaut corps. This position was unlike that of the first teacher in space, Christa McAuliffe, who trained briefly as a guest spaceflight participant for a single specific mission but was not a NASA astronaut. The daughter of science and math teachers and a geology major with fieldwork experience, Dottie taught high school earth science and astronomy for five years and coached the Science Olympiad and cross-country running. She was an Academic All-American in cross country and track during her college years.

Metcalf-Lindenburger was the first astronaut to have attended Space Camp at the US Space and Rocket Center in Huntsville, Alabama. That experience in 1990, when she was fourteen, and the wonders revealed by the Hubble Space Telescope, launched at the same time, inspired her to want to work at NASA, whether as an astronaut or in some other position.[31] After being selected as a candidate she stayed at NASA for ten years, during which she flew on a shuttle mission and commanded a NEEMO mission. She also had assignments in the ISS and station operations branches of the Astronaut Office and worked as a Cape Crusader for the final three space shuttle missions in 2011.

Her fifteen-day shuttle mission, STS-131 (2010), resupplied the International Space Station. The crew delivered more than thirteen tons of supplies and equipment and returned with more than three tons of hardware, science results, and trash. Metcalf-Lindenburger served as shuttle flight engineer, assisting the commander and pilot during launch, entry, and return. She also operated the shuttle remote manipulator system arm and supported all three spacewalks as the choreographer and backup EVA crewmember.

Two other women, NASA's Stephanie Wilson and Japan's Naoko Yamazaki, were on this flight, the last one to include three women, the last one with a seven-person crew, and the last one to include any rookies on their first mission. The three women shared duties operating the robotic arm to inspect the vehicle's surfaces for any signs of damage, support three spacewalks by the men on board, and transfer cargo between the shuttle and space station. When the shuttle docked with the ISS, where Tracy Caldwell Dyson was a resident crewmember, four women were in space together for the first time.

In 2012, Metcalf-Lindenburger welcomed her assignment to command the twelve-day NEEMO-16 mission as an opportunity for leadership and responsibility.[32] Its focus was a simulated exploration of a near-Earth asteroid, then being considered as a possible NASA mission. The aquanauts tested techniques and protocols for such a remote mission, and they operated with a communication time delay keyed to the longer distance of interplanetary spaceflight. Because communication delays will be a significant factor in a Mars mission, crews will be forced to be more autonomous in responding to problems.

For fun while at JSC, Metcalf-Lindenburger was a lead singer with the all-astronaut rock band Max-Q, which performs at NASA and astronaut events, weddings, and parties, mostly in the Houston area. Composition of the group changes as astronauts are available to perform.

After leaving NASA in 2014, Metcalf-Lindenburger moved to Washington State, where her family could pursue their enthusiasm for outdoor activities. She also earned a master's degree in applied geosciences and joined an environmental consulting firm. She remains active as a public speaker, serves on nonprofit boards such as the Challenger Center and the Museum of Flight in Seattle, and supports other organizations to advance STEM and STEAM education. She looks back at her astronaut years with humility and gratitude for the opportunity: "I was not perfect, but I did a very good job."[33]

Flights: **1** on space shuttle • **42nd** US woman astronaut in space • **53rd** woman worldwide • Time in space: **15** days, **2** hours (**362** hours)

Shannon Walker

Shannon Walker has accumulated a number of astronaut "firsts" in her career, and it hasn't ended yet. She is the first astronaut born, bred, and educated in Houston, home of NASA's Johnson Space Center and the astronaut corps. She is the first astronaut to have earned her bachelor's, master's, and doctoral degrees from Houston's Rice University. A physicist, she was the first woman to launch into space from US soil since the space shuttles were retired. When SpaceX was ready to transport crews, Walker and crew were launched from Kennedy Space Center on its first operational flight in late 2020, after nine years of NASA's dependence on Russia for rides to and from orbit.

Walker grew up with NASA and astronauts in the background of her daily life. She was four years old when Americans landed on the Moon, and her parents took her outside to gaze up, telling her that people were up there. She thought it "sounded like the best thing ever" and decided to go to the Moon someday, too.[34] As a member of the Artemis team, she may indeed have a chance.

Walker applied five times in ten years and when selected as an astronaut had accumulated so much relevant experience that she was hardly a novice.[35] Upon graduating college in 1987, she worked at JSC with NASA's space operations contractor, Rockwell, as a robotics flight controller for space shuttle missions. She took a three-year break for

Shannon Walker wearing the Russian Sokol suit, ready for her Soyuz launch to the ISS, 2010. *NASA*

graduate school in space physics and returned to JSC in 1995, this time as a NASA employee working on development of robotic systems and general problem resolution for International Space Station missions. She then moved to Moscow for a year to work with the Russian Space Agency and its teams on avionics and ISS problem solving, basically ensuring that hardware and software provided by each of the partner agencies worked together seamlessly. When she returned to JSC, she moved into technical leadership roles for ISS operations and engineering until her selection in 2004 as an astronaut candidate. Her assignments after basic astronaut training included ISS crew support and CAPCOM for both shuttle and ISS missions.

After training as a backup crewmember for two ISS expeditions, Walker flew her first mission to the space station as Soyuz pilot and as an Expedition 24/25 crewmember in 2010. During this time, the ISS broke the record previously held by the Russian space station *Mir* for longest continuously occupied facility in space. The crew also celebrated the tenth anniversary of a continuous human presence in space on ISS.

Ten years later, Walker was assigned to her second mission as a crewmember on the first operational flight of the SpaceX Crew Dragon spacecraft (SpaceX Crew-1) as part of NASA's Commercial Crew Program. This flight made her part of ISS Expedition 64/65 in 2020/2021, with Kate Rubins also aboard for the Expedition 64 segment. This expedition occurred during the height of the worldwide coronavirus pandemic, which forced the crew to take strict precautions against exposure to the virus and do much of their training remotely. Walker served briefly as commander of Expedition 65, the third woman to be in charge, but her tenure was the shortest, only eleven days before her departure.[36] On both ISS missions, Walker was fully engaged with the scientific research program and station operations.

Between the two space station expeditions, Walker commanded NASA's NEEMO-15, a two-week underwater mission to simulate exploration on the surface of an asteroid. The aquanaut team tested ways to anchor a spacecraft, equipment, and spacewalkers without the aid of gravity, which is negligible on small asteroids. They also evaluated strategies for using instruments and moving around to collect data.

Upon returning from space, Walker served as chief of at least three branches in the Astronaut Office, supervisor for the 2021 astronaut class, and deputy chief of the Astronaut Office. As she awaits her next assignment, she has inspired two unusual honors. SpaceX named one of its Dragon support vessels *Shannon* in her honor, and the Houston Public Library named a new branch near her childhood home the Dr. Shannon Walker Neighborhood Library.[37]

Flights: **2** on International Space Station, via Soyuz and SpaceX • **43rd** US woman astronaut in space • **55th** woman worldwide • Time in space: **330** days, **13** hours (**7,933** hours)

The Class of 2009

The 2007 announcement for applications was the first to indicate that long-duration stays on the International Space Station and missions to the Moon were possible.[38] There was no mention of the space shuttle, nor any mention of pilots or mission specialists, although educators were still encouraged to apply. This was the first recruitment to implement the updated skills and experience criteria that were developed under Ellen Ochoa's leadership since 2003 for long stays on the ISS and expeditions to the Moon and beyond.

Along with the standard press release announcement, NASA issued a companion "Help Wanted: Astronauts" ad on its website announcing that "NASA HAS SPACE FOR YOU!"[39] Mimicking the lingo of military recruitment posters with "America needs *YOU* at the frontiers of space and technology!," it called for "a few men and women who want to fulfill their dreams and be a part of the next generation of explorers." This ad spelled out the scope of technical training and duties astronauts must master for ISS expeditions in preparation for missions to the Moon in the Constellation program.

More than 3,500 applications arrived, from which NASA selected nine new astronaut candidates.[40] The 2009 class included three women: Serena Auñón, Jeanette Epps, and Kathleen Rubins. Five international astronauts, two from Canada and three from Japan, brought the class size to fourteen. This class signaled the transition from spaceflight on the shuttle to future exploration of space, with preparation on the International Space Station.

Serena Marie Auñón-Chancellor

Serena Auñón, later Auñón-Chancellor, first worked with NASA as a flight surgeon before selection as an astronaut candidate. After earning a bachelor's degree in electrical engineering, she graduated from medical school in 2001 and then completed residencies in internal medicine and aerospace medicine and a master of public health degree. From 2006 to 2009, as a NASA flight surgeon, she was involved in medical activities for the Astronaut Office and flight crews and supported crews training in Russia. She also practiced medicine at a free clinic. Her published research focuses on astronauts' exposure to radiation.

As part of her astronaut training, Auñón-Chancellor participated in specialized exploration training on a two-month Antarctic Search for Meteorites (ANSMET) expedition and on two underwater NEEMO missions during which she operated a submersible and worked as an aquanaut. She then was a flight engineer on International Space Station Expedition 56/57, traveling via Soyuz and staying for 197 days, to become the first Hispanic woman resident on the space station. During that time, she and crewmates installed a new life sciences glovebox for expanded research in a sealed work area and worked on experiments in other disciplines. Since returning, she has been a management astronaut working in the CAPCOM branch and handling medical issues for the ISS operations branch of the Astronaut Office. She also took a part-time university position as an associate professor of clinical medicine working with residents in internal medicine and was on duty in a hospital during the peak of the COVID pandemic.[41]

At some point, Auñón-Chancellor or NASA and the media ceased using the accents in Auñón, her Cuban American father's surname, and the pronunciation shown on her official biography fact sheet is anglicized as ON-un.[42] She is an avid amateur radio operator who enjoyed making calls from the space station. A Colorado native, she enjoys hiking in the mountains, and since living in Galveston, Texas, during her medical training, jet-skiing along the coast.

Auñón-Chancellor has noted a connection between engineering and medicine: "Both require that you examine problems from all angles and reason through multiple solutions."[43] Of her career thus far, Auñón-Chancellor says, "The decisions I made in my professional career were not toward a specific goal but because I loved what I was doing at the time."[44] Those decisions led her directly into the post-shuttle era of space exploration.

Flights: **1** on International Space Station via Soyuz • **45th** US woman astronaut in space • **61st** woman worldwide • Time in space: **196** days, **17** hours (**4,721** hours)

Jeanette Jo Epps

A physicist and aerospace engineer, Jeanette Epps worked in engineering research for ten years before she applied to be an astronaut. Her graduate studies focused on uses of shape-memory alloys, or "smart materials." She then went to Ford Motor Company's scientific research laboratory as a research engineer investigating applications of those materials in the automative industry, earning a patent for her work. From Ford she was recruited into the Central Intelligence Agency as a technical analyst to "reverse engineer" technology assets of US adversaries.[45]

Epps had long considered being an astronaut but humbly doubted whether NASA would select her. However, her admiration of Sally Ride, Guy Bluford, and Mae Jemison, and the encouragement of her astronaut friend Leland Melvin, finally persuaded her. At age thirty-eight, she decided it was "now or never" and had to try. To her amazement but no one's surprise, she was selected on her first application.

From childhood, Epps had been an excellent student, especially in science and math, and she was surrounded by encouraging family, teachers, and other adults who mentored her. She credits her graduate school advisors for opening doors for her into the engineering profession and a couple of astronauts for guiding her into the corps. Always grateful for the encouragement she received along her journey, Epps is committed to "paying it forward" by encouraging young people to pursue STEM interests.

After Epps completed initial astronaut training and technical assignments, in 2017 she was assigned to the crew of ISS Expedition 56/57 and a Soyuz roundtrip. She completed two years of training in Russia and elsewhere for the mission and was just weeks away from launch in 2018 when NASA inexplicably announced a crew change and substituted her backup, Serena Auñón-Chancellor, in her position.[46] Epps would have been the first African American to serve on a resident ISS crew, a first later accomplished by Victor Glover and Jessica Watkins. Epps seemed as ready as possible for her first spaceflight, having completed not only mission-specific training for Soyuz and ISS but also the NEEMO-18 underwater mission as an aquanaut. Epps said that she had no medical, training, or family problems to warrant removal from the Soyuz-ISS mission and claimed that she herself did not know the reason and could not speculate.[47] That decision remains a mystery, since neither NASA nor Epps has revealed what happened, despite a flurry of speculative comments posted on the internet in the aftermath. Epps returned to Houston to resume duties in the Astronaut Office, working on the Orion project and serving as an ISS CAPCOM. In 2019, she completed a European Space Agency CAVES expedition, an otherworldly analog to a lunar or planetary mission, doing scientific research with a team and finite resources in the isolated environment of remote caves.

A year later, NASA assigned Epps to the first operational mission of Boeing's new *Starliner* spacecraft to the ISS, then anticipated to fly in 2021. She trained for that mission, but the flight tests and the first mission were delayed by technical issues. She also began cross-training on the SpaceX Crew Dragon spacecraft. In mid-2023, NASA assigned her to the SpaceX Crew-8 crew rotation mission for ISS Expedition 70/71, and in 2024 she launched and completed this mission after waiting almost fifteen years to

Jeanette Epps suited up for SpaceX mission to the ISS, 2024. *NASA*

reach space. At that time, she set a record as the African American who had spent the most time in space: 235 days.

Epps recalls being in high school and the profound impression of seeing Mae Jemison celebrated on the cover of *Jet* magazine as the first African American woman astronaut. "It made me think . . . 'If she can do this then maybe I can do this as well.'"[48] That inspiration and the consistent encouragement she received along her journey toward space has made Epps passionate about being a mentor and role model: "Giving something back is huge for me. How can I not do that?"

Flights: **1** on International Space Station via SpaceX • **54th** US woman astronaut in space • **92nd** woman worldwide • Time in space: **235** days, **3** hours (**5,643** hours)

Kathleen Hallisey Rubins

A microbiologist whose interests range from disease-causing bacteria and viruses in humans to signs of microbial life in ancient rocks and lava, Kate Rubins holds degrees in molecular biology and cancer biology and has done virology fieldwork in Africa to study pox viruses, Ebola, and Lassa fever.[49] Before coming to NASA, Rubins gained government grants to establish her own lab of fourteen researchers at the Whitehead Institute for Biomedical Research at MIT. Since joining NASA, she has completed the European Space Agency's Pangaea geological training course, an analog for lunar and planetary surface exploration that is conducted in otherworldly terrains in Germany, Italy, and Spain.[50]

Rubins has completed two International Space Station missions, a four-month stay in 2016 on Expedition 48/49 and a six-month stay in 2020–21 on Expedition 63/64. Her biomedical research expertise and interest in genomics is a natural fit for many of the thousands of investigations conducted on the space station. On her first mission, she became the first person to sequence DNA in space, and she also cultured beating heart cells in a far-reaching study of changes in the heart in microgravity. Rubins is also a veteran spacewalker, having performed two maintenance and installation EVAs on each mission, and she has operated the space station robotic arm to capture arriving automated resupply spacecraft. After returning from the ISS, Rubins became chief of the EVA and robotics branch of the Astronaut Office, and she is involved in the development of new space suits for missions to the Moon.

Sequencing DNA in space has practical relevance as well as basic research value, especially on long, remote missions. It could enable astronauts to diagnose an illness or identify microbes in their environment. Spacecraft are routinely sanitized but are not sterile, and the human body is populated with myriad bacteria; despite precautions, it is possible that microbes presenting a health threat could proliferate. About that potential, Rubins says, "There are some interesting places to take a look at the microbiome of the space station."[51]

Rubins has a long association with the US Army Research Institute of Infectious Diseases and the Centers for Disease Control and Prevention. In 2021, after completing her second spaceflight during the COVID-19 pandemic, she decided to join the US Army Reserve to continue pursuing her interests in public health. She "really wanted to give back" for the opportunities she has had and also sharpen her leadership and operational skills.[52] Major Rubins serves with the 75th Innovation Command in Houston.

Flights: **2** on International Space Station via Soyuz • **44th** US woman astronaut in space • **60th** woman worldwide • Time in space: **300** days, **1** hour (**7,201** hours) • EVA time: **26** hours, **46** minutes

The Class of 2013

NASA's next recruitment was announced in 2011, the year of the last space shuttle missions.[53] This announcement clearly indicated that the next class of astronauts would be preparing for long-duration International Space Station missions and future deep space exploration missions beyond low Earth orbit, including a possible mission to an asteroid. NASA was surprised by the extraordinary response—more than 6,300 online applications, the second-highest number yet—through its first use of the USAJobs website. Longtime manager of the Astronaut Selection office Duane Ross attributed the high number to a major recruitment effort on social media and the internet to counteract public perception that NASA went out of business when the shuttles were retired.[54] Better communications extended NASA's outreach to spread the message that exciting exploration programs lay ahead and being an astronaut was still a promising career.

From that large applicant pool, NASA culled the smallest selection since 1978—eight candidates introduced as the Class of 2013. For the first and only time to date, the class included an equal number of men and women, a fact noted with approval but more coincidental than intentional. Parity was not a policy, and subsequent classes again included more men than women. In introducing this group, NASA Administrator Charles Bolden raved that "these new space explorers asked to join NASA because they know we're doing big, bold things here—developing missions to go farther into space than ever before."[55] The women in this class were Christina Hammock (later Koch), Nicole Mann, Anne McClain, and Jessica Meir.

Christina M. Hammock Koch

Christina Koch can't remember a time when she didn't want to be an astronaut.[56] She prepared well for this career: attending space camp five times, graduating from a math and science high school, receiving an astronaut scholarship to study engineering and physics, attending NASA Academy for collaborative research training, and in 2002 taking her first postgraduate job at NASA's Goddard Space Flight Center. She earned bachelor's and master's degrees in electrical engineering and another bachelor's degree in physics.

During the ten years between completing her studies and applying to become an astronaut, Hammock, as she was then known, alternated between engineering work and scientific research in two-year intervals, augmenting her experience in various kinds of exploration. She said that she had always set her sights on working with NASA, "but I didn't want to get there by checking the usual boxes, like learning to fly and scuba dive. I was passionate about science and the next frontier."[57] As an electrical engineer at NASA's Laboratory for High Energy Astrophysics and the Johns Hopkins University Applied Physics Laboratory, she developed instruments for space science and remote sensing. After each of these jobs, she did scientific and engineering fieldwork in remote environments, including a full year in Antarctica with the US Antarctic Program, seasonal stays in Greenland, and more fieldwork in Antarctica. When she returned from those expeditions in 2012, she went to work for the National Oceanic and Atmospheric Administration (NOAA) at research bases in Alaska and American Samoa. Hammock arrived at NASA in 2013 already well experienced as an explorer living and working in isolated, harsh environments.

(*left to right*) Jessica Meir and Christina Koch in ISS *Quest* airlock, preparing power tools and spacesuits for their EVA, 2019. *NASA*

After six years of training, Koch went to the International Space Station via Soyuz as a crewmember on Expedition 59/60 in 2019. Classmate Anne McClain was already on board to greet her. While there, her stay was extended into 2020 on Expedition 61 as an opportunity to collect more data on the physiological effects of longer duration space-flight. Koch set a new record for the longest single spaceflight by a woman—328 days, just five weeks shy of a full year. No woman has yet spent a complete year in space, and only one American man thus far has done so.[58] While on the ISS, Koch conducted micro-gravity research in protein crystals, biomedical studies, 3D printing of biological tis-sues, and a variety of other disciplines. She accomplished six spacewalks, including the first EVA by a pair of women with partner Jessica Meir.[59] The women paired again on two additional spacewalks.

Speaking from the space station on Women's Equality Day in 2019, Koch noted, "I have truly been inspired by the struggle and triumphs of so many women that dare to break new ground in all aspects of society, and those who have pushed the bound-aries of human imagination. And now I am both privileged and challenged to have the opportunity to be a part of that story. . . . I say to the record-breakers, 'Thank you.'"[60]

After her flight, Koch returned to serve as a branch chief in the Astronaut Office and rotated into the JSC Director's Office as a technical assistant. In 2023 Koch was named to the four-person crew of the Artemis II mission, now planned to launch in 2026 as a lunar flyby rehearsal for a subsequent landing mission. Until then, she is in training for the complex mission.

Flights: **1** on International Space Station via Soyuz • **47th** US woman astronaut in space • **64th** woman worldwide • Time in space: **328** days, **14** hours (**7,786** hours) • EVA time: **42** hours, **15** minutes

Nicole Aunapu Mann

Nicole Mann is the first Native American woman in space, registered with the Wailaki of the Round Valley Indian Tribes in Northern California, where she was born and grew up. Her original surname, Aunapu, reflects her Estonian heritage as well.

A distinguished graduate of the US Naval Academy, where she was a star and cap-tain of the women's soccer team and earned a slew of scholar-athlete and All-American honors, Mann was commissioned into the Marine Corps in 1999.[61] Upon earning a mas-ter's degree in mechanical engineering, she completed basic training and flight train-ing, picking up her call sign "Duke" and earning her wings of gold as a naval aviator in 2003. She was assigned to pilot the F/A-18 Hornet fighter/attack jet in the aircraft car-rier fleet and was deployed twice. Mann flew almost fifty combat missions before reporting to the Naval Test Pilot School, which she completed as an honor graduate. She was an F/A-18 Hornet and Super Hornet test pilot and project officer when selected to train as an astronaut. Since joining NASA, Mann has advanced in rank from major to colonel.

Before her first spaceflight, Mann served as the T-38 safety and training officer and led the astronaut corps' developmental work on NASA's Orion spacecraft, Space Launch System heavy-lift booster, and exploration ground systems for missions to the Moon. Mann then commanded SpaceX Crew-5 for ISS Expedition 68 from late 2022 into early 2023. During their five-month stay, the crew participated in hundreds of experiments and technology demonstrations. Mann completed two spacewalks doing maintenance on the solar arrays and other tasks, and she operated the robotic arm for two other EVAs. Upon returning, Mann was assigned to the test flight crew for Boeing's Starliner spacecraft, still in development for roundtrip crew transportation for International Space Station expeditions.

Mann didn't contemplate becoming an astronaut until she considered her career path after combat duty. She missed engineering. Her decision to become a test pilot, and an opportune tour of NASA, put her on a course that never seemed possible in her youth. She understands that she is now an example and inspiration, especially for Native American young people, whom she encourages to dream big. "I hope that young women can connect with me and my journey and maybe see a little bit of themselves in me," she says. "Maybe that will give them the inspiration and the courage that they need to follow their dreams."[62]

Flights: **1** on International Space Station via SpaceX • **51st** US woman astronaut in space • **79th** woman worldwide • Time in space: **157** days, **10** hours (**3,778** hours) • EVA time: **14** hours, **2** minutes

Anne Charlotte McClain

Anne McClain is the first female graduate of the US Military Academy at West Point to enter the astronaut corps. Commissioned as an Army officer in 2002, she earned her wings as a pilot in the OH-58D Kiowa Warrior scout/attack helicopter, claimed the call sign "Annimal," and flew 216 combat missions. She later became qualified in other rotary, fixed-wing, and multiengine aircraft and graduated from the Navy Test Pilot School. Colonel McClain is a Master Army Aviator, experimental test pilot, instructor pilot, and experienced leader in command at the platoon, company, battalion, troop, and detachment levels. She came to NASA after more than a decade of military service.

McClain earned undergraduate and master's degrees in mechanical/aeronautical and aerospace engineering. She holds two additional master's degrees in international relations and strategic studies. She attended the Army War College and completed the Command and General Staff College. During some of her postgraduate years, she also played rugby on the USA Women's National Team and captained a USA Women's Rugby All-Stars team, and she played rugby in England while studying there as a Marshall Scholar. McClain applied to NASA in 2009 and was not accepted, but she reapplied after graduating as a test pilot in 2013 and was selected.[63] In the astronaut corps, McClain has been a branch chief and a CAPCOM, EVA, and robotics instructor.

McClain flew roundtrip on a Soyuz spacecraft as a crewmember for ISS Expedition 58/59 (late 2018 to mid-2019), where she was a flight engineer, researcher, and spacewalker. She completed an EVA and was scheduled to do one with crewmate Christina Koch, the first time for a pair of women spacewalkers. However, McClain recommended that she not participate in that EVA because the large suit she would have worn did not fit her as well as a medium size.[64] She and Koch both needed to wear a medium size suit and only one was ready; assembling and checking out another would put the crew off schedule. For safety and schedule reasons, she thought it wiser to change places with her male EVA partner, who normally wore the large suit, for the spacewalk with Koch. All agreed, and McClain later did a second EVA in the medium suit. This instance highlighted a spacesuit size and fit issue that also had affected other women astronauts' ability to do EVAs.

Speaking to students at her hometown alma mater, McClain said that she wanted to be an astronaut since she was three years old. That is all she ever wanted to be, and she told her mom that her reason to go to school was to learn to be an astronaut. "I didn't have any special gifts or talents or advantages or disadvantages," she said, but "where you end up is a series of decisions you make, one after another. It's not a giant leap; it's making little decisions each day" to push yourself toward your goal, because "Our dreams don't live in our comfort zones." She suggested some affirmations that helped her along the journey from school to space: "I am brave" and "I can do hard things."[65] McClain's next assignment was to command the SpaceX Crew-10 mission in 2025.

Flights: **1** on International Space Station via Soyuz • **46th** US woman astronaut in space • **62nd** woman worldwide • Time in space: **203** days, **15** hours (**4,887** hours) • EVA time: **13** hours, **8** minutes

Jessica Ulrika Meir

Jessica Meir drew a picture of an astronaut standing on the Moon when she was five years old to show her class what she wanted to be when she grew up. At age thirteen, she attended a space camp at Purdue University. As a high school senior, she wrote in her yearbook that she intended to do a spacewalk. Her friends called her "space girl."[66]

Meir took a path less traveled to become an astronaut; instead of studying physics or engineering, she chose to follow her passion, biology, with graduate degrees in space studies and marine biology.[67] She pursued interests in marine mammals and birds in extreme environments, focusing on the physiology of oxygen deprivation in animals that dive below the ice in Antarctica and a species of geese that fly above the Himalayas. Between her master's and doctoral studies, Meir worked at the Johnson Space Center supporting human physiology research on the shuttle and space station, which led to research flights on NASA's KC-135 parabola-flying aircraft and the undersea NEEMO-4

(2002) mission as an aquanaut. She also participated in Smithsonian diving expeditions to the Antarctic and Belize. When selected as an astronaut candidate, Meir was an assistant professor at the Harvard Medical School and Massachusetts General Hospital.

After completing NASA astronaut training, Meir was a crewmember on a European Space Agency space analog CAVES mission in 2016, and she served as lead CAPCOM for an ISS expedition. On her first flight as an astronaut, she traveled via Soyuz to be a flight engineer on the International Space Station Expedition 61/62 in 2019–2020. During her stay she participated in the first three all-woman spacewalks with ISS crewmate Christina Koch, recognizing their significance in the long history of women striving for equal access and opportunity in space.[68] As she handles management responsibilities in the Astronaut Office, Meir remains eligible for future missions to the Moon and Mars or back to the space station.

Meir is a first-generation American whose parents immigrated from Israel and Sweden to the United States, where she was born. She has dual citizenship in the United States and Sweden. She is also a multi-instrument musician and holds a private pilot's license. Asked about her encyclopedic interests and abilities, Meir responded that they stem from curiosity and seeking to understand what she doesn't know, the same goal that propels exploration on or off the planet.[69]

Flights: **1** on International Space Station via Soyuz • **48th** US woman astronaut in space • **65th** woman worldwide • Time in space: **204** days, **15** hours (**4,911** hours) • EVA time: **21** hours, **44** minutes

The Class of 2017

In 2015 NASA announced that it was preparing to select another group of astronauts. This time the notice was a friendly, direct appeal: "Be an Astronaut: NASA Accepting Applications for Future Explorers.... NASA is looking for the best candidates to work in the best job on or off the planet."[70] This announcement touted the news that astronauts soon would fly on US spacecraft on missions to the International Space Station, Moon and Mars. They would have opportunities to fly on four different spacecraft: the International Space Station, SpaceX Crew Dragon, Boeing's Starliner, and NASA's Orion.

The educational requirement changed slightly in specifying at minimum a bachelor's degree in engineering, biological science, physical science, computer science, or mathematics, with an advanced degree desirable with at least three years of related, progressively responsible professional experience. Pilots still needed at least a thousand hours of pilot-in-command time in jet aircraft, and everyone interviewed had to pass the NASA physical for long-duration spaceflight. For the first time, applicants were directed to complete an online self-assessment in addition to the application form on the USAJobs website.

Six women astronauts among the 2017 class: (*front*) Zena Cardman, Loral O'Hara; (*middle*) Jennifer Sidey-Gibbons (Canada), Jasmin Moghbeli, Jessica Watkins; (*rear*) Kayla Barron. *NASA*

A record-setting more than 18,300 people applied, 10,000 more than responded to the 1978 call for astronauts. NASA interviewed 120 and selected twelve, of whom five were women: Kayla Barron, Zena Cardman, Jasmin Moghbeli, Loral O'Hara, and Jessica Watkins.[71] The new class of twelve US candidates and two from the Canadian Space Agency reported for training in August 2017. It is customary for each new class to adopt, or be given, a nickname. This group became the Turtles, because shortly after their arrival Hurricane Harvey flooded the Houston area, stranding turtles and other wildlife. Four years later, class member Kayla Barron used a stuffed turtle as the zero-gravity indicator, which drifted weightless when the SpaceX-3 crew reached space.[72]

The Turtles were the first class to graduate under the Artemis program.[73] The new curriculum for their two years of training had five main courses: flying T-38s, ISS systems, robotics, spacewalking, and Russian language, as well as training in expeditionary skills such as leadership, followership, team care, and communication. Upon graduation, they were eligible for assignments to the International Space Station, the Moon, and eventually Mars.

Kayla Jane Sax Barron

A distinguished graduate of the US Naval Academy, Kayla Barron earned a bachelor's degree in systems engineering and a master's in nuclear engineering as a Gates Scholar abroad at the University of Cambridge. She was inspired by the events of 9/11, when she was fourteen years old, to serve her country, and after graduating from the academy she was in the first group of women admitted into submarine service. She was trained and assigned to be a submarine warfare officer and then completed three tours of duty at sea before joining NASA as a lieutenant. Barron says that after serving on a submarine she had a random conversation with astronaut Kay Hire about the International Space Station, and she suddenly realized that it seemed much like a submarine. She talked to her Navy mentor about whether she might be able to qualify as an astronaut. The mentor responded, "You know how to become an astronaut? You apply! You put yourself out there."[74]

Barron's first spaceflight was on SpaceX Crew-3 to the International Space Station for six-month duty on Expedition 66/67 in 2021–2022. Barron served as mission specialist on SpaceX and as flight engineer on the station, completed two spacewalks and operated the robotic arm for another, and was mainly engaged in onboard research and routine activities. Promoted to lieutenant commander after her mission, she was named to the Artemis team to support development of technologies and operational concepts for upcoming lunar missions. She is eligible for a future ISS or Artemis flight assignment. She thinks that "There is no better preparation for spaceflight than serving aboard a submarine."[75]

Flights: **1** on International Space Station via SpaceX • **49th** US woman astronaut in space • **72nd** woman worldwide • Time in space: **176** days, **2** hours (**4,226** hours) • EVA time: **13** hours, **26** minutes

Zena Maria Cardman

A doctoral candidate in geosciences when selected by NASA, Zena Cardman holds degrees in biology and marine science, with minors in chemistry and creative writing. Her research interests are in geobiology and extreme life forms (extremophiles) in subsurface environments—caves, ice, hydrothermal vents, and other places that might be analogous to environments off the planet. She has studied ecosystem changes and microbial systems in different locales from the Arctic to Antarctica, and she has participated in several NASA-supported research projects relevant to scientific planetary spacewalks. Cardman says that, at NASA, "You want to be as versatile as possible. You want to be a payload who's worth your weight. You want to be able to be that scientific Swiss Army knife in the field."[76]

Cardman was assigned to command SpaceX Crew-9 for ISS Expedition 71/72 in 2024–2025, but plans changed when NASA decided to use that flight to return the first two Starliner astronauts from the space station after their spacecraft developed worrisome technical problems.[77] Crew-9 launched with only a commander and pilot and two empty seats for the Starliner pair's return. Cardman was reassigned to command SpaceX Crew-11 for an ISS expedition in 2025–2026.[78]

Jasmin Moghbeli

Jasmin Moghbeli arrived in America as a one-year-old, born to Persian parents who had fled Iran during the 1979 revolution, first to Germany, where she was born, and then to the United States. She became a US citizen when she was seven years old and is the first person of Middle Eastern descent to be a NASA astronaut. Going to space was her ambition since she presented an in-costume book report in sixth grade about Valentina Tereshkova.[79] She attended the Advanced Space Academy (a higher-level Space Camp) in Huntsville, Alabama, as a teenager, and then charted her path to space through degrees in aerospace engineering and becoming a military pilot. She was a multisport athlete as an undergraduate at MIT, where she was captain of the women's basketball team and also played lacrosse and volleyball. She says that experience was significant in her application, interviews, and training as an astronaut because "sports are one of the great ways" to learn teamwork.[80]

Moghbeli was commissioned as an officer in the Marine Corps in 2005, completed basic and pilot training, flew the AH-1W Super Cobra attack helicopter, deployed overseas three times, and tallied 150 combat missions. She completed US Naval Test Pilot School in 2013 as both the honor graduate and outstanding student of her class and became a helicopter test pilot. Along the way, she acquired the call sign "Jaws." Moghbeli now holds the rank of lieutenant colonel.

Until she began training for her first spaceflight, Moghbeli worked on development of the human landing system for future missions to the Moon. She then commanded the SpaceX Crew-7 mission for a six-month stay as flight engineer on the International Space Station, where she lived and worked in 2023–2024 on Expedition 69/70. Classmate Loral O'Hara joined her aboard the ISS for the Expedition 70 segment. They performed their first EVA together, working outside for almost seven hours to replace components in a variety of maintenance tasks. After serving on the space station, Moghbeli remains eligible for other assignments, and possibly a mission to the Moon or eventually Mars.

Flights: **1** on International Space Station via SpaceX • **52nd** US woman astronaut in space • **86th** woman worldwide • Time in space: **199** days, **2** hours (**4,778** hours) • EVA time: **6** hours, **42** minutes

Loral Ashley O'Hara

Growing up in a Houston suburb, Loral O'Hara was no stranger to NASA when she was selected to become an astronaut. She was always aware of space exploration and decided at age eight that she wanted to be an astronaut.[81] She made almost every decision in her life with that goal in mind. En route to earning her undergraduate and master's degrees in aerospace engineering, aeronautics and astronautics, she participated in three NASA internships and research programs for students. She then spent eight years at Woods Hole Oceanographic Institution as a research engineer working on the engineering and operations of both human-occupied and remotely operated submersibles, good preparation for space missions and spacewalking. She is a veteran of eleven scientific cruises aboard research vessels and is also a private pilot and certified emergency medical technician. She applied twice to be an astronaut before being selected on her third attempt, but in the interim she developed skills and qualities of an explorer while discovering the connections between space and ocean exploration.

O'Hara served as the Astronaut Office's director of operations for crew training in Russia before her first spaceflight in 2023. She traveled via a Soyuz spacecraft to the International Space Station to join the crew of Expedition 69/70. With classmate and ISS crewmember Jasmin Moghbeli, O'Hara completed her first spacewalk, working outside for almost seven hours as the second pair of women on EVA. After serving on the space station, O'Hara remains eligible for other missions in the Artemis program.

Flights: **1** on International Space Station via Soyuz • **53rd** US woman astronaut in space • **87th** woman worldwide • Time in space: **203** days, **15** hours (**4,887** hours) • EVA time: **6** hours, **42** minutes

Jessica Andrea Watkins

Jessica Watkins became the first Black woman to complete a long-term mission on the International Space Station in 2022 and, with a total of 170 days, was the African American who had spent the most time in space until surpassed by Jeanette Epps in 2024.[82] Other Black women have visited the ISS on shuttle missions, but Watkins was the first to be a member of a resident crew. As a flight engineer for six months on ISS Expedition 67/68, she engaged in Earth and space science, biology, biomedical research, and station operations with European Space Agency astronaut Samantha Cristoforetti, one of her SpaceX Crew-4 teammates.

Watkins earned degrees in geology and environmental sciences while also playing intercollegiate and international rugby on both the Stanford University Division I national champion team and the USA Eagles Women's World Cup team. She also trained with the Olympic team. As a student, she held internships supporting Mars exploration programs at NASA's Ames Research Center and planetary projects at the Jet Propulsion Laboratory. She has worked on the Mars Science Laboratory and *Curiosity* rover science team, the Mars *Perseverance* rover, and a planned Mars sample-return mission. She has participated in two analog missions at the Mars Desert Research Station and the underwater NEEMO-23 mission, all testing technologies, science objectives, and operations for lunar and deep space missions. At the time of her selection as an astronaut candidate, Watkins was doing research in geological and planetary sciences as a post-doctoral fellow at the California Institute of Technology (Caltech).

Watkins says she was inspired to become an astronaut at age eight or nine when attending an after-school program at the Judith A. Resnik elementary school in Maryland and learning about America's second woman astronaut. Sally Ride and Mae Jemison also inspired her. Having dreamed of spaceflight since childhood, Watkins says that realizing a dream is "just putting one foot in front of the other on a daily basis. If you put enough of those footprints together, eventually they become a path toward your dreams."[83]

Flights: **1** on International Space Station via SpaceX • **50th** US woman astronaut in space • **75th** woman worldwide • Time in space: **170** days, **13** hours (**4,093** hours)

The Class of 2021

In 2020, NASA announced "Explorers Wanted" for missions to the Moon and Mars.[84] In a uniquely personal recruitment appeal, NASA Administrator Jim Bridenstine invited applicants to "join us in this new era of human exploration that begins with the Artemis program to the Moon. If you have always dreamed of being an astronaut, apply now," adding that "it's an incredible time to be an astronaut." He vowed, "We will send the first woman and next man to the lunar South Pole by 2024, and we need more astronauts to follow suit on the Moon, and then Mars."[85]

At that time, there were forty-eight active astronauts, most involved with the International Space Station. The ISS had reached twenty years of continuous operation with at least one American, and usually two or three (but occasionally more) aboard, and that program had no definite end in sight. To support both ISS and the coming Artemis missions, the agency needed to expand the corps and prepare younger astronauts for missions beyond the 2020s.

The 2020 announcement made some notable changes that elevated the basic requirements for the job. Now a master's degree in a STEM field was *required*, whereas previously a bachelor's degree was enough, but a graduate degree was *desired* or *preferred*. Alternatively, two years of work toward a technical PhD, a completed medical degree, or completion of test pilot school with at least a bachelor's degree in a STEM field would satisfy the requirement. Acceptable STEM fields were engineering, physical sciences, biological sciences, computer science, and mathematics, and, for the first time, certain disciplines in psychology (clinical, physiological, or experimental). Candidates also needed to have at least two years of related, progressively responsible professional experience and be able to pass the long-duration spaceflight physical. In addition to technical training, candidates could expect expeditionary behavior training in leadership, followership, and teamwork skills.

Four women in the class of 2021 (*left to right*): Nichole Ayers, Jessica Wittner, Deniz Burnham, Christina Birch. *NASA*

This recruitment yielded more than twelve thousand applications. NASA selected just ten candidates, four of them women—Nichole Ayers, Christina Birch, Deniz Burnham, and Jessica Wittner.[86] Of the women, three were military officers. This class graduated from training in 2024 and became new members of the Artemis generation, eligible for assignment to the International Space Station or deep space missions on the Orion spacecraft to the Moon.[87] Upon their graduation, NASA concurrently announced a new recruitment with applications due in March 2024 for a class to be selected in 2025.[88]

Nichole Stilwell Ayers

Nichole Ayers was by age thirty-two already an experienced T-38 Talon and F-22 Raptor instructor pilot and combat aviator, and she had led the first all-woman F-22 formation in combat. Her call sign is "Vapor" for light steam that may form inside the F-22 in hot, humid weather. Inspired when she saw the Air Force Thunderbirds flight demonstration team perform over her Colorado hometown, she wanted to attend the Air Force

Nichole Ayers conducting research on the ISS, 2025. *NASA*

Academy, be a pilot, and follow that path to become an astronaut. She applied to NASA as soon as she had the required number of flight hours in addition to her master's degree. "I always had an affinity for the sky and for space," she says, and "the little explorer in me" who camped and hiked throughout Colorado informed her desire to explore space.[89]

A graduate of the US Air Force Academy who majored in mathematics, she earned a master's degree in computational and applied mathematics from Rice University. At the academy she took advantage of intensive language training and studied abroad in Kazakhstan and Kyiv to minor in Russian, knowing that fluency in Russian language and culture would be an asset in spaceflight. Ayers played Division I volleyball all four years at the academy and later competed internationally on the US Armed Forces team. She held the rank of major upon selection as an astronaut candidate. Her new goal is to get to the Moon. Even before she began astronaut training, Ayers remarked, "Everybody else has paved the way for us to become part of the NASA team. Hopefully, we can start to build the way for others . . . to the Moon and long-term spaceflight."[90] Upon completing astronaut candidate training in 2024, Ayers began training to pilot a SpaceX mission to the International Space Station in 2025 as the 55th US woman astronaut.[91]

Christina Marie Birch

Chris Birch holds an undergraduate degree in mathematics, biochemistry, and molecular biophysics and a doctorate in biological engineering. While in graduate school at MIT, she took up competitive cycling from 2009 through 2015, entering twenty-five races in one year alone.[92] She then pursued a professional cycling career, aiming for the elite level and supporting herself by teaching at the University of California, Riverside, and Caltech. She found that teaching and training full-time was not sustainable, so she took a sabbatical in 2018 to become a professional track racing cyclist on the US National Team and qualified for the 2020 Olympics in Tokyo, postponed until 2021 due to the global COVID pandemic.[93] With teammates, Birch has earned three World Cup medals, twice participated in world championships, and is a gold medalist in the Pan American Games.

She realized while doing lab research that her work was similar to work on the International Space Station and that she had the right skill set, so she applied to NASA. Upon selection as an astronaut candidate, this dedicated athlete began working on upper body, shoulder, and grip strength to prepare for spacewalk training. Among her first assignments after graduation from initial training was to practice and assess procedures for ocean splashdown and recovery in the Orion spacecraft, as if it had just returned from the Moon. These exercises in the Pacific Ocean involved a large team of personnel in helicopters and water craft in what Birch called "an amazing learning opportunity."[94] Asked about going to the Moon, she replied, "I don't need to be the first, I just want to be a part of this program."[95]

Deniz Melissa Burnham

Deniz Burnham has perhaps the most unusual work experience background among all the astronauts. With a degree in chemical engineering, she entered the energy industry in 2007 and managed oil and gas drilling rig projects in the Arctic, Alaska, Canada, offshore and in the continental United States for more than a decade, usually as the boss and the only woman on the rig. She earned a master's degree in mechanical engineering during this time.

Burnham's work schedule usually was two weeks on duty and two weeks off, with regular travel between rig sites and her home in Alaska. Fascinated with aircraft since childhood, she had ample time between jobs to develop a hobby in aviation. Burnham became a private pilot with fixed wing, helicopter, and instrument ratings. She also skydives and hang-glides; flies paramotors and paragliders, seaplanes, and sailplanes; and enjoys riding motorcycles. She trained to fly on her own as a private citizen paying for lessons and renting aircraft at flight schools during her weeks off. She says that flying is her opportunity to learn new skills, build confidence in her competence, and take a break from the isolation and man's world of oil and gas rigs. Flying keeps her life balanced between hard work and pleasure. Burnham mostly flies solo and sees herself as a cautious pilot, wary of paying for mistakes with her life.[96]

In 2017 Burnham was commissioned into the Navy Reserve and trained as an engineering duty officer. As a lieutenant she served as the executive officer of a shipyard operations unit in Alameda, California.[97] Her work in the drilling industry and the Navy gave her the operational experiences needed to become an astronaut, as well as problem-solving skills, adaptability, discipline, and teamwork skills. She says, "I always wanted to be a NASA astronaut, but I didn't quite know the path I would take to get there. . . . Engineering is what spoke to me."[98]

Jessica Wittner

An aerospace engineer, naval aviator, and graduate of the US Naval Test Pilot School, Jessica Wittner began her naval career as a teenager in the Naval Sea Cadet Corps. Entering astronaut training at the rank of lieutenant commander, she had enlisted in the Navy right out of high school in 2001 as an aviation machinist mate/aircraft mechanic. She earned her commission as an officer after graduating from college in 2009 and then entered Navy flight training. Wittner flew F/A-18 fighter jets, serving in two Strike Fighter Squadrons, and worked as a test pilot and project officer in air test and evaluation, leading engineering teams working on improvements to the F/A-18E-F Super Hornet. She also attended the Naval War College.

As a girl, Wittner played with rockets. She cites her submariner grandfather for inspiring her to join the Navy and working on motorcycles with her dad for her early interest in engineering.[99] She has always wanted to know how and why things work as they do. During astronaut training, she turned her curiosity to a new interest: geology lectures and fieldwork through the European Space Agency's Pangaea course. She now feels prepared to explore the lunar surface should she be assigned to an Artemis landing mission. "I would love to go to the Moon," she says. "That's like the next frontier right now. We know what questions need to be answered."[100]

The Artemis Team

Named for Apollo's twin sister, Artemis both harkens back to the first era of lunar exploration and signals that women will go to the Moon this time. NASA committed to send "the first woman and next man to the Moon," a mantra that soon became "the first woman and first person of color." Apart from demonstrating innovative technologies and techniques for exploration, NASA wanted the public to understand that the new era of exploration would be inclusive and representative of the diversity of America.[101]

In December 2020 NASA named eighteen astronauts to form the Artemis team for missions on and around the Moon.[102] Nine of them were women, a historic milestone in equity and parity. Eighteen months later, the agency announced that all current astronauts would be considered for Artemis missions, thus expanding the roster to forty-two astronauts with a total of sixteen women as the Artemis generation.[103] Most were members of the 2013 and 2017 classes, but several had been in the corps longer. When the 2021 astronaut class graduated from training in 2024, they joined the Artemis generation eligible for flights, adding ten more to the total and increasing the number of women to twenty. As of the end of 2024, the active-duty astronaut corps numbered forty-seven, almost half of them women. Six decades after women were dismissed from consideration as astronauts, five decades since American men journeyed to the Moon repeatedly, and four decades since women were admitted to the US astronaut corps, women finally have equal standing, if not equal number, to reach the Moon and eventually venture on to Mars.

The twenty women of the Artemis generation are Nichole Ayers, Kayla Barron, Christina Birch, Deniz Burnham, Zena Cardman, Tracy Caldwell Dyson, Jeannette Epps, Christina Koch, Nicole Mann, Megan McArthur, Anne McClain, Jessica Meir, Jasmin Moghbeli, Loral O'Hara, Kate Rubins, Shannon Walker, Jessica Watkins, Sunita Williams, Stephanie Wilson, and Jessica Wittner. The most senior among them are Stephanie Wilson and Sunita Williams, both selected in the late 1990s. They are the most experienced women, and most experienced women of color, in the astronaut corps. Most of these women are veterans of a long-duration stay on the ISS, with a few of them awaiting their first assignments to an ISS expedition. The wait for an Artemis flight may be a long one, though, since only one lunar mission per year has been scheduled. In 2024 NASA issued another call for applications, with astronaut candidate selection anticipated in 2025.[104]

Space for Women

Since 1959, NASA has selected 360 astronauts, 61 of them women. As of the end of 2024, with twenty women among the forty-seven active astronauts ready for flight assignments and another three women astronauts serving in management roles, a total of twenty-three women were on duty as NASA astronauts. If frequency of flights is a measure of success, the 1980s through 2011 were prime time for US woman astronauts. With a peak corps of 150 astronauts, a total of 135 space shuttle missions flying a total of 852 crew-members on an average pace of five flights a year over thirty years, astronauts had a good chance of going into space every two or three years. Some flew as many as five, six, or seven times. Flight frequency slowed after the shuttles retired and crews were smaller for the International Space Station, leaving about four opportunities a year for assignment to an ISS expedition. Even so, more than thirty US woman astronauts so far have served on space station crews, several more than once. Artemis flights to the Moon will be even less frequent; once annually is planned. Those missions will be exotic and thrilling, but the wait to fly will inevitably be long.

The astronaut corps approached gender parity in four selections (Groups 20–23, 2009–2021), but overall men still predominate. Women astronauts on the job now make up about 42 percent of NASA's working astronaut corps, whereas they made up 18 to 20 percent of the corps when it was larger. This is notable progress since 1978 when women were first selected, but the numbers remain low for people of color.[105] Thus far, twenty-one astronauts have been Black, and fewer than that were of Hispanic/Latin, Asian, or Native descent. There has not yet been a recruitment class of astronauts that was majority women, and except for the large 1996 class, which included five persons of color, there have rarely been more than two or three in a class. In contrast, the Russian space program, which has orbited more than 135 cosmonauts, has flown only three more women since making headlines with Valentina Tereshkova and Svetlana Savitskaya decades ago.

Private or commercial spaceflight, sometimes called space tourism, accounts for an increasing number of women in space who are able to pay the fare or travel as invited guests of wealthy sponsors. Iranian American entrepreneur Anousha Ansari was the first woman to buy a ticket for roundtrip transportation and a ten-day stay on the International Space Station in 2006. SpaceX has scheduled customers' private space-flight into orbit, and one flight in 2024 looped round the Moon. Blue Origin and Virgin Galactic have offered suborbital spaceflights in rocket-powered aircraft that give pas-sengers about four minutes of weightlessness during a half-hour flight. Such private flights have attracted celebrities and enthusiastic adventurers, notably Wally Funk from the Lovelace Women in Space program. As noted earlier, SpaceX also provides commercial transportation for its NASA and Axiom customers. At least one company is trying to develop a space hotel, the ultimate destination for wealthy vacationers, which may increase the population of men, women, and maybe children in space.

Since 2000, the number of women in space who are not professional astronauts has increased markedly, mainly on brief Virgin Galactic and Blue Origin suborbital flights or private SpaceX orbital flights during the 2020s. By the end of 2024, their number exceeded thirty. A smaller number of professional astronauts from other coun-tries flew during the first quarter of this century: Julie Payette of Canada, on her sec-ond shuttle mission; Claudie Haigneré of France, on her second ISS expedition; Naoko Yamazaki of Japan, on a shuttle mission; China's taikonauts Liu Yang, Wang Yaping, and Wang Haoze on the Tiangong space station; Russian cosmonauts Yelena Serova and Anna Kikina on ISS expeditions; and ESA astronaut Samantha Cristo-foretti of Italy on two ISS expeditions.

The number of women in space is likely to increase steadily with the growth of national space programs, commercial spaceflight ventures, space tourism, and private initiatives. For example, the Association of Spaceflight Professionals and other commer-cial groups have selected astronaut candidates for commercial flights, not affiliated with any national or international space agency. As space becomes more accessible, such enterprises may open paths for more women to fly in space. Meanwhile, the space agencies of other nations continue to recruit and train astronauts, including women, for partnership flights with the United States or commercial providers.

At the same time, the number of professional NASA astronauts may reach a steady state for a while because of limitations on the residential capacity of the International Space Station and the developing lunar base infrastructure. ISS resident crews typically arrive four times a year for overlapping six-month stays with a head count of six or seven occupants per expedition. Visitors may come for a few days several times a year, but that doesn't add up to many flight opportunities, and traffic at the ISS must be managed to permit automated resupply craft to come and go on a regular basis. Another consideration is that as of 2024, the United States has committed to use the ISS only until 2030, which, without an extended term, would certainly affect future space station operations and reduce flight opportunities for NASA astronauts. Artemis missions with four crew members are planned at the rate of one a year, beginning in 2026. Until missions to the Moon occur at a higher rate, there may be little need to increase the size of the professional astronaut corps beyond maintaining an essential capability. In that case, increases in the number of people traveling in space will occur primarily through commercial and private spaceflight, and through independent space programs of other nations, rather than NASA and its partner space agencies.

Counting both shuttle and ISS missions by the end of 2024, fifty-four US women have flown a cumulative total of 134 times, and most of them have flown two or more times. That seems a good return on the investment of preparing for, training, and serving in the Astronaut Corps. With few exceptions, these women worked as astronauts for ten to thirty years, making it the heart of their careers. They took pride in being equals to the men, doing everything alike and expecting no concessions. US women astronauts have undeniably made their place in space and excelled in the profession. The one thing they wish everyone to realize is that they are astronauts, period. They were never separate or classified as different from the men, so they do not identify themselves, or want to be identified, as women astronauts or female astronauts. Their record of skills and accomplishments, the respect they earned doing the job, and the inspiration they have become earns them the title: Astronauts.

CHAPTER 5

Astronauts at Work: From the Ground Up

Astronauts spend far more time on Earth than in orbit, much of it in training for flight assignments and even more working on technical assignments at their desks, in labs, or at remote locations. They may be involved in developing new technologies, assessing procedures, testing software, evaluating tools and equipment, leading study teams within the Astronaut Office or elsewhere within NASA, learning Russian for international work, participating in missions in an underwater station, doing a rotation at another NASA center or at NASA headquarters in Washington, DC, or any number of tasks that expand their expertise and serve the needs of the space program.[1] They usually do several jobs at a time. A call for astronaut applicants in 2024 gave a summary of expected duties on the job, but it barely scratched the surface of the multiple roles that astronauts hold concurrently.

Many of the women chosen to be astronauts have spent ten or more years of their careers at NASA. Shannon Lucid and Anna Fisher, from the 1978 class, had the longest tenures as of 2024: thirty-four years for Lucid and thirty-nine years for Fisher. A few others have marked twenty to thirty years with NASA. Lucid worked during the full duration of the space shuttle era and was a CAPCOM for the final shuttle mission, STS-135 in 2011, before she retired in 2012. Fisher spent her last twenty years focused on the International Space Station until she retired in 2017. These two astronauts exemplify ways that women successfully managed careers with long workdays, stressful assignments, frequent travel, and the rhythms of family life. Other women moved on from NASA after much shorter stays to pursue other career options. Mae Jemison left to found her own company and foundation a year after her 1992 flight. Sally Ride left in her tenth year at NASA and then spent twenty-five years in a second career in physics and science education.

As of 2024, the active astronaut corps had forty-seven members, twenty of whom were women, and three more women are among twelve astronauts serving in management roles. NASA anticipates recruiting every four years or as needed to maintain a cadre of about fifty people eligible for spaceflight. As civil servants, astronauts

At Mission Control Center, 2017: (*left to right*) Megan McArthur, Christina Koch, Ellen Ochoa, Nicole Mann, Tracy Caldwell Dyson, Anne McClain, Anna Fisher, Marsha Ivins, Serena Auñón Chancellor, Jessica Meir, Sunita Williams, Stephanie Wilson, Kate Rubins, Shannon Walker, Ellen Baker, and Karen Nyberg. Peggy Whitson on screen. *NASA*

are not as highly paid as one might think, given the exceptional expertise they have and the risks they face in spaceflight, but thousands of applicants seek the job at every recruitment.

Life in the Astronaut Office

The daily life of astronauts on the ground revolves around the Astronaut Office at Johnson Space Center (JSC), their home base, where they carry out a variety of roles. It is organized into various branches by program or function, which change in name and number over time as the size and priorities of the astronaut corps fluctuate. The office typically has at least five branches for such functions as extravehicular activity (EVA), robotics, CAPCOM, shuttle (formerly), International Space Station (ISS), commercial crew, safety, software testing, crew equipment, crew mission support, payloads and habitability, and others. When they are not training for a specific mission, astronauts in each branch work on projects networked to engineering offices and labs around JSC to ensure that crew perspectives and expertise are factored into all flight-related projects. These representatives keep the rest of the Astronaut Office informed about activities in all relevant technical and policy areas and guarantee that astronauts work closely with other experts throughout the center to avoid becoming "siloed" in specializations.

For example, the EVA branch engages in planning, problem-solving, and developing anything to do with spacewalks, from conducting training runs in the neutral buoyancy lab and virtual reality simulations to resolving suit design issues, providing input into new suit concepts, and mentoring new astronauts in necessary skills. EVA works closely with other parts of the NASA organization, such as the crew equipment office, the training and simulation office, the virtual reality lab, and the neutral buoyancy lab. The robotics branch members interact with the training and simulation and robotics offices at JSC while working with the remote manipulator arms used in orbit and also innovative technologies for missions to the Moon and Mars. The crew equipment branch works with engineers and other specialists to develop and evaluate the multitude of tools used in space, such as standard manual and power-drive hand tools and custom-designed tools for unique tasks, such as a container to hold bolts and screws captive to keep them from floating away in space. Flight clothing is considered crew equipment; after space shuttle astronauts wore the original standard blue pants, shorts, and polo shirts for five years, astronauts wanted more choices. Rhea Seddon and Mary Cleave worked on crew equipment in the late 1980s and managed to retire the solid blue clothing in favor of khaki pants and shorts and win a decision that each mission crew could choose shirts in a variety of styles, colors, and patterns.[2] That change made a difference in crew morale and team identity.

Rhea Seddon described a crew equipment liaison as one who deals with everything from clothing to the galley to the toilet—anything onboard for the crew's use.[3] She noted that if the crew needs a special pair of gloves to handle an experiment, the crew equipment liaison will make it happen. This fellow astronaut reviews options before presenting them to the crew for approval. Once the crew decides what they want, others responsible for stowage, operations, and maintenance decide where the new items will be stowed, how they will be used, and what the crew needs to know about equipment failure or repair. There is always more process involved than simply adding a piece of equipment.

Kathy Sullivan recalled her work in mission development, a role that brings crew perspective into early stages of mission concept planning before an actual crew is assigned. At that stage, the planning team decides what payloads will fly on a particular mission, how they will share space and onboard resources, and how the crew will interact with the experiments, satellites, or other payloads. As the early consultant to a cargo group putting together a flight, Sullivan "brought the crew perspective into all their planning efforts . . . and made sure that each flight crew, when they came along, got a fairly mature engineering and checklist package to work from. So I managed the [STS] 41-G cargo . . . for six to nine months. I was charged with getting it laid out right, laying out the experiment objectives, and writing the initial checklist."[4] As a result, she knew the logic and technicalities of the payload plan and was selected to be a crewmember on that mission, during which she became the first American woman to do a spacewalk.

An astronaut may be assigned to more than one branch during her career and may be tapped to be a branch chief managing a group of astronauts. The branch chiefs also assist the chief of the Astronaut Office with administrative tasks as needed, such as preparing reports and briefings or chairing a task group on a new policy. For some of the women astronauts, becoming a branch chief is a first step toward higher management, although no formal ladder exists. It is an opportunity to exercise leadership and connect more broadly with NASA's engineering and administrative offices in the arena of human spaceflight. Others among the women choose to stay focused on technical work and readiness for flight assignments. Management roles can take an astronaut out of flight rotation, in effect postponing the next opportunity to fly. Not everyone wants to risk losing her position in the queue.

Rhea Seddon was among the first women to request management experience rather than wait for it to be offered.[5] In the 1980s there were only fifteen women in the astronaut corps and a limited number of leadership positions, with more than one hundred men vying for stature. Seddon negotiated an opportunity to work as an assistant to Carolyn Huntoon, head of the Life Sciences Directorate. Seddon aspired to work on upcoming Spacelab life science research missions and thought helping to lead the life sciences lab would be good preparation. Huntoon was amenable. However, working as staff of a unit other than the Astronaut Office hadn't been done before, and Seddon proceeded with care not to be seen as a complainer or troublemaker when presenting a new idea. When she proposed her plan, the chief of the Astronaut Office balked at the idea. Instead, he asked her to be an assistant to the flight crew operations director (FCOD) in the umbrella organization that included the Astronaut Office. Seddon was the first woman to assume that role. She worked with JSC leadership in the headquarters building, learned how FCOD was managed, handled or delegated tasks assigned by the director to ensure their timely completion, and worked on return-to-flight projects after the *Challenger* tragedy. And despite never working in the Life Sciences Directorate as she had hoped, she was assigned to both of the Spacelab Life Science missions flown in 1991 and 1993, serving in leadership as payload commander.

Seddon reflected that if she expressed interest in a certain technical assignment, she had a good chance of getting it. As a physician, she helped put together the onboard medical kits and procedures, train the designated crew medical officers, and figure out how to do CPR in weightlessness. Her first assignment seemed more random—working on the galley design, nutrition, and food products—an assignment that some thought was sexist or second-rate compared to working on spacesuits or the robotic arm. The unspoken code was to do what you were assigned to the best of your ability and not complain, as if each assignment were a test of your mettle for a possible flight assignment. Seddon was satisfied with her assignments, noting that "if there was something that you wanted to work on, something that you found really interesting, a lot of times you would say, 'I want to go work on this,' and they'd say, 'Okay, your next assignment will be that,' or 'Add that to your list of assignments.' . . . So you got to work on things that you were interested in and that you could help with, or that you could learn from if you knew nothing about that at all. You had to be brave enough to say, 'I don't know anything about that; I need to work on that for a while."[6]

No one seems to recall precisely which of the early women astronauts was first to become a branch chief, but several held the role in the 1980s, and it became common from the 1990s on. Bonnie Dunbar and Kathy Sullivan had turns as mission development branch chief. Shannon Lucid and Marsha Ivins headed the Cape Crusaders team at times. Ellen Ochoa was chief of the space station branch and then the CAPCOM branch. Ivins headed the Constellation branch when that was the upcoming post-shuttle program. Astronauts moved into the branch chief role as they grew in experience and expertise, but they rarely held it more than three years so they could stay eligible for flight assignments. For women, being in a management role could conveniently coincide with pregnancy at a time when they were not allowed to be on a mission, although they could be assigned to training for a future flight. Anna Fisher recalls how satisfying it was to lead the ISS branch during the planning for assembly of the space station in the late 1990s.[7] It was a time of intense, complex work that included much negotiation and coordination with international partners to decide how and where crewmembers would be trained and how in-orbit responsibilities would be divided or shared. Fisher held that position long enough to consider it a pinnacle of her NASA career. Janet Kavandi also led the ISS branch and considered it one of her favorite roles for its broad exposure to all issues of human habitation in space.[8]

When training for a flight, the assigned crew moves into a shared office until the mission is completed. This arrangement furthers the bonding, camaraderie, and mutual support essential to their teamwork. They also participate in team-building experiences, such as a National Outdoor Leadership School wilderness expedition when they spend a week outdoors honing leadership and followership skills, sharing responsibilities, problem-solving, dealing with adversity, and developing trust in one another. An office housing a three-to-seven-person crew may not seem ideal, but it is good practice for sharing the close quarters and lack of privacy in orbit. It is also a good way to learn how each other thinks and, as Peggy Whitson noted, "see how they react to things, see how to interact with them well."[9]

More experienced astronauts often share offices with newer ones to mentor and foster fellowship for a tightly knit astronaut corps. The work is serious and challenging, but astronauts also have a lot of fun together participating in team sports with other NASA units, an annual chili cook-off, an all-astronaut rock band (Max-Q), party skits, pranks, and good-humored teasing. Astronauts tend to live in neighborhoods close to NASA, so families become friends, children attend school and play together, parents socialize, and a strong sense of community is prevalent. It is common for astronauts to say that they feel part of an extended family.

Astronauts always serve as chief and deputy chief of the Astronaut Office. In almost sixty-five years through 2024, sixteen men and one woman have held the chief astronaut position. Alan Shepard served twice for almost eight years, and John Young served for thirteen years, but the average tenure is about three years. An astronaut cannot be assigned to a flight while serving as chief, and astronauts generally prefer to stay in the flight rotation, so they limit their term length or never aspire to be chief at all. Most chief astronauts have been military pilots, more often from the Navy than other services, and have commanded missions. Only one scientist and one educator astronaut, both mission specialists, have served as chief thus far.

Until recently, mission specialists (scientists, engineers, and physicians) were not even considered for the chief astronaut position. For years, the Astronaut Office hierarchy operated much like a military squadron. In a scheme familiar to them, pilots *assumed* leadership. They already had the benefit of formal leadership training and experience through their military service and had held the highest position on spaceflight missions. They were well versed in spacecraft systems and mission operations. Furthermore, the main responsibility of the chief astronaut is to assign crews to missions and ensure that all astronauts are ready to fly, which means deciding on their technical assignments and training opportunities. There was no one better suited to make such judgments than senior pilots and experienced commanders, they reasoned.

The military pilots gradually accepted mission specialists—at first, males—holding the deputy chief astronaut position. Linda Godwin broke through as the first woman in 1992 and served as deputy chief for five years. Peggy Whitson followed in 2002–2005. Janet Kavandi was deputy chief in 2006–2008, succeeded by Suni Williams in 2008–2009, and then Megan McArthur took the position in 2017–2020.

It was not until 2009 that a woman, civilian, or scientist became chief of the Astronaut Office, breaking the lock on a position always held by a military pilot. Mission specialist and biochemist Peggy Whitson was the first and, as of 2024, only woman to head the Astronaut Office. More important to her colleagues, she was the first chief who was a scientist and civilian, not a military pilot. In the quasi-military hierarchy of the Astronaut Office this was quite a change. She had previously served as the ISS operations branch chief and as deputy chief of the Astronaut Office. She had flown two tours on the International Space Station, once as commander, and she held the chief position until 2012, when she was ready to return to flight and command a second ISS expedition.

NASA usually issues a press release to announce a new chief of the Astronaut Office, but Whitson's position change was buried, perhaps inadvertently, in a notice of new crew assignments.[10] NASA didn't herald this historic moment, and the national media didn't pick up on it, but the astronauts certainly took notice, and Whitson detected some shock among her colleagues. Some women astronauts think Whitson's elevation as a scientist was more important than her being female, because the pilots had such a long record of leadership of the astronaut corps. Others think the timing for her appointment was perfect given the increasing focus on the ISS research facility, and Whitson had more experience than anyone in that realm. The end of space shuttle operations was only a few years away, and Whitson's elevation signaled transitions.

Whitson herself felt that not being a pilot was more significant to her appointment than being a woman.[11] Some wondered how different her leadership style might be. They had seen her in action as deputy to the chief astronaut and deputy to the director of flight crew operations, who oversees the Astronaut Office, but not as the head of an organization, and she was well prepared to move into the job. The office needed to be reorganized and downsized for ISS missions and the post-shuttle era. In fact, the last space shuttle missions flew in 2011, and the program was phasing out during her time as chief. Amid uncertainty and low morale during a period of change, Whitson communicated well, stayed optimistic, and helped her peers make decisions about whether to stay or leave NASA. By the end of her term as chief, the astronaut corps shrank from almost eighty to about fifty members, a hard loss of one-third of the astronauts to industry, academics, or retirement. Orchestrating that reduction called for extraordinary finesse.

One of the most challenging of the chief's responsibilities is to project mission and crew requirements up to five years ahead, taking into account planned missions, necessary skills, training periods (2.5 years for an ISS mission), astronaut attrition, the list of who has not flown, and uncertainties like delays or cancellations, medical issues, and other complications. Putting crews together is a complex job made more so by a dynamic environment. During her term as chief, Whitson had another challenge—a review by the National Research Council, triggered by the Office of Management and Budget, to evaluate management of the astronaut corps. The purpose of the review was to evaluate whether the astronaut corps was properly organized, right-sized, and ready for staffing the space station in the post-shuttle era.[12] She deftly managed the six-month process in an exhaustive team effort with a productive outcome. Reflecting on her time as chief of the Astronaut Office, Whitson remarked that she was pleased to be in the right place at the right time.

Jobs on the Ground

In addition to the various technical assignments arising from the Astronaut Office, some jobs take astronauts out of the office on special duty. For these assignments, the astronaut works in another facility on the JSC campus (Mission Control, for example, or the headquarters building) or at another NASA location (Kennedy Space Center or NASA headquarters, for example) or outside the United States, as at the cosmonaut training complex in Star City, Russia, or another of the international partner sites. Some of those on-the-ground jobs follow.

The capsule communicator (CAPCOM) is the flight crew's representative in the Mission Control Center. CAPCOM is the direct contact through whom the mission control team and the in-flight crew communicate orally; only the CAPCOM speaks to the crew to ask questions, relay updates, and give guidance as needed. Information flows to the CAPCOM from flight controllers and "backroom" technical support teams through the flight director, the leader or "boss" of the Mission Control Center team. CAPCOMs cultivate a calm, dispassionate, brief, and clipped style of communication; there is no chatter or wordiness, just efficient factual communication of instructions or requests for information. There are rare exceptions to this protocol—a few words of good wishes before launch and upon reaching orbit, a welcome home greeting upon landing, and perhaps a bit of humor or congratulations when some action in orbit goes well. Otherwise, it's strictly business.

CAPCOMs are assigned to specific shifts on specific missions and become integral parts of the mission team through all the preflight simulations and the actual flight. Astronauts rotate through CAPCOM training and duty, so everyone serves in this crucial position. CAPCOMs essentially live the mission minute by minute with the crew, know what they are doing, understand any problem they may have, and interact with them as peers. It is almost like being a member of the in-orbit crew. Rhea Seddon claimed, "There was no better way to be in the thick of things than to be a CAPCOM."[13] Serving as CAPCOM involves so much problem-solving and decision

CAPCOM Shannon Lucid on duty for the last space shuttle mission, 2011. *NASA*

making that it is the best way to understand how a flight crew is supported by teams on the ground. Anna Fisher recalled asking for CAPCOM experience before her first flight, intuiting that it would help her be a more effective crewmember. Her mission commander was initially skeptical, fearing that it would distract from her required training, but he relented, and she found that being CAPCOM for the first Spacelab mission (STS-9) was invaluable preparation for her flight.[14]

Mary Cleave served as CAPCOM for STS-7. She and Sally Ride made the first space communications between two women but didn't even notice. They were just doing their jobs, as they had done them during training and simulations. The media noticed, though, and wanted to report it as a momentous occasion, although Cleave and Ride declined to support that messaging. Being CAPCOM, said Cleave, "You live the mission with the people up there [and] know exactly what's happening."[15]

CAPCOMs learn about other missions in depth, gaining a breadth of view beyond one's own missions and understanding the extent of ground support that is essential for mission success. The crew in space and the mission control team on the ground are so well synchronized that everyone experiences the mission together for days on end. It's easy to lose track of the day of the week and time of day on the ground (real time) while working together in mission elapsed time (MET) from launch to landing.

CAPCOM duty has lighter moments, despite the strict protocols for brevity and clarity in communications. When Kathy Sullivan was CAPCOM for STS-26, she had an idea for a different sort of wake-up music, which is traditionally suggested by family members.[16] She and a Houston area producer arranged for comedian Robin Williams to tape a shuttle-themed version of his radio monologues in the 1987 film *Good Morning, Vietnam*. It was a complete surprise to everyone in mission control and in orbit, a real coup.

Astronauts often say that serving as CAPCOM is their favorite assignment other than being in orbit. "The next best thing to being in space is being in Mission Control as a CAPCOM, but the view isn't as good," says Kathy Thornton. Astronauts agree that being the voice of Mission Control-Houston for the crew in space is as close as one can be to actually flying the mission.[17]

Playing off the comic-book nickname for the superhero Batman, "Cape Crusader" is a term coined by the astronauts for their more formal title, Astronaut Support Personnel (ASP). It, too, is a job that astronauts especially enjoy because it keeps them close to an upcoming mission and has them working with engineers and technicians in other specialties. During the space shuttle era, Cape Crusaders were on intermittent duty at the Kennedy Space Center (KSC). Their job was to be the flight crew's representative during preparations of the vehicle and payloads, tests of equipment and procedures, launch countdown rehearsals, launches, and landings.

For space shuttle missions, the Cape Crusader team of two to five astronauts was responsible for the configuration of the orbiter interior for launch. They set cockpit

Cape Crusader Marsha Ivins at Kennedy Space Center, 1995. *Photo by Carolyn Russo, National Air and Space Museum* (NASM.2014.0025-bx050-fd007_011)

switches in the correct positions, and someone stayed in the orbiter to monitor all entries and final installations to ensure that nothing changed accidentally. When launch countdown started, the Cape Crusaders verified communications links between the vehicle and the control centers. Before the crew boarded, they set up the seats, ensured that flight checklists and other crew aids were accessible, and made whatever adjustments the crew had requested. The lead member of the team helped the close-out crew strap the crewmembers into their seats and do their comm-checks, perhaps with some humor, and then wished them Godspeed on the mission. They were the last ones out of the vehicle and also among the first ones inside at landing to greet the crews and get them ready to exit. Cape Crusaders were called the crews' eyes and ears at KSC, and at the same time they gained knowledge about what to expect for their own launch experience.

Anna Fisher served as a Cape Crusader for the third and fourth shuttle missions and as the lead crusader for Sally Ride's STS-7 mission. She recalls spending the night inside the shuttle orbiter while eight months pregnant, determined that no one would touch anything inside *Challenger* before the crew boarded.[18] Kathy Sullivan remembers being focused on payloads as they were processed, checked out, and installed.[19] Marsha Ivins recalls the camaraderie with the NASA contractor close-out crew, who were responsible for the safety of everyone on the launch tower and ensured that the astronauts were fully settled before closing the hatch and vacating the area.[20] Pamela Melroy served as a Cape Crusader many times and says that she "loved the intimacy of spending the last couple of moments on Earth with the crew."[21] She was on duty as a Cape Crusader for the final *Columbia* mission, STS-107 in 2003. As the prime ASP, she was assigned to stay at KSC after *Columbia* disintegrated during its return from space and to lead the effort to identify recovered debris of the crew cabin and crew equipment—a heartbreaking task for any astronaut.

When NASA and the Russian Space Agency Roscosmos became partners in the International Space Station program, NASA established a liaison office near Moscow in Star City, home of the Yuri Gagarin Cosmonaut Training Center. Teams of engineers and managers were deployed there to coordinate the myriad details of hardware design and operations of the Soyuz spacecraft and *Mir* space station, training regimens, and preparations to fly in space as partners. Astronauts assigned to support their colleagues who were training for flights in Star City called themselves Russian Crusaders and represented NASA crew interests in technical meetings and problem solving so the crewmembers could stay focused on their training. Being a Crusader was a good way to learn how the Russians operated and what cultural differences between the two space agencies would need to be settled if they were to work together effectively.[22] It was also good preparation for an astronaut aspiring to be assigned to a Russian flight. Books written by the American astronauts who served on *Mir*, including Shannon Lucid, give insight into the experience of living and training in Russia.[23]

Family escorts and casualty assistance and calls officers (CACO) are assigned to every mission. Each mission crew selects two astronauts to be family escorts for launch and landing. They help spouses with travel arrangements, shepherd them to briefings and receptions, accompany them to the viewing sites, make sure that children are accommodated, and generally take care of details so that everyone has a good experience.[24] On two tragic occasions, the family escorts and other astronauts attended to grieving family members. Each astronaut selects a trusted fellow astronaut to care for his or her family in case of an accident or death while on duty. This role, derived from military tradition, is a service born of friendship and compassion, but there are also guidelines to ensure that everyone receives the appropriate level of care. The CACOs are called to duty in the worst-case scenarios—the *Challenger* and *Columbia* tragedies, when each time seven families suffered the sudden, horrific death of their beloved astronaut. CACOs provide comfort and practical assistance with funeral arrangements, insurance filings, communicating with extended family, and helping with household routines. After a tragedy, some stay especially close to the family for life. Among the women who have accepted this trusted role, Janet Kavandi, Lisa Nowak, and Stephanie Wilson bore somber responsibilities for members of the STS-107 *Columbia* crew. Mary Cleave, a good friend of *Challenger* crewmember Judy Resnik, took on escort duty at the NASA memorial service attended by President Reagan at JSC.[25]

It's not an on-the-ground job, of course, but flight proficiency is a requirement for all astronauts, all of whom are trained to fly in the supersonic T-38 jets that are the signature aircraft of the astronaut corps. Pilot astronauts must keep up their proficiency by logging a certain number of flight hours in these jets by flying over Texas and the Gulf of Mexico and also by flying T-38s on their business trips to other NASA locations and contractor sites. For space shuttle missions, they flew around KSC to gauge the wind conditions for launches and flew as chase planes for landings in Florida and California to monitor the spacecraft's configuration and capture in-air photographic documentation. Kathy Sullivan drew chase plane duty on the second shuttle mission and remembered how thrilling it was to watch and photograph the shuttle from that vantage point.[26]

When astronauts who are not designated pilots go flying, a pilot astronaut or trainer is always in command. Nonpilot astronauts also need to log flight hours in the back seat where they handle communications and navigation. Most women who enter the astronaut corps from the military services are experienced pilots, but unless NASA designates them as pilot astronauts, they are back-seaters in the T-38s like the scientists and engineers. Occasionally someone who has flown comparable jets as pilot-in-command in the military has been approved to pilot the T-38s, but it is an exception to the rule.

During the space shuttle years, pilot astronauts also stayed proficient in approach and landing skills by flying the shuttle training aircraft, a Gulfstream II jet that was modified to "feel" and fly like the shuttle. They also spent hours upon hours in the shuttle training simulator learning all the possible malfunctions and failures that could occur and practicing how to respond quickly and correctly. They had to build the "muscle memory" and confidence to stay calm and know exactly what to do in any flight emergency.

Yet another opportunity to fly was assignment to the NASA B-57 (WB-57F) High Altitude Research Aircraft. Originally a high-altitude bomber and reconnaissance military aircraft, the modified twin-engine airplanes hosted a varied program of scientific research projects. Kathy Sullivan was assigned to this aircraft program early in her astronaut career. Flying in this aircraft required wearing a full-pressure suit. As a result, she was the first of the women astronauts to be certified for full-pressure suit operations, which gave her a head start toward neutral buoyancy training in the EVA pressure suit, about which more later.[27]

Management is an opportunity that some astronauts seek and others prefer to avoid, but eventually almost all advance into leadership positions. Occasionally an astronaut may request to be taken off the active flight roster and assigned to a management position elsewhere in NASA. There is no hard-and-fast rule about how many missions an astronaut may fly, but informally or customarily, three to five flights are considered enough, because other astronauts in the queue await assignments, and it is important for all to gain spaceflight experience. After their last flight, some astronauts move into management within NASA, some decide to leave NASA, and others transition into interesting management jobs outside NASA. Bonnie Dunbar, for example, moved up into the JSC director's office to carry out special projects such as building research relationships with universities. Shannon Lucid spent a year as NASA chief scientist at headquarters in Washington, DC. Yvonne Cagle spent most of her NASA career in flight medicine and medical research at the Ames Research Center in California. Nancy Currie went to Cleveland, Ohio, as chief engineer in the NASA Engineering and Safety Center that was established after the 2003 *Columbia* tragedy. Janet Kavandi moved to Ohio to become director of the Glenn Research Center. Pamela Melroy retired from NASA, worked at two other federal agencies, and then returned to NASA twelve years later in 2021–2024 as deputy administrator, the fourth woman confirmed by the Senate to hold that senior position.

For those who seek management experience directly related to the astronaut corps, the next level up from the Astronaut Office is the Flight Crew Operations Directorate (FCOD) at JSC. It includes both the Astronaut Office and Aircraft Operations at NASA's Ellington Field airport. The director and deputy director are responsible for policies and decisions about the astronauts' flight opportunities and advancement and the need for and selection of new astronaut candidates. George Abbey wielded great influence over the astronaut corps as the flight crew operations director for the first fifteen or so years of the space shuttle era. After he moved up to NASA headquarters and then became

JSC director, a path opened for some astronauts to move into management in flight crew operations.

Women who moved into the Astronaut Office deputy chief role were Bonnie Dunbar, Linda Godwin, Peggy Whitson, Janet Kavandi, Sunita Williams, and Megan McArthur. In time, Godwin, Kavandi, and Ellen Ochoa, who had served as deputy director, were appointed to leadership of the Flight Crew Operations Directorate, now the Flight Operations Directorate. Rhea Seddon served a period as assistant to the FCOD director before returning to flight status.

The most common leadership role among astronauts is mission commander. Two military women pilots held that role on shuttle missions, and both civilian and military women have commanded ISS expeditions and SpaceX missions. Pilot astronauts Eileen Collins and Pamela Melroy were the only women to command space shuttle missions. Thus far only three women have served as commander of International Space Station expeditions; they were mission specialist astronauts Peggy Whitson (Expedition 16 in 2007–2008), Sunita Williams (Expedition 33 in 2012), Whitson again (Expedition 51 in 2017), Shannon Walker (Expedition 65 in 2021), and Williams again (Expedition 72 in 2024–2025).

There are only four formal leadership positions at the top of the Astronaut Office and Flight Crew Operations Directorate, which are usually held for three-year terms, so astronauts eager to lead either had to wait their turn or seek other leadership opportunities. For mission specialists on shuttle crews, the one chosen to be payload commander led the onboard scientists and engineers and was responsible for scheduled research and tests. These payload command positions held during the training period and flight were important opportunities for women who sought to fulfill their potential as managers. Kathy Sullivan, Bonnie Dunbar, Rhea Seddon, Linda Godwin, Ellen Ochoa, Tammy Jernigan, and Kathy Thornton served as payload commanders and welcomed this leadership role on their missions.

For International Space Station expeditions, astronauts are no longer designated as pilots and mission specialists. Rather, there is one commander, and everyone else is a flight engineer. All ISS crewmembers essentially share the same jobs, tending to space station systems and maintenance, housekeeping, and scientific investigations. Crewmembers experience some division of labor for specific tasks, but basically everyone is capable of doing everything under the general management of the ISS commander and the timeline masters on the ground. There was a brief interlude in 2006 when NASA decided to designate one member of the crew as the ISS science officer, and Peggy Whitson was first and last to hold that title.[28] It evidently proved unnecessary and redundant, given the way that ISS crew share the workload.

The new commercial spaceflight missions to the ISS since 2020 opened another, seemingly quicker path for women to become commanders. On a four-person crew, one person serves as pilot, another as commander, and the others as flight engineer. As of

2024, seven NASA women have flown on SpaceX crew missions to the space station, with two—Nicole Mann and Jasmin Moghbeli—as commanders of SpaceX Crew-5 and SpaceX Crew-7, respectively. Anne McClain and Nichole Ayers were assigned to be commander and pilot of SpaceX Crew-10 in 2025, the first time for two women in those positions on the same mission. Zena Cardman was scheduled to follow as commander of SpaceX Crew-11 in 2025–2026. In addition, retired NASA astronaut Peggy Whitson, working as director of human spaceflight with commercial space company Axiom, flew as commander of the Axiom-2 mission in 2023 and was scheduled to command Axiom 4 in 2025.

When asked about any perceived differences in leadership styles between men and women, some women astronauts remarked either that they had left the Astronaut Office before Peggy Whitson took charge or that they had encountered too small a sample of women in management to make a judgment. Their perspective was that personality rather than gender made a difference, or that the noticeable differences were generally between military and civilian leaders rather than males and females.[29] The main trait the women astronauts said they appreciated in managers was collegial communication about decisions rather than issuance of edicts.

NASA Extreme Environment Mission Operations (NEEMO) underwater missions are simulations, or "analogs," of space missions. Since 2001, NASA and its partner space agencies have assigned astronauts to participate in underwater research missions as part of their preparation for a space station mission.[30] An analog mission involves spending time isolated with a crew in harsh conditions somewhat like the space environment at undersea, Antarctic, and desert sites. The crews conduct research, test equipment and procedures, and follow such mission practices as timelines and communications protocols with their remote mission control team. A preferred analog facility is the underwater Aquarius Reef Base located in the Florida Keys. It is shared by marine researchers from various institutions, and since 2001 NASA has booked it for more than twenty NEEMO sessions. Along with more than forty astronaut men, fifteen of the women astronauts—one quarter of them—have completed NEEMO missions, with six serving as mission commanders.[31] An all-woman crew carried out the NEEMO 23 mission led by Italian astronaut Samantha Cristoforetti. NEEMO missions are an effective training experience for missions and leadership in space.

Each NEEMO mission, which may last from a few days to three weeks, after a week of intense topside training, has research objectives for the aquanaut crew to accomplish. Often these include underwater extravehicular activities (EVAs) in scuba or helmet dive gear while wearing weighted suits to permit walking on the ocean floor. Crewmembers may be assigned to a NEEMO mission before a spaceflight to gain operational experience living and working with a team in confinement or after a flight to keep their skills fresh and gain experience as a commander. Four-member crews sharing the habitat often include international astronauts and engineers or scientists from other agencies or NASA units, as well as two support technicians. As in space, men and women serve together on NEEMO missions.

It is advantageous to problem-solve and try out new technology and procedures closer to home than in space. Tests can be done more quickly and less expensively, and any trial-and-error solutions can be refined long before flights. Analog missions are also a way to evaluate team dynamics and any behavioral effects of confinement, such as tension, depression, or boredom, that might affect morale and effectiveness. If these should occur, they can be addressed and remedied before spaceflight. Analogs also allow for mimicking unfamiliar conditions, such as the forty-minute round-trip communications delay that will be experienced on Mars missions. Crews must become more self-sufficient in planning their days and dealing with emergencies without near-instantaneous voice and email transmissions. NEEMO missions are reminiscent of the Tektite II series of underwater research expeditions that NASA treated as space analog missions in 1970.

Activities on these missions have included assembling space station–like structures underwater, testing tools and robotic crew aids, evaluating telemedicine techniques for remote surgery, testing high-tech breathing helmets and features of spacesuit design, simulating moonwalks and sample collecting, setting up devices on the seabed, operating a small rover, navigating and tracking outside the habitat, taking photos to document work, dealing with the unexpected, monitoring biomedical functions, conducting audio and video media interviews and educational chats with students, making meals and performing housekeeping chores, unpacking and packing up—activities much like those carried out on spaceflight missions. The distinctive NEEMO 15 mission simulated a full human exploration mission to an asteroid using suitable tools, techniques, timelines, and submersibles.

Summing up their experience, the NEEMO 12 crew led by Heide Stefanyshyn-Piper wrote in their mission journal, "How neat it is to use the ocean to help launch our next return to the moon!" and "We are amazed at all of the analogies that we have with space and lunar exploration, but also how the ocean floor is a very unique and wonderful place."[32]

Astronauts also may participate in analog programs offered by the European Space Agency (ESA) at sites in Europe. One, called CAVES, carries out exploratory expeditions in complex cave systems. Teammates start with classroom training and then move deep underground to live, do scientific research, and adhere to safety protocols in the dark, hazardous environment from which there is no quick escape. They also monitor and adjust their reactions under duress to improve communications and problem-solving behaviors. Practicing these skills is valuable preparation for spaceflight. Jeanette Epps and Jessica Meir participated in three-week CAVES expeditions before their missions on the space station. The other ESA analog program, Pangaea, is a training

Aquanaut Megan McArthur Behnken (*lower right*) and crew of NEEMO-21 expedition at Aquarius Reef, 2016. *NASA*

course for geology and astrobiology fieldwork in preparation for surface exploration of the Moon and Mars. Astronauts explore craters, deserts, and other lunar and Martian analog sites to hone their observation and sampling skills and the related technical vocabulary. Kate Rubins and Jessica Wittner completed the Pangaea course in advance of possible Artemis missions.

Training, training, and more training: that is what astronauts do. Astronauts often jest that their initial basic training as candidates is tantamount to earning a master's degree in "Advanced NASA." They might well add that as they progress through mission training they are earning the equivalent of doctoral degrees in systems engineering and the research fields represented by the investigations they conduct in space. The earliest astronauts selected before the space shuttle was flying hardly knew what to expect. They found out quickly, as if drinking from a firehose, that in one year they had to learn everything about spacecraft systems, orbital mechanics, astronomy, and computer software. Whatever their background, they had to learn engineering through hours spent in classroom lectures and then studying at home. Mary Cleave said, "We were all pretty shocked at the amount of coursework."[33] Astronauts always want to leap into mission-related training and technical assignments, but first they go through generic in-the-field training in parachuting, parasailing, water survival, land survival, parabolic KC-135 aircraft flights for familiarization with weightlessness, and flying in the T-38. They go on field trips to other NASA centers to become familiar with their projects and facilities. In labs scattered about the JSC campus, they gradually become familiar with the remote manipulator system (robotic arm), exposure to underwater neutral buoyancy, and the EVA suit. Then they learn how to prepare space meals, use space toilets, work out on space exercise equipment, set up sleep restraints, use a variety of cameras and tools, and practice everything related to spaceflight. Absorbing all the necessary technical knowledge and skills is mentally and physically demanding, sometimes bruising, as astronauts sometimes discover when they undress after a day of grueling activity.

Once assigned to a shuttle flight, astronauts underwent further intensive mission-specific training for a year or longer. Training for space station missions, however, lasts two to three years, partly in the United States and partly in Russia, Japan, Europe, and Canada. Crewmembers learn everything about the mission profile and payloads, as well as their own roles and responsibilities. Then they are cross-trained to back up others on the mission in case they need to share or pick up tasks to stay on schedule. They travel to labs elsewhere in the world to learn the science and operations of onboard experiments provided by researchers abroad. The designated RMS operators and the EVA teams train intensely on their tasks. Other crewmembers train for and carry out EVA "prep and post," getting the suits and tools ready, helping the EVA pair suit up, and then reversing the process when they return inside. Training is typically not a nine-to-five job but a total commitment in preparation for around-the-clock work in space, with training sessions often held at night when competition for the facilities is not as heavy. Training is the path to teamwork. "Making the team the best that it can be" is a very important soft skill, according to Peggy Whitson.[34]

Beyond classroom training, astronauts participate in different kinds of simulations from following on-paper procedures for individual tasks to hands-on work with hardware, integrating tasks into a full procedure, and rehearsing portions of the mission timeline. Some simulations focus on routine operations, malfunctions, and repair procedures for various spacecraft or payload systems. Culminating simulations are full-team rehearsals of eight-hour to three-day segments of a mission with astronauts as well as the complete mission control center and technical support teams. Simulation supervisors insert devious malfunctions and unexpected crises to force the teams to resolve problems and practice responding to situations that may arise in flight. Debriefings and analysis after these full-scale simulations ensure that mistakes are addressed and build confidence that everyone is fully prepared and ready for the mission. Simulations are so realistic that participants feel fully immersed in the mission rather than a rehearsal, and astronauts often remark that actual missions seem easy after such rigorous simulations.[35]

Training evolves from theory to practical application. As Peggy Whitson describes it, you have to understand what a system or payload or experiment is meant to do and then learn how you actually use this information on orbit. "You have to be pretty proficient on all of it."[36] Much of one's preparation for a stay on the space station is solo training in assigned roles. Crews train as a group for emergencies—fire, depressurization, toxic atmosphere, or medical crisis—learning all the hardware and procedures and understanding the logic of responses so they can follow the process under stress.

Training for robotic arm operations and extravehicular activity is the most intensely rehearsed, apart from piloting maneuvers when the space shuttle was in operation. These two roles require exceptional concentration, precision, and endurance, with many hazards to be avoided. Astronauts assigned to these roles spend hours upon hours practicing with the mechanical arm trainer or in EVA suits underwater, doing tasks repeatedly until they build up "muscle memory" so their bodies as well as their minds are trained. Executing the task becomes almost Zen-like; as they concentrate, their arms and hands learn what to do so that it becomes habitual. They also rehearse through virtual reality simulations, wearing headgear screens and motion sensors on their hands, that visually place them in highly realistic space scenarios. Ideally, all of this training means there are no surprises on orbit. Time in space is a precious commodity, so tasks must be carried out efficiently but not hurriedly. An astronaut motto, "slow is faster," means that deliberate action is less prone to time-consuming mistakes. Practice makes perfect. Astronauts have periodic refresher training for RMS and EVA operations so they do not lose their skills between missions.

The Neutral Buoyancy Lab is the water facility—a large, deep, indoor pool—where astronauts train for two kinds of extravehicular activities: scheduled and contingency. Scheduled EVAs are unique to a mission, such as Hubble Space Telescope servicing tasks or ISS assembly tasks. The mission is organized around accomplishing those well-rehearsed EVAs. Contingency EVAs are planned for but not scheduled; they occur when problems arise that must be addressed quickly, like removing and replacing a critical hardware component that directly affects the space station power supply or stability, or in times past, failure of the shuttle's payload bay doors to close for reentry—which fortunately never happened. Astronauts trained for contingency tasks are on call should those contingencies arise. If all goes well, they never have an opportunity to do an EVA, which means a nominal mission but probably a personal disappointment.[37]

Basic physical training is part of an astronaut's daily life. Many are runners or work out regularly in the gym; others are active in sports such as tennis, squash, or racquetball to maintain an overall fitness level. For long-duration missions, astronauts adhere to a specific regimen before and after a stay on the space station to boost their fitness before and their adaptation after months in weightlessness. EVA astronauts do enhanced strength training in their shoulders, arms, hands, and legs to prepare for working in the EVA suit, which can be quite taxing, since the body has to work against the pressure that stiffens the suit and gloves.

As if all this training isn't demanding enough, astronauts heading to the International Space Station study the Russian language for years, one of their most challenging tasks. Although English is the common language for ISS crews, Russian trainers, the Russian mission control team, and cosmonauts speak Russian among themselves. Non-Russian astronauts must be fluent enough to communicate with the Russians in technical conversations. A shared vocabulary and understanding are necessary for safety reasons. Russian language study continues indefinitely for anyone who wants to serve on the space station or as the NASA liaison for crew support in Star City.[38]

When Russia became a partner in the space station program in 1992, the agreement set in motion a series of crew exchanges between the space shuttle and Russia's *Mir* space station that, in turn, set in motion new requirements for crew training. Russian crewmembers flew on eleven shuttle missions to *Mir*, and seven American astronauts rotated through residencies on the Russian space station. These joint missions laid the groundwork for eventual joint occupation of the International Space Station by giving the two space agencies experience in working together to jointly manage missions and crews. Given differences in language, expected levels of proficiency, philosophies of space operations, training facilities, perceived national stature, and more, much negotiation ensued to reach agreement on what training was required and where it would occur. Cosmonauts were to spend time in the United States for certain kinds of training at JSC, and astronauts would be sent to Star City to prepare for stays on *Mir* and possible flight on the Soyuz spacecraft. As the International Space Station came into being, that plan continued, with use of the Soyuz becoming standard practice.

To oversee training in Russia and support the astronauts selected to fly on Shuttle-*Mir* missions as space station residents, NASA set up a liaison office in Star City. Bonnie Dunbar was the first US woman astronaut sent to train in Russia as a backup to the first man chosen for a stay on *Mir*.[39] She spent thirteen months training there, learning the language while attending classes and taking oral exams entirely in Russian. At the time there were no women cosmonauts, and as an anomaly in a staunchly male program, Dunbar encountered sexist disrespect but soldiered through it to complete her training. In 1995 she was certified for long-duration flight on *Mir* and, if necessary, on Soyuz. Negotiations for her assignment to *Mir* faltered, so she returned to work at JSC without being assigned as a *Mir* crew member. The Russian space agency evidently was not yet ready to have a woman peer on board, although it had allowed a woman visitor, Helen Sharman of Great Britain. Instead, Dunbar served on the shuttle side of two Shuttle-*Mir* missions: STS-71, for the first docking and crew exchange with *Mir* in 1995, and STS-89 in 1998 to deliver the last American resident on *Mir*. On both missions she was engaged in scientific research. She also had ongoing responsibilities for coordinating the training and readiness of *Mir* crews.

Wendy Lawrence served in Star City as NASA's director of operations in 1995 and the next year began training for assignment as a resident *Mir* crew member.[40] Unfortunately, she was disqualified when the Russians determined that she was too small to fit properly in their Sokol and Orlan spacesuits worn during Soyuz launch and return and on EVAs. Although she would have traveled on the shuttle to and from *Mir*, any emergency evacuation would have meant using the Sokol suit and Soyuz spacecraft. Likewise, she would have had to wear an Orlan suit if assigned to do EVA from *Mir*. Her disappointment and NASA's over disqualification was somewhat mitigated by Lawrence's crew assignment on two Shuttle-*Mir* missions, STS-86 in 1997 and STS-91 in 1998.

Shannon Lucid became the sole woman to reside on *Mir* as a crewmember, an assignment for which she unexpectedly set a world record for continuous time in space and was awarded the Congressional Space Medal by President Bill Clinton upon her return. Lucid gamely volunteered for a *Mir* mission and years later published a memoir of her experience that records much of the hardship and frustration she experienced while living and working in a culture so different from the United States and NASA.[41] She also recounted unexpected pleasures and humor found in her spartan circumstances. Lucid is known for being unflappable and practical in responding to the unexpected, and true to form she organized her life in Russia and on *Mir* to her satisfaction. She encountered sexism, but her two Russian crewmates, who did not speak English, developed a respectful and affectionate companionship with her, and they got along well in their long confinement. Lucid boarded *Mir* in March 1996 and returned home in late September, spending 188 continuous days in space, 179 of them on *Mir*. She returned in good health, determined to walk off the shuttle unassisted to show that she was physically strong after months in weightlessness, as the American and woman who at that point had spent more time in space than any US astronaut.

The Shuttle-*Mir* program of training and long-duration spaceflight enabled NASA and the Russian space agencies, astronauts, cosmonauts, engineers, mission control teams, and others to work out the agreements and protocols for joint operations in space. They were able to address many problems before assembly and occupation of the International Space Station, enabling that program to proceed more smoothly if not perfectly. To this day, there are still some tensions and disagreements at the agency level, but the crews who actually serve on the space station consider themselves a team and see their cooperation as an illustration of international harmony.

Software testing and verification has been an ongoing process for the highly computerized space shuttle and space station systems. Many astronauts rotated through the Shuttle Avionics Integration Laboratory (SAIL) at JSC as one of their first technical assignments after graduating from initial training. Software for spacecraft systems was often updated and improved, and astronauts put it through its paces to drive out bugs and ensure its reliability. The heart of the SAIL was a fully functional cockpit simulator that replicated all of the shuttle's command and control electronics from the general-purpose computers that ran the vehicle to all the specialized avionics and control units for various systems. This high-fidelity electronics simulator was used to test and prove software and procedures for all phases of flight and for malfunctions and contingencies, especially during launch and landing, the two most critical portions of a mission. Physician Rhea Seddon recalled that software testing "was one of those jobs that I knew absolutely nothing about, so I had to just jump in and figure it out. . . . I had to work really hard to understand what was going on, but it was absolutely essential [to] do that."[42] Every shuttle mission was flown in advance in the SAIL. The facility was retired but preserved as a historic site when the shuttle program ended. A similar facility organized around a full-scale functional mockup of the International Space Station serves the same purpose. Astronauts complete more than two hundred training courses in all operations, procedures, and malfunctions.

Selecting new astronauts is a recurrent job during each astronaut recruitment cycle. Many astronauts see service in the astronaut selection process as a privilege. Some participate in the initial winnowing of thousands of applications to a few hundred most qualified by reviewing all applications, weeding out the least qualified, and forwarding the most qualified as finalists for further consideration. Others serve on the nine-member Astronaut Selection Board with representatives of the JSC personnel, medical, flight operations, science, and engineering offices. Their role is to conduct interviews and evaluate the finalists during their week at JSC, then select the best fits to be reviewed by the Astronaut Office. At this stage the board looks for nuanced qualifications in leadership, attitude, and commitment, and they try to ensure that each finalist gains a good understanding of what the job of astronaut entails. This process consumes about three months and becomes a full-time job. Selection board members say it is an amazing experience to see the impressive levels of talent and enthusiasm that applicants present. Most of the women astronauts have served on this panel at least once.

Jobs in Space

On the space shuttle, the commander and pilot flew the vehicle and were responsible for the safety of the crew and success of the mission. They ruled the flight deck during launch and return and performed all vehicle maneuvering in orbit. Typically, the commander had already flown at least once, and the pilot was on a first or second flight. The pilot had certain roles, such as dropping the landing gear and deploying the drag chute at the exact moment, and assisted the commander throughout the flight. They trained together constantly before the mission so they knew each other and their roles so well that they could act as one. Only NASA pilot astronauts held these positions; mission specialists were responsible for operating many of the spacecraft systems but did not fly the shuttle. The commander and pilot supported the rest of the crew during complex activities, such as EVAs, when everyone had a role to play, but they were not allowed to go on EVAs and only rarely operated the RMS arm. The commander was the ultimate decision maker about any mission-critical matters. When the shuttle was flying a two-shift around-the-clock science mission, one of the pilot astronauts was on duty for each shift.

Eileen Collins, Susan Still (Kilrain), and Pamela Melroy were the only women designated as pilots in the astronaut corps during the space shuttle era. Collins piloted two missions and commanded two missions as the first woman in each role. Still piloted two missions, and Melroy was pilot twice and commander once. Collins flew four different types of missions. Her first as pilot was a rendezvous with *Mir* and her second docked with *Mir*. Her first mission as commander was to deploy the Chandra X-ray Observatory, and her final command was the return-to-flight mission after the *Columbia* tragedy. On that mission she executed a new shuttle maneuver, a backflip in close proximity to the ISS, so the space station crew could photograph the entire vehicle in search of any launch-related damage. Still had the unique experience of flying the same mission twice; the first time the planned sixteen-day Spacelab microgravity science laboratory mission launched, a technical problem forced the shuttle home on the fourth day. NASA scheduled a repeat flight three months later, and the full research mission was completed. All three of Melroy's missions were for ISS assembly; her crews delivered and installed major elements of the ISS: the *Harmony* node and portions of the truss structure that supports the solar arrays and radiators.

On the space station, the pilot role is not needed because the station has automated and limited maneuverability. The commander role is still necessary to orchestrate the crew and their activities to accomplish mission goals, and also to ensure their safety, especially in response to any emergencies that may arise. ISS crews have ranged from two to nine people but typically six, one of whom is the designated commander. The role of commander alternates among the international partners, so the position may be held by an American, Russian, European, Canadian, or Japanese astronaut who may have a piloting, engineering, or scientific background. Astronauts' stays on the ISS usually last six months, with half the crew rotating in or out in three-month expeditions or increments. Commanders also rotate on a three-month cycle. A commander usually serves for one expedition after first serving as flight engineer in the previous expedition. An exception occurred when Peggy Whitson was on the ISS for nine months in 2016–2017; she was flight engineer for Expeditions 50 and 52 and commander for Expedition 51.

To date (2025) three US women have commanded ISS expeditions: Peggy Whitson, a scientist, was first on Expedition 16 in 2009 and was commander again on Expedition 51 in 2016–2017. Sunita Williams, an engineer, commanded Expedition 33 (2012) and Expedition 72 (2024–2025). Both entered the astronaut corps in 1996 as mission specialists and flew on the shuttle and an ISS expedition before assignment as commanders. Commander Shannon Walker led Expedition 65 in 2021 after being a flight engineer on Expedition 24/25 in 2010. Selected in 2004, she flew all of her missions to the ISS without being on the shuttle, traveling instead on Soyuz and SpaceX spacecraft.

Whitson's first stay on the ISS (Expedition 5, 2002) convinced her that she wanted to be a commander. "[I knew] I could do that job. I wanted to do that job. . . . Being first was not my goal, being the commander was my goal."[43] To prove herself, she took an assignment as commander on a NEEMO underwater mission before returning to the space station.

On shuttle missions, the flight engineer was a mission specialist (designated MS-2) who supported the commander and pilot during launch and return. Sitting behind and between the two with a view of the cockpit displays, the flight engineer kept an eye on the checklists and called out steps in their procedures. Of course, the pilot astronauts knew what they were doing, but it was standard practice to have another set of eyes and ears involved in these critical phases of flight to ensure that nothing was missed. The flight engineer also kept an eye on certain displays while the pilots' attention was directed to other displays. Many of the women astronauts served as flight engineer on shuttle missions in addition to their other onboard duties.

On the ISS, everyone is a flight engineer. The ISS position is more generalized with responsibilities for monitoring, operating, and maintaining all systems, making repairs as needed, installing new equipment, reconfiguring and relocating hardware, and keeping the station shipshape as it requires constant operational attention. ISS flight engineers also tend to all the onboard science facilities and investigations, operate the robotic arms, participate in or support spacewalks, manage arrival and departure of supply ships and visitors, and engage in public outreach communications with schools and media around the world.

One of the most demanding positions in space is operating the Remote Manipulator System, known as the RMS or robotic arm. The RMS is actually a telerobot operated by a human, not an independently operating device. The space shuttle had one arm from the outset, and a long extension was added after the *Columbia* tragedy to permit visual inspection of the complete vehicle. The ISS has two long arms to permit access to almost all parts of the space station. The fifty-foot-long arms serve as cranes for moving large equipment such as satellites, space station modules, the Hubble Space Telescope, large experiments, and equipment such as solar arrays, relieving astronauts of such inherently risky movements. The arms also serve as mobility devices for EVA astronauts, moving them into position across tiring distances or placing them precisely where they need to be on a stable platform so they can reach and handle items for installation, removal, or repair. The arms have rotating joints and can be moved in any direction as commanded by the operator.

All astronauts are exposed to the RMS during their initial training. Those who show aptitude continue with advanced training and are designated prime or backup operators for missions. Most of the women astronauts have been RMS operators, a role that requires extreme concentration and precision. One has to master the characteristics of the arm itself, its control systems, geometry, and spatial awareness. The job depends on exceptional hand-eye and fine-motor coordination and an intuitive sense of where the arm is in imaginary three-dimensional space. The RMS is an indispensable device for extravehicular activity, so the RMS operators closely coordinate with EVA plans and crews.

More than half of the women astronauts have operated the robotic arms on the space shuttle or space station. Among the first women considered especially proficient in operating the RMS were physicist Sally Ride and electrical engineer Judy Resnik, who worked on it while it was still under development and used it well on their missions in 1983 and 1984. Rhea Seddon had an impromptu challenge using the arm to try to flip an activation switch on a satellite that did not deploy properly on her 1985 mission.[44] Stephanie Wilson, called "Madame Robotics Expert" by her crewmates, guided the arm through some of the most difficult maneuvers on missions in 2006 and 2007. She was the first person to use the extended arm and sensor system to conduct an inspection of the entire space shuttle orbiter exterior, and she operated the ISS arm at its full length to precisely position an EVA crewmember as he repaired the solar array at the far end of the truss structure.

Unlike extravehicular activity, where a small person may be at a disadvantage, body size doesn't matter for robotic arm operations. The operator works at a control station using two hand controllers and watching data displayed on a screen, as well as images from cameras on the shoulder, elbow, and wrist joints of the arm. One's strength or reach aren't factors; the operator stands at the controls, watches the screens, uses both hand controllers to translate and rotate the arm with minute hand motions and a gentle touch. The operator must block all distractions, concentrate on delicate, precise motions, and "feel" the arm as an extension of one's own while ensuring that it doesn't touch anything it shouldn't and doesn't violate the geometry of its workspace.

That ability comes from hours upon hours of practice, much as a musician practices to control every aspect of the sound from an instrument. One could not succeed in operating the arm in space without dedicated training and simulations in advance. Even so, the arm in space operates slightly differently than the training version, more quickly yet more steadily, so every bit of finesse is needed. Women astronauts have excelled in this crucial role.

Every astronaut hopes for an exciting, challenging mission assignment. They say that every mission is a good mission, but some are better. Sooner is better than later; longer is better than shorter; higher is better than lower. And an EVA is best of all. It is no secret that most astronauts consider extravehicular activity the plum role. It is physically taxing and dangerous but exhilarating work. It is the essence of spaceflight: to be outside in one's own personal, protective spacecraft (the spacesuit) with an unlimited view of

Nancy Currie at the shuttle RMS controls, Hubble Space Telescope servicing mission, 2002. *NASA*

the heavens and Earth and to perform superhuman feats, such as assembling a space station from massive parts using only your hands and small tools.

EVA is not yet for everyone, though, for the most trivial reason. Not everyone fits the existing spacesuits. For economic reasons, NASA decided decades ago to forgo custom-fitted spacesuits for standard sizes—small, medium, large, and extra-large.

Unfortunately, the smallest women astronauts did not fit safely or comfortably in any suit, because the proportions were off. Even in a small suit, the shoulders were too broad, the girth too large, and the legs and arms hard to size properly. Women are not small men; their proportions and curves are different. When the small suit was retired in the mid-1990s, women were forced to make do in a larger suit if they could add padding to maintain the best posture inside the extra volume. Some women—maybe as many as twenty—who might have been quite capable in EVA were thus precluded for lack of a suit that fit. Whether intentional or not, the message they heard was "You are expendable; we don't need you in EVA."

The problem was compounded when NASA decided that anyone assigned to a space station expedition must be certified for EVA to do whatever maintenance and repairs were required and to work on any "outside" emergencies that might arise. That requirement further precluded most small women from tours of duty on the ISS and again sent a message of exclusion. Meanwhile, tall, broad-shouldered, long-boned men reaped most of the plum assignments to service the Hubble Space Telescope and build the space station. NASA has promised that new suits for missions to the Moon and Mars will fit everyone to remedy such inequity.

Fortuitously, another twenty or so women were just enough taller or larger framed to be able to wear the medium-size EVA suit effectively. They collectively performed more than fifty EVAs since Kathy Sullivan first ventured out of the shuttle's airlock in

Kathy Thornton on the RMS arm installing a new optics unit into the towering Hubble Space Telescope, 1992. *NASA*

1984. Peggy Whitson holds the current record with ten EVAs, all from the space station. One or more of the younger active astronauts may one day surpass ten. Sunita Williams and Christina Koch are already in range with nine and six, respectively. Among the most thrilling EVAs by women, two occurred on the first Hubble Space Telescope servicing mission that breathed new life into the observatory by exchanging old components for new: Kathy Thornton served on that 1992 mission and was the only woman spacewalker on any of the five Hubble servicing missions. Later, in 2002 and 2009, Nancy Currie and Megan McArthur supported Hubble servicing EVAs by operating the RMS arm. Susan Helms, the first woman resident on the ISS (Expedition 2 in 2001), and her male EVA partner set the world record for longest spacewalk, lasting almost nine hours, as they installed hardware on the exterior of the *Destiny* laboratory module. In 2019, ISS Expedition 61 crewmates Christina Koch and Jessica Meir became the first female pair to do a spacewalk, logging more than seven hours working outside the ISS. They did two more all-woman EVAs in 2020, and Jasmin Moghbeli and Loral O'Hara paired up for an EVA in 2023. None of these all-woman EVAs was contrived to make history but happened because the women were qualified and available.

Some women passed general training for EVA and were qualified for contingency EVA on missions but never had an opportunity to go outside because contingencies didn't occur. They were not scheduled for the primary EVA tasks, but if some unexpected situation arose they would have gone outside to resolve it. As an example, the shuttle had a communications antenna that stowed automatically so the payload bay doors would close for reentry. Had the antenna failed to stow or had the payload bay doors not latched closed, the contingency EVA pair would have risen to the occasion. Cady Coleman, Janet Kavandi, Mary Ellen Weber, and several others, were designated for contingency EVAs that were not required.

On shuttle missions, the payload commander was the chief scientist responsible for seeing that all research objectives were met. This position was assigned only on Spacelab missions, when the shuttle carried far more scientific experiments than usual. The payload commander kept an eye on the experiment timeline to ensure that all experiments were under way as scheduled, assisted the science team in adjusting the plan as problems arose, and served as the decision maker for any major changes to the plan. Experiments had competing needs for power and crew time, all of which was worked out in advance, but if problems arose the payload commander decided what was best for the overall scope of the mission.

Kathy Sullivan was the first person designated as payload commander, a position she held on the ATLAS atmospheric science mission STS-45 in 1992. Bonnie Dunbar served as the next payload commander on the US Microgravity Laboratory mission, STS-50, also in 1992. Thereafter, Rhea Seddon, Linda Godwin, Ellen Ochoa, Tammy Jernigan, Kathy Thorton, and Bonnie Dunbar again, as well as several male astronauts, were payload commanders on science missions. Rhea Seddon found it to be her favorite assignment.[45] The last payload commander to fly was on STS-107, the all-science *Columbia* mission in 2003.

Peggy Whitson was the first person to be named ISS science officer when she joined Expedition 5 in 2002. The space station was then in its rudimentary configuration, not yet fully equipped for scientific research, but enough experiments were on board to command a crewmember's attention. Designating a science officer called attention to the fact that the ISS was being constructed for research.[46] Whitson's role was to activate and operate science experiments and evaluate how suitable the onboard provisions were for scientific research. She was the primary point of contact with investigator teams as their representative before and during the expedition, and she debriefed them upon her return. She made it a highly collaborative role. Whitson liked her new title but admitted that "many supportive friends have sent an incredible amount of Star Trek/Mr. Spock email."[47]

All astronauts are trained to use a variety of still and motion cameras, and over time they transitioned from film to digital cameras as imaging technology advanced. NASA acquires the most reliable cameras from the commercial market, adapting them for spaceflight as needed. Hasselblad, Canon, and Nikon are the primary brands used for Earth observation and crew activities. Missions include lists and schedules for targeted photography and allow the crews leeway for spontaneous photography. Experiments often require use of specialized scientific cameras for studies of the eye or monitoring of fluid physics or crystal formation. Almost every astronaut ends up taking photos and videos in space, whether assigned to do so or not. It's almost irresistible to record the views and activities in space.

The most challenging camera used in space was the large format IMAX camera flown on a variety of missions to accumulate footage for documentary films. It, too, evolved from 1D to 3D format for even higher-fidelity imaging. The footage shot in space during several missions was edited into narratives, scored with original music, and often narrated by a celebrity actor. Presented as feature films in museums and commercial theaters, these beautiful IMAX films brought the experience of spaceflight down to Earth, where they were well received by audiences and critics alike. The crews who were trained to do the shoots enjoyed their role in these productions, although the large camera was something of a beast to handle. Even the astronauts were moved by the realism and beauty of an IMAX film; it recalled their experiences vividly and made them feel as if they were in space again.

Marsha Ivins took photography to heart and made it her mission to organize the camera collection and procedures to ensure the efficiency of all astronauts tasked with photography. The variety of cameras, lenses, and accessories were a photographer's dream, and she had a hand in keeping the inventory up to date. She also became responsible for reviewing all in-flight photographs and helping NASA public affairs officers select the best ones for public release.[48] This was no small task, since each mission produced thousands of photos as general or scientific documentation.

Usually, two astronauts per mission are trained to be onboard paramedics in case a crewmate becomes ill or is injured. Their role ranges from conducting physical exams and dispensing medication as needed to cleanup of scrapes or irritated skin. They are also trained to administer CPR, use a defibrillator, and perform minor surgery, but it has not yet been necessary to use these procedures in space. The onboard medical kits contain many of the usual first aid materials and medications for headaches, nausea, sinus and respiratory discomfort, digestive problems, some medical emergencies, and any regular prescriptions an astronaut may have. In addition, there are instruments for minor surgery, like suturing a wound. An astronaut with or without an MD degree may be trained for the role of crew medical officer. Astronauts who are physicians usually serve as crew medical officer on their missions. That group includes Serena Auñon-Chancellor, Ellen Baker, Laurel Clark, Anna Fisher, Mae Jemison, and Rhea Seddon, but others among the women astronauts have also served in this role, as have male astronauts.

Many crewmembers are trained to draw blood required for life science experiments or general health monitoring, and some can perform eye and dental exams. An ultrasound device and a portable echocardiogram (Holter monitor) are available for health monitoring and diagnosis. NASA's flight surgeons on the ground are always available for consultation.

Before flight, all astronauts receive a thorough medical checkup to ensure they are healthy and have no emerging medical issues that would compromise their health or the mission. Great care is taken to keep everyone onboard well. While health matters are confidential, no flights in the US space program have been terminated due to crew health issues.

On shuttle missions to the International Space Station, one person with good organizational skills, often a woman, was the designated loadmaster. (On the ISS, everyone shares that role.) The job is to manage the transfer of supplies and equipment coming up and equipment, experiment samples, and trash going down. The space shuttle could carry tons of stuff each way, but today's cargo capacity on Progress, SpaceX, and expendable small vehicles is limited. It took days to unload and reload the shuttle; now it can be done more quickly, but it still takes careful planning, strategy, and choreography to pack, load, unload and stow hundreds of items of various sizes and mass. The goal is to fit the maximum amount of mass without altering a vehicle's center of gravity. Sequencing is part of the process: what must be unloaded first has to be loaded last, and things must be loaded in a way that the loadmaster has access to all the storage space. Food, clothing, equipment, tools, experiment hardware, and other materials are prepacked into standard-sized soft containers and bags for easy handling and stacking.

Trash, broken devices, and dirty clothes are now loaded into expendable vehicles that burn up in the atmosphere. Experiment hardware and samples are returned to Earth via crewed spacecraft, such as Soyuz and SpaceX.

Once materials have been unloaded from the delivery craft, the next part of the job is to unpack and stow everything in its proper place on the ISS so it can be located when needed. Items are barcoded and inventoried, and their locations tracked. Nothing is more frustrating than not finding something where it is supposed to be. No one wants to lose track of essential items. The logistics of keeping an orbital space station well supplied and removing what is no longer needed is an essential function that most people don't consider, and the loadmaster's job is a crucial although less well-known role.

US astronauts began to fly on Russia's Soyuz spacecraft during the Shuttle-*Mir* program, and Soyuz has been a reliable mode of transportation to and from the International Space Station ever since. Even before the space shuttle program ended, some US crewmembers arrived at the ISS and departed on Soyuz, just as some Russian ISS crewmembers came and went on the shuttle, as negotiated by the space agency partners. The same was and still is true for European, Canadian, and Japanese astronauts, who can travel on any available spacecraft.

The Soyuz spacecraft is a three-person capsule, with seats in a tight row and almost no wiggle room for the pressure-suited crew. The commander is always a cosmonaut. The other two are either a copilot, flight engineer, or passenger. US astronauts have served as copilot assisting the commander and ready to take control if necessary, and as flight engineer with limited duties in the highly automated craft. The flight time from launch to docking at the ISS ranges from three or four hours to three days, depending on the orbital mechanics of rendezvous.

Likewise, crew positions are similar on SpaceX Crew Dragon and Boeing Starliner commercial spacecraft. For the test flight of each, two crewmembers flew as commander and pilot. For operational service the SpaceX crew includes a commander, pilot, and one or more mission specialists. Megan McArthur served as pilot of a 2021 SpaceX mission, Nicole Mann commanded a 2022 mission, and Jasmine Moghbeli commanded a mission in 2023. Two women—Anne McClain and Nichole Ayers—were assigned as commander and pilot of the SpaceX Crew Dragon-10 mission in early 2025. Shannon Walker, Kayla Barron, Jessica Watkins, and Jeanette Epps were mission specialists on other SpaceX flights, monitoring timelines, telemetry, and consumables. Peggy Whitson, a retired NASA astronaut, flew as an Axiom Space Company astronaut and commander on a SpaceX Crew Dragon spacecraft and is set to command another Axiom-SpaceX flight in 2025. In 2024, the Boeing Starliner made a crewed test flight with Suni Williams as pilot but has not yet flown an operational mission.

Jobs Wherever You Are

Being an astronaut is a 24/7 commitment. Some days have a normal schedule, and then there are long periods of training and travel when the clock doesn't matter. The work to be done must be done then. One job can be performed wherever you are: being a public figure and role model. The Astronaut Office wisely regulates public appearances to avoid interfering with primary work responsibilities and schedules but still expects all astronauts to do a share of public outreach, typically at least once a month but more often right after a flight, when crews make the rounds of international partners' home countries and space agencies.

Astronauts are considered to be the "face" of NASA and ambassadors for space exploration and STEM education. They represent NASA wherever they are, whether on the ground or in space, in the United States or abroad, and in public or private settings. Although they are widely admired, like other celebrities they can be subject to public criticism if they misstep. NASA has always been sensitive to their public image and has prepared them for media attention, which they learn to handle adroitly.

The NASA Public Affairs Office receives and responds to thousands of requests for astronaut appearances. Astronauts are typically scheduled for at least one public appearance a month, which can be at the White House or before Congress, at a convention or on a television program, at a school or a science camp, at civic club meetings or ceremonies, on a radio program or a documentary film shoot. Most public appearances involve travel and time away from work and home. Astronauts may be called upon to explain the rationale for particular NASA programs, advocate for space exploration, describe their upcoming or just-completed mission, inspire students to study hard and dream big dreams, or make goodwill visits around the world.

Since the explosion of social media, most astronauts have websites and write blogs and posts that reach people electronically with an informal, in-person intimacy. Through social media platforms, astronauts actively broadcast their activities and thoughts about their work. While in space, they do frequent media interviews and audio and video broadcasts, and through downlinked video and ham radio transmissions they engage in live conversations with students. Astronauts mark special occasions, tell stories, explain what they are doing, answer questions, demonstrate weightlessness, and connect with people—all of which spark a positive impression of the space program.

Not everyone is initially comfortable with so much public exposure, but gradually they learn to handle and usually enjoy it. Of the 1978 class of astronauts, Rhea Seddon remarked that there was wide variation in comfort level with the press. "Some people didn't want to talk to the press at all, and other people were like, 'Here's my life' or 'Hey, that's part of the job.'"[49] Kathy Sullivan and Judy Resnik enjoyed doing radio and TV commentary with network news anchors for the early shuttle missions, talking about their roles without using technical jargon in a way that helped the public understand what was happening. Both were on duty for STS-1, with Sullivan doing commentary for the ABC network and Resnik joining host Tom Brokaw on NBC *Today*.[50]

Media exposure and public appearances are key to astronauts being seen as role models, and the women astronauts have grown to be comfortable with that image. Said Mary Cleave of public relations duties, "I like the role model part of it."[51] She was concerned about math anxiety and girls' losing interest in math and science during their early teen years. Cleave and other women have been keen about their potential to prompt girls and young women to consider doing something they wouldn't have thought of before and sharing the message, "If I can do it, so can you." Many carry that enthusiasm for outreach into retirement, when they can devote far more time to public appearances and extend their reach as role models. Nicole Stott, for example, travels widely, almost full-time, as a retired astronaut, using art as an entree to think about exploring space and taking care of spaceship Earth.

The women interviewed for this book who have moved on from NASA are clear that being an astronaut is a highly satisfying career. The work is varied, intense, challenging, and fuels their drive to excel, often taking them outside their knowledge comfort zones. They consider it a hard-earned privilege and are grateful for the many opportunities that come with the job. Most tellingly, they would do it all over again if they could, and they now focus on encouraging younger people, especially girls and young women, to aim high, too.

CHAPTER 6

Balancing Acts: Risks and Rewards of Being an Astronaut

America's women astronauts often say how fortunate they feel to have been chosen for this career. To be among the seven hundred or so people to date who have left the planet to live and work in space—about one hundred of them women—is a rare privilege. Although the number of people venturing into space is increasing with other nations' space programs and the growth of commercial spaceflight, the sixty-one US females are the ones who really proved that women belong in space. They went from standing out as women to fitting in as full-fledged astronauts, not a distinctive group different from the men. They train, work, and fly together as equals. They are astronauts, period. Yet astronauts who happen to be women occasionally encounter circumstances where they *are* different or are treated differently for physical or medical reasons, and at times some people think that they *are* treated differently, perhaps unfairly. Some individuals suspect that managers do not fully recognize their potential, and they must advocate for themselves. The tension between fitting in and standing out is a balancing act. So, too, is making trade-offs between the risks and rewards of being astronauts.

All astronauts, men and women, accept the inherent risk of catastrophic death in training or spaceflight. Seventeen spaceflight crewmembers have lost their lives in US spacecraft—Apollo 1, *Challenger*, and *Columbia*—and at least five others died in T-38 or commercial aircraft accidents while on duty. This is not a trivial risk. In addition, there have been near-misses in flight when technical problems in the spacecraft or space-suit could have become catastrophic—an electrical short, a leak in air pressure or fluid flow, a tiny crack or metal shaving somewhere. There are also risks related to parachute and survival training activities, simulated high altitude and space environment in hypo-baric and vacuum chambers, and underwater neutral buoyancy training. Risks are managed and mitigated to make the work environment as safe as possible, but risk of accidental injury or death is never zero.

Astronauts also face risks that are less obvious and immediate but may gradually affect their physical and behavioral health. Both spaceflight and the return to Earth are stressful as astronauts adapt to microgravity and readapt to normal gravity environments. So far, only a few physical stresses and changes seem to have potential for long-lasting, detrimental effects. Long spaceflight may provoke emotional and behavioral stress through isolation from family and friends, confinement, a taxing work schedule, poor sleep, little privacy, even the quality of food. Countermeasures, such as regular exercise, varied menus, and frequent email and video contact with family and friends help to reduce these risks. Frequent travel, long absences, erratic work hours, time in preflight quarantine—simply not being there—also pose risks to astronauts' relationships and family life.

Offsetting risk considerations are many rewards of being an astronaut: opportunities to see the beautiful Earth from above, magnificent sunrises and sunsets, the deepest black of space punctuated by an infinity of bright stars; the satisfaction of executing all the tasks of a challenging mission, often doing something never done before; the sense of contributing to something bigger than oneself; participating in the history of exploration, expanding knowledge, doing research that will have benefits on Earth; working with brilliant, capable, and dedicated colleagues in the spacefaring enterprise in the United States and elsewhere in the world; and being an inspiration to others of all ages. There is also the satisfaction of pushing oneself past previous boundaries and achieving one's full potential. Women astronauts interviewed for this book cited all of these reasons for taking the risks.[1]

It is unlikely that any astronaut, after working so hard and aspiring for so long, would sacrifice such rewards to avoid the inherent risks of spaceflight. To date, four men have resigned for various reasons while in astronaut training and another three left after qualifying but seeing no upcoming flight opportunities.[2] No woman has walked away from this career before flying in space. When astronauts leave NASA, it is typically because they have new opportunities to pursue, know that they will not fly again, want to prioritize their family, or are ready to retire.

General Health Risks of Spaceflight: Bodily Responses to Microgravity and Radiation

Some risks of being an astronaut are obvious and immediate, such as the possibility of death in an aircraft or spaceflight accident or injury from a training accident. Others—radiation exposure, for example—are latent, and effects may not appear until years later.

All astronauts have exceptionally good health and fitness. When astronaut applicant finalists report to the Johnson Space Center for interviews and evaluation, they undergo detailed physical and psychological exams to assess any health-related risks that might affect the individual or future mission success. The applicants' detailed medical histories also are reviewed. Applicants may be disqualified for hundreds of reasons, among the most common being visual, cardiovascular, kidney, and thyroid abnormalities, and psychological/ behavioral disorders.[3] Applicants who pass this rigorous assessment may then be selected for astronaut training. The admission standards are the basis for annual evaluations of continued health and fitness for spaceflight. These are itemized in a standards document issued by the NASA chief medical and health officer.[4]

Most members of the astronaut corps participate in a longitudinal (lifetime) study of astronaut health, with annual evaluations and monitoring for issues that might be associated with their time in space. This study originated in 1992 but includes data from astronauts since 1959 and is continually updated.[5] In addition to annual medical evaluation and recertification for flight, astronauts also have pre- and postflight physical exams, medical consultations during flight, and medical debriefings after flight. Many participate in biomedical experiments during flight, providing a wealth of health data available for study.

Decades of experience with humans traveling, living, and working in space, from the briefest flights to increasingly longer duration missions, have yielded evidence of the profound effects of microgravity (weightlessness) on the human body, from transient system-level disruptions (or adaptations) to changes at the cellular level. The longer the time in space, the longer the recovery time may be.

A number of effects on the body in microgravity have been recognized for years: fluid shifts from the lower to upper body, decreased bone density through loss of calcium and other minerals, loss of muscle strength and mass, changes in red and white blood cell production and in immune response, disturbance of the neurosensory system related to balance and body position, and cardiovascular deconditioning, among others. Most of these changes are reversible naturally or with in-flight countermeasures and rehabilitation upon return to normal gravity on Earth. These physiological findings are recognized as normal adaptive responses to the microgravity environment that typically do not pose an immediate or long-term health risk. With longer-duration missions on the space station, additional changes have been noted. For example, some astronauts have experienced changes in visual acuity related to swelling of the optic nerve and other factors. There is currently no evidence of long-term cardiopulmonary, neurological, or muscular-skeletal health consequences if exercise and dietary protocols are followed on missions lasting up to six months.

NASA experts periodically compile research findings in space biology and space life sciences to summarize current knowledge about health risks of spaceflight and what happens to the body in space. Astronauts on Skylab, the first US space station in the early 1970s, monitored and tested their bodies during missions lasting one to three months, and the results gave the first comprehensive account of bodily changes in space.[6] *Space Physiology and Medicine*, first published in 1982 and updated intermittently, has become a standard reference for system-by-system explanations of the body's responses to microgravity, as well as discussions of radiation, toxic hazards, psychologic and psychiatric considerations, and various other topics.[7] The most recent edition appeared in 2016 with intended use as a textbook for clinicians and researchers. A comparable compilation and textbook, *Principles of Clinical Medicine for Space Flight*, edited and in part written by two NASA physician-astronauts and most recently issued in 2019, likewise explains the nature of the spaceflight environment, the bodily systems that respond to spaceflight, and the medical systems and procedures in place to deal with adverse physiological or psychological issues that may arise. While addressing the many hazards and changes to human beings in space, the book nevertheless indicates that "the expectation of healthy well-adapted crew-members . . . for standard 6-month missions has been firmly cemented."[8] Adaptation to spaceflight and readaptation on Earth are predictable, normal, and do not impair astronauts' ability to work productively.

Of greater concern than routine adaptations to spaceflight are changes to the body that may not be noticed immediately but appear over time and are noticed anecdotally or through examination of longitudinal study data. In a report published in 2004 after the first ten years of such data collection, three trends were noted: cataracts, cancer, and thyroid dysfunction.[9] However, there was no significant difference in occurrence among astronauts as compared to matched non-astronaut populations. The thyroid issue was readily resolved and corrected when it was traced to the high iodine content of the space shuttle's drinking water. Cataracts and cancer tend to be associated with radiation exposure and aging and have not been attributed to microgravity. The cancers diagnosed at that time included malignant melanoma, leukemia, lymphoma, Hodgkin's disease, and kidney and gall bladder carcinomas. There were four instances of prostate cancer and one instance of breast cancer in a total of fourteen cancer cases among more than three hundred astronauts.

At that time, the primary cause of death among astronauts since the beginning of spaceflight was accidents and injuries, with cancer and cardiovascular disease accounting for multiple, but fewer, deaths. The available evidence then did not suggest that these diagnoses were outside the range of normal compared to the general population, nor was there evidence of a different propensity for these medical conditions between men and women (except prostate and breast cancer). In the twenty years since the original study, there have been other accidental deaths, surpassed by an increasing number of natural deaths by disease and aging, but the basic conclusions about astronauts' health have not been challenged by a similar review of NASA's longitudinal health data. Research has not indicated a correlation between normal astronaut mortality and spaceflight.[10]

Radiation exposure is the most daunting risk of spaceflight because of its known correlation with various types of cancer on Earth. Astronauts are exposed to ionizing radiation and cosmic rays when they are beyond the protective atmosphere, and radiation in the space environment penetrates spacecraft and spacesuits. Astronauts wear dosimeter badges while in space, and their cumulative radiation doses over time are carefully tracked. The level of prior exposure may become a factor in crew selection for a particular mission, along with mission duration and whether the mission requires spacewalks, to avoid exceeding permissible exposure limits. An astronaut may meet the radiation exposure lifetime limit by flying three or four space station missions. The possible long-term effects of radiation exposure may include cataracts, early onset dementia, and increased risk of cancer and circulatory diseases. There is no indication yet that astronauts or cosmonauts have a higher-than-normal mortality rate from radiation exposure, but conservative limits on exposure are accepted as ethically responsible protection.[11]

Certain body tissues are highly sensitive to radiation. Formation of cataracts seems to be associated with aviation and spaceflight. For women, cancers of the thyroid, breast, ovaries, uterus, and lungs are of concern with radiation exposure because they are common cancer sites among Western women in general. Any such cancers might occur many years later with aging, so it is important that astronauts be informed and reasonably protected from this risk. Until recently, female astronauts were restricted to a lower lifetime level of cumulative radiation exposure, which in effect meant that they could not fly as often or as long as males.

In 2021, following recommendations by the National Academies Institute of Medicine, NASA proposed to revise its radiation exposure standard to a common dose-based limit rather than an age- and sex-based calculation.[12] This revision erased the more conservative dose standard for females and set a single new standard for both sexes, regardless of age, while keeping all astronauts below a 3 percent increase in probability of death by radiation exposure sometime during their lives. This universal standard eliminated an astronaut's sex as a factor in making mission assignments to the space station or beyond. NASA implemented this policy change in 2022. Both scientific data and a commitment to equal access to space, as well as closer alignment with the standard used by international partner space agencies, informed NASA's decision to adopt this new protective standard for maximum lifetime radiation dose for spaceflight in low Earth orbit. As research continues, the standard may be adjusted again for longer, more distant expeditions to the Moon and Mars, where astronauts will be exposed to a much harsher radiation environment beyond Earth's protective magnetic field.

Much that is currently known and unknown about the body's responses to both microgravity and radiation exposure in space will be research priorities in preparation for future exploration missions beyond Earth. Longer missions may entail greater risks. Exposure times to radiation and microgravity sometimes will be logged in years instead of weeks and months, and the body may respond differently at much longer durations. Current knowledge is based on spaceflight experience in low Earth orbit and modeling based on experiments with animals, simulated space environments, and analog data from survivors of nuclear accidents, Hiroshima, and Nagasaki. Risks related to flying on missions of very long duration beyond low Earth orbit on the Moon and Mars, and any additional physical and psychological stresses that may entail, will require study and possibly further protection.

Researchers are also interested in possible interactions between exposure to microgravity and exposure to radiation, which may become more pronounced on long expeditions in deep space. Astronaut Scott Kelly's yearlong stay on the International Space Station in 2015–2016 revealed several concerning changes in his genes and chromosomes, eyes and optic nerves, carotid arteries, and immune response, as well as slight cognitive declines in periodic testing. Only some of these changes reverted to normal after his return from space.[13] His experience is helping to set the agenda for new and ongoing research to ensure the health and safety of long-duration expeditionary crews. Younger members of today's astronaut corps and those yet to be selected for long expeditions may face increased risks to their health. Advances in shielding technologies, countermeasures, and medicines may afford future crews additional protections not yet available.

Differences in the Effects of Spaceflight on Men and Women?

Since women joined the astronaut corps, researchers have wondered whether there are any differences between males and females in short- or long-term health risks from exposure to microgravity and space radiation. That question could not be addressed for some time, because there were too few female astronauts for adequate data or detection of trends. It is still a challenging research question, because not as many women as men have spent six to twelve months in space. However, evidence is gradually becoming available for assessment of some, but not all, health risks through the lens of male/female differences. Since the 1980s, several studies focused on possible sex-related differences in the body's response to spaceflight have reached similar conclusions that differences are minor, usually with the caveat that there is insufficient data to identify or predict really significant differences.

One of the earliest papers to present a medical perspective on women in space was published in 1984 when only three women—Valentina Tereshkova, Svetlana Savitskaya, and Sally Ride—had flown in space.[14] Its author was a NASA flight surgeon trained in psychiatry, Patricia Santy, who was interested in physiological and psychological considerations for women in space. She reviewed the then-known physical adaptations experienced by astronauts and posed some speculative medical and social issues that might uniquely affect women. Anthropometric differences in body size, reach, and strength would need to be accounted for in spacesuit and equipment design. Management of menses would need to be addressed. No data yet existed about gynecological and hormonal function in space, topics ripe for investigation. She noted some psychological and social concerns about stereotypes of masculinity and femininity that might affect crew cohesion, particularly with women in leadership and command roles. None of these were reasons to restrict women in space but were suggestive areas for research. Santy surmised from earthbound studies that slight but insignificant differences in physiological adaptation to spaceflight might manifest.

In general, women astronauts were not keen to seek out differences that might be used to restrict their participation, preferring to treat gender as a nonissue unless it somehow affected performance.[15] One true gender issue was accommodation for urination while strapped in their seats during launch and entry or while on a spacewalk, when a toilet wasn't accessible. Male pilots and astronauts used a urine collection device that was a condom or cuff attached to a small plastic drain tube attached to a collection bag, all worn under the flight suit. Relief was effortless unless there was a leak. That device obviously would not work for women, but it took time and false starts before a solution was reached. NASA developed and the women astronauts evaluated a few unsatisfactory options.[16] No one wanted to risk irritation or infection by using an invasive catheter. An alternative was a close-fitted fabric undergarment like a girdle or biking shorts lined with absorbent padding that became a gel when wet, but it was uncomfortable and expensive to make. Another was a small elongated funnel with an attached drain tube to a collection bag; the funnel was to be worn snug against the woman's body inside her underwear. It leaked, overflowed, and was uncomfortable. The final solution was inspired by disposable baby diapers. The adult version for incontinence was soft, comfortable, highly absorbent, disposable, and inexpensive, and several pull-on or adjustable styles were commercially available. Relief was effortless and not messy. The women liked this option so well that the men eventually adopted it too for launch, reentry, and extravehicular activity. To use the toilets on the shuttle and space station, everyone had a personal funnel, with a slightly different shape for males and females, for urination into a drain hose.

By 2000, when women had been flying on missions for almost twenty years, interest in possible gender-related differences related to spaceflight was growing to ensure that appropriate countermeasures and health care could be provided.[17] NASA enlisted the National Space Biological Research Institute to conduct a study on gender-related issues in spaceflight research and health care. The institute convened a two-day workshop in 1999, chaired by physician-astronaut Rhea Seddon.[18] Participants included another physician-astronaut, Ellen Baker, and at least ten researchers from medical schools and health agencies who were invited to engage in a review of the knowledge base and the status of research. The group reported that as yet there were only a few instances of measurable gender differences, such as postflight orthostatic intolerance (dizziness or fainting from insufficient blood flow to the head), but that there was not enough data to identify or predict significant differences. They recommended more research involving more women, aiming for a better understanding of differences, and then development of feasible countermeasures if necessary to protect women astronauts' health and safety.

This workshop report also identified several unique health care concerns for female astronauts, mainly their ability to prevent or plan pregnancy and to rely on highly effective contraception. These needs were crucial to their flight availability and family planning. There had been no effort yet to investigate any microgravity effects on hormonal activity, the efficacy of birth control methods, or the efficacy of contraceptive medications used for other purposes such as maintaining bone density or reducing the risk of certain gynecological issues—issues that should be studied. Finally, the workshop noted the possibility of performance differences between men and women due to technology, such as spacesuit design, and task procedures. The report clearly stated, "It should be the goal that all people selected to be astronauts be able to perform all tasks associated with the astronaut job regardless of size or gender." The unstated point was that some equipment was indeed hampering the performance of some women astronauts, and such design issues should be remedied.

Ostensibly meant to identify priorities for research into gender-related issues, the workshop had another purpose. At the time, NASA was toying with the idea of an all-female shuttle mission.[19] There was a flurry of NASA and press commentary on the possibility of an "unmanned" crew. The NASA administrator then, Daniel Goldin, thought it would be inspirational. The life sciences chief thought it would be valuable for gathering more data. Both agreed that such a mission needed a strong science justification and hoped the workshop would provide that. However, the group's report concluded, "None of these recommendations require that all-female crews be flown." In addition, most of the women astronauts wanted no part in a mission that could be viewed as a publicity stunt; it seemed too political, and new research findings from one brief mission seemed unlikely. A scientific program for a series of missions to study specific women's health issues might have gained more interest and legitimacy. Some reports in the media reflected the same concerns.[20] Although all-male crews were common, the women didn't want to be assigned to a sex-segregated mission; they wanted to be assigned on the basis of their abilities, not their anatomy. Seeing no good rationale for proceeding, NASA dropped the idea, and an intentional all-woman mission has not yet occurred.

In 2001 NASA authors published an overview of known and potential gender-related differences in physiological responses to spaceflight.[21] At the time, the only data-supported difference was postflight orthostatic intolerance, a transient condition that women were significantly likelier than men to experience upon returning to Earth's gravity, but it was not a clinical problem. Some male astronauts experienced cardiac dysrhythmias in flight, but there was no comparative data yet from in-flight cardiovascular monitoring of female astronauts. No gender-related difference in bone density by mineral loss was reported; regular exercise in space was credited for that result. Nor did any evidence indicate gender-specific risks for formation of renal stones during spaceflight. There was no indication that spaceflight altered the immune system differently for males and females, or evidence of any significant gender differences in neurosensory-motor functions. Gender-related differences in drug metabolism had been recognized on Earth but not yet studied in spaceflight; such differences could be important for medical treatment in space, a reason for exploring this issue more thoroughly. Although the average woman has less muscle mass than the average man, in the physically fit astronaut population strength training reduces these general differences, especially in microgravity, which largely negates them. At the time, there was not much data to evaluate women's deconditioning and changes in strength after spaceflight, so no comparisons with men could be made. This report's conclusion was that "individual differences in physiological responses within genders are usually as large as, or larger than, differences between genders."[22]

With both the space shuttle and space station in operation after 2000, individual astronauts were flying more often and staying longer in space, the number of women veterans of spaceflight was increasing, and more data was becoming available for study, but researchers still could not address the question of sex-based differences with certainty. In 2011 the National Academy of Sciences stressed the need to understand the influences of sex and gender on human responses to spaceflight. In response, NASA and the Space Biomedical Research Institute undertook another review of relevant research and priorities, establishing six workgroups to summarize the current body of data and publishing a set of reports in 2014 as The Impact of Sex and Gender on Adaptation to Space.[23] Each workgroup reported on a particular area of health—cardiovascular, immunologic, sensorimotor, musculoskeletal, neurosensory, reproductive, and behavioral—and made recommendations for new research to fill knowledge gaps.

Some differences were noted between women and men, but their causes and significance remained uncertain. For example, male astronauts are more likely to experience visual impairment and hearing loss in space than female astronauts, and females have a more robust immune response in space (as on Earth) than males. Women have a greater loss of blood plasma volume than men. Some men and women develop kidney stones after, but not during, spaceflight. Female astronauts report more urinary tract infections in space than their male colleagues do, a difference possibly related to the design of the waste management system. These conditions are not major health problems, and they may not differ appreciably from incidence in the population at large. Measurable or statistical differences do not necessarily mean clinical significance.

While the summary report argued that "it is imperative to examine and understand the influences that sex and gender have on physiological and psychological changes that occur during spaceflight," it also acknowledged the difficulty of drawing conclusions from a still small sample size of female astronauts.[24] Nevertheless, this review indicated that progress had occurred in the effort to understand how spaceflight affects males and females. Recommendations to reduce the disparity in data included assigning more women to space missions, encouraging more female and male astronauts to participate in research studies, including sex and gender factors in experiment designs, and pursuing more detailed research into reported differences.

The motive for investigating and understanding differences between the sexes is to ensure astronauts' safety and appropriate health care as they encounter the risks of spaceflight. As the understanding of space physiology continues to deepen, the consensus holds that while individuals vary, there are few if any significant differences in the ways that male and female bodies react to spaceflight. The authors of the most recent edition of *Principles of Clinical Medicine for Space Flight* present no definitive findings of significant differences between male and female physiological adaptations.[25] The human body in general seems to have a common and predictable range of responses to microgravity and radiation, and individuals vary within that range regardless of their sex. No alarming preponderance of males or females has been noted for any of the conditions studied. Minor differences do not affect ability to do the job and do not portend different health consequences. The first generation of women astronauts may have been wise not to worry about sex-related differences unless they became a problem; no such problems in health or performance have arisen.

This consensus has been fifty years in the making since the first data reviews were compiled. Its consistency over time is reassuring, despite lacking as much data from women astronauts as from men. Early women astronauts who were concerned about health risks of spaceflight had to trust the findings derived from men's experiences in space, absent any compelling evidence to the contrary. Now they may have greater confidence that they are at no greater risk than men in spaceflight. Historically, inadequate data from women participants is not unusual; it has been noted in pharmaceutical and other medical research as a gender data gap that may mask subtle but clinically important differences between the sexes until more women are included in research studies and clinical trials.[26] As more women spend more time in space, researchers will welcome additional evidence of normal or sex-related differences in astronauts' health.

As of 2024, only three active female astronauts had reportedly experienced conditions that could affect flight status or become chronic health issues. Two of them had transient conditions that were treated successfully and resolved. Neither condition was among those that NASA had planned to remedy in space. Another developed cancer late in her career. None of these conditions was alleged or verified to be triggered by spaceflight or a sex-related difference.

Of the three, one female astronaut on the space station discovered an external jugular vein thrombosis during an ultrasound exam as part of a research study. This was the first time this type of blood clot was detected in an astronaut while in space. Treatment was improvised using the onboard medications, and she was monitored for the remainder of the mission with ultrasound exams and consults with physicians on Earth.[27] The in-flight treatment was successful in reducing the clot, and postflight monitoring indicated that the clot had dissolved, so no further treatment was required.

She remained in the astronaut corps, but not on active flight status. The incident raised an important question: was blood clot formation related to changes in blood flow in microgravity? That question is now under study, and the answer could affect all astronauts. Another woman developed a kidney stone after flight; had it occurred in space, it would have been considered a medical emergency.[28] After treatment and no further occurrence of stones, the affected astronaut was cleared to fly again, and indeed did. A third woman developed breast cancer and died of that cause at age fifty-five in 2012.[29] She had flown on five missions and spent a total of forty-nine days in space. Her last flight was in 2000, her diagnosis occurred in 2002, and she worked in management roles thereafter.

As of 2024, only seven of NASA's sixty-one female astronauts had died, three in the *Challenger* and *Columbia* accidents, one fatally injured in an aircraft accident, two from cancer, and one of a stroke.[30] There is no published evidence that the nonaccidental deaths were related to time in space, and there is no evidence that NASA's women astronauts have an increased risk of breast or gynecological cancer.[31]

Reproductive Health

The primary physiological and anatomical difference between the sexes is obviously the reproductive system. In this realm, some risks and health issues are unique to the female astronaut population. Speculative concerns about gynecological and reproductive issues predate women going into space but were usually mentioned only in passing because there was no factual evidence to analyze. The 1999 workshop report on gender-related health differences recommended research into women astronauts' reproductive health, and in 2000, the first systematic review of gynecological issues and spaceflight appeared in a medical journal.[32]

The coauthors of this review were two NASA physicians, flight surgeon Richard Jennings and astronaut Ellen Baker. Jennings, who was trained in gynecology as well as aerospace medicine, had a long affiliation with the Johnson Space Center Flight Medicine Clinic and shuttle crews. He counseled women astronauts on birth control, managing their menstrual periods for spaceflight, pregnancy and related conditions, childbirth, and fertility, and he followed them through their pregnancies but did not deliver their babies.[33] The women under his care were fortunate to have their reproductive health evaluated by a flight surgeon in the context of spaceflight conditions. Baker, with a specialty in internal medicine, had been a medical officer in the same clinic before becoming an astronaut. She had flown on three shuttle missions; later, after her flights, she became the head of flight medicine at JSC. Both participated in the 1999 gender and health workshop led by physician-astronaut Rhea Seddon. Both have since written, separately and collaboratively, about astronauts' reproductive health.

Their first article expressly addressed gynecological and reproductive issues for women in space over time from their selection as astronauts and their medical certification for flight, then training and mission operations, and on to postflight fertility and pregnancy.[34] A spectrum of gynecological conditions could impact a woman's participation in spaceflight or her ability to bear children. Jennings and Baker argued that the best health strategy was based on preflight prevention of such problems, early diagnosis of any abnormalities, and conversion of surgical conditions to medically treatable conditions. Conditions such as menorrhagia (excessive bleeding), painful cramps (dysmenorrhea), benign fibroid tumors, ovarian cysts and uterine polyps, recurrent infections or inflammation or other complications, and malignancies should not arise during spaceflight but be prevented or treated on the ground in advance. Endometriosis, once disqualifying for flight because it could be exacerbated by backed-up menstrual flow or radiation exposure, could be tolerated if it were mild and manageable. Many conditions could be resolved with medication or minor surgery. The authors concluded that for healthy women "no medical or gynecological problems have developed that cannot be addressed," and with appropriate reproductive health care there are no constraints on women's participation in space exploration. This review allayed many of the earlier speculative concerns about possible adverse gynecological effects of spaceflight.

After this review article was published in 2000, female reproductive health in relation to spaceflight became a topic of wider interest for research and reporting. Earlier publications on space physiology either had not mentioned the reproductive system or had done so only in passing. Another review article published in 2001 by a group of NASA coauthors that included Jennings and astronaut Peggy Whitson offered a section on reproduction that discussed menstruation, fertility, and unknown effects of gravity and radiation.[35] By 2011 the National Academy of Sciences had recognized reproductive health as a research priority in investigations of the influences of sex and gender on bodily changes that occur in spaceflight. One among several workgroups convened by the academy focused on reproductive health concerns for both male and female astronauts, the results of which were published in 2014 in the *Journal of Women's Health*.[36] The 2019 edition of *Principles of Clinical Medicine for Space Flight* includes a comprehensive chapter on gynecologic medical standards, operational concerns, and reproductive considerations, coauthored by Jennings and Baker.[37]

Over the course of forty-plus years, health issues unique to female astronauts moved from unmentionable to open discussion, from speculation to limited understanding, and from little or no data to enough data to reach some reassuring conclusions. However, there are still many gaps in knowledge. Important questions remain about hormone functioning, radiation exposure, fertility, and other concerns, especially for women on long duration expeditions. The incidence of pregnancy complications and infertility after prolonged spaceflight is not known, and it may be difficult to identify effects of microgravity and radiation exposure versus the natural effects of aging.[38] Research into spaceflight's effects on the reproductive and endocrine systems continues.

Astronaut Gynecology

Gynecology and reproductive health encompass menstruation, contraception, fertility, pregnancy, and their related abnormalities and disorders. Of concern for spaceflight is any disorder of the female reproductive system that might interfere with an astronaut's health or performance of her duties or that might become a medical emergency in space.[39] NASA Standard OCHMO-STD-100.1A (2021), issued by the office of the chief health and medical officer, includes the current list of ten gynecological disorders and abnormalities that are of concern for women astronauts.[40] On the list are disorders of the uterus, cervix, and ovaries; abnormal bleeding; recurrent pelvic infections or inflammation; and any menstrual abnormality or related hormonal dysfunction. Pregnancy is disqualifying for spaceflight until complete postpartum recovery.

During the astronaut selection process, women finalists undergo a pelvic examination, cervical cancer screening, abdominal ultrasound screening, and a breast exam and mammography to confirm that their female anatomy is normal and healthy and to rule out any disqualifying conditions. If treatable conditions are found, finalists may undergo procedures to remedy them. Once in the astronaut corps, women continue to have annual exams, so any problems that may arise can be promptly diagnosed and treated. The purpose of initial screening and annual examination is health maintenance, prevention, and treatment of conditions that may progress and affect health and flight certification.

Women apply to become astronauts during their childbearing years, and most have not yet been pregnant. Occasionally, a woman is pregnant at the time of finalist interviews. She is not disqualified from selection, but she cannot complete the screening exams until after childbirth and thus may have to postpone her exam until the next astronaut selection cycle. Likewise, a woman originally disqualified for a gynecological reason may reapply and be reconsidered when the medical condition is treated and corrected. Screening before selection is so effective that "no active female US astronaut has been permanently grounded because of a gynecologic condition that developed after selection."[41]

Managing Menstruation

The main practical consideration for women astronauts is the natural process of menstruation. Many of the women astronauts choose to manage it, for convenience, by suppressing their monthly periods with hormonal contraceptives while in space. Those who choose not to interfere with their menstrual cycle report that it is not a problem in space and occurs no differently than on Earth.[42]

Recalling the Lovelace Women in Space research program of 1960–1962, in which thirteen female pilots passed the same physical and psychological exams that were used to select the first male astronauts, the sole published medical research report appeared in the *American Journal of Obstetrics and Gynecology* in 1964.[43] The authors' commentary focused on possible problems associated with the female reproductive system if women were to become astronauts sometime in the future.

The authors, both medical doctors, posited that an astronaut's menstrual cycle could interfere with the mission schedule: "The intricacies of matching a temperamental psychophysiologic human and the complicated machine are many and, obviously, both need to be ready at the same time. . . . Menstruation may complicate the use of the female astronaut in an environment of time tables and rigid schedules needed for a perfectly manned space voyage." They cited flawed studies that women are "more prone to show a lack of attentiveness and [be] more accident prone" during menstruation, and that there is an increase in mental illness and suicide during menstrual flow. They noted that the degree of incapacitation due to menstrual cramps should be ascertained. On the other hand, their more reasonable observations included the fact that menstruation and its waste products would be "a challenging problem for disposal" and that delaying menstruation by using oral contraceptives "may prove to be of merit." The report concluded with a statement that research was needed on women during the entire menstrual cycle to evaluate whether they could serve effectively in space. Such research was never deemed necessary.

Apparently unknown to these physicians, in the 1940s an Air Force physician, Nels O. Monserud, had studied the performance of pilots in the Women Airforce Service program (WASP) before, during, and after menstruation.[44] Previously, it was a widely accepted belief that women pilots should be grounded during their menstrual periods because they were less stable and tended to become emotionally upset, even to the point of crashing their airplanes. Monserud found that the WASPs, like some earlier female racing pilots, flew without regard to menstruation and without suffering any adverse effects or causing any crashes. He found that menstruation "did not interfere with either training or dependable job performance" and concluded that "menstruation is not a handicap to flying."

Menstruation in spaceflight is now considered to be a normal female process, not a debilitating medical or occupational problem, and one to be managed. NASA's women astronauts have active agency to manage menstruation for reasons of health or convenience, making their own choices, in consultation with their physicians, about letting their periods occur naturally during spaceflight or suppressing them for the duration.[45] This decision is a balancing act of weighing pros and cons, risks and benefits. Many women astronauts choose to suppress their periods in space by taking hormonal oral contraceptives (birth control pills) continuously until they are ready to have another period. Advantages to this method may include reducing loss of bone density and red blood cell volume, avoidance of anemia and iron deficiency, and prevention of heavy or irregular bleeding. However, oral contraceptives may carry long-term risks of blood clots or cancer. As yet, hormonal contraceptives have not been linked to any specific effect of spaceflight.[46]

Women who choose not to suppress their periods during spaceflight have access to pads and tampons and anti-inflammatory medications, such as ibuprofen and acetaminophen, to relieve any menstrual discomfort. Because onboard mass and volume are consequential, menstrual hygiene products affect calculations for stowage and waste management. This relatively small mass must be accounted for in relation to everything else. The amount of trash, such as food packaging, is carefully controlled, so disposal of sanitary products must be factored into waste disposal capacity. Menstruation is a consideration in the design and function of the human waste management system (toilet) and the water reclamation system that processes urine and moisture from sweat and breathing into drinking water. Because menstrual blood and secretions may be released during urination, the design of filters and processing methods must take that into account. While suppression of menstruation during spaceflight reduces such engineering requirements, there is no pressure on women to do so. The impact of menstruation on engineering design of spacecraft systems is not trivial, but properly designed systems can and do function effectively.[47]

This story has been told elsewhere and is now legendary.[48] In 1983, when the crew equipment team put together Sally Ride's personal hygiene kit for her initial six-day mission without consulting her, they calculated the number of tampons she might need. When she inspected everything that had been prepared for her needs in space, she was astonished to find a hundred tampons, neatly tied together like sausages so they wouldn't float away in weightlessness. "Is one hundred the right number?" the men asked. "No," she replied, "that would not be the right number." Thereafter women astronauts specified the products and quantities needed.

Contraception

The ability to manage their reproductive health by using contraceptives to manage their periods and schedule their pregnancies has been a fundamental right for women since they entered the astronaut corps.[49] If they do not wish to become pregnant, it is their responsibility to manage that decision. Because pregnant women are precluded from spaceflight, most women astronauts use the most effective contraceptive methods available. If they do wish to become pregnant, they can plan an optimal time. Some of the women find it prudent to postpone pregnancy until the less hectic, more settled time after initial training. Others may wait until they have flown at least once before starting or growing a family. Some plan their pregnancies between missions when they can work on the ground in other assignments in the Astronaut Office. All contraceptive methods are available for each woman's choice.

Pregnancy

Women are selected to become astronauts during their prime childbearing years, between the ages of twenty-six and thirty-five, and most have not yet borne children. Some have no plans to have children, and others foresee eventual pregnancies. Childbearing decisions may become another balancing act for women astronauts, as they consider such trade-offs as postponing pregnancy to advance their career and fly in space, possibly missing the window of their prime fertility, or choosing not to have children. Those who desire to have children often postpone pregnancy until they have flown in space once or twice.[50] The downside of delay is that becoming pregnant when older may be more difficult, with increased risks of infertility, miscarriage, complicated pregnancies, or birth defects.

Pregnant astronauts are temporarily disqualified from spaceflight and from many training activities that pose a risk to a woman or fetus.[51] After the first trimester, pregnant astronauts are not allowed to fly in T-38 jets and other aircraft with ejection seats in case of accidental decompression or forceful ejection. They may not participate at any time during pregnancy in KC-135 aircraft parabolic flights, altitude and vacuum chambers, and underwater neutral buoyancy training to avoid exposure to high g-forces, decompression, and hyperbaric environments. Nor can pregnant women participate in water survival and parachute training to avoid possible injury. Astronauts are expected to self-report pregnancy and are not routinely tested, but they do take a pregnancy test forty-five days, ten days, and two days before their upcoming mission. A pregnant woman would be removed from a mission.

If a woman becomes pregnant, she can continue her work outside these restrictions. Work assignments are adjusted to activities without unusual risks, and she remains a productive member of the astronaut corps by assuming one or more of the many technical duties in the branches of the Astronaut Office. As federal employees, women astronauts are able to use sick leave, vacation leave, and family medical leave for their preferred maternity leave duration, usually six weeks to three months. Some women are able to arrange with the chief of the Astronaut Office for part-time work during their maternity leave. Astronauts willing to share their experience reported that they were treated with respect and understanding during pregnancy and childbirth, and so were the women who adopted children. Supervisors and peers were supportive in meeting their needs and adjusting their work assignments as appropriate.

Rhea Seddon and Anna Fisher, both medical doctors, were the first astronauts to become pregnant. Seddon gave birth first, in 1982, an event that made national news and earned her son the nickname "astrotot," bestowed by *Time* magazine. Fisher was the first to fly after giving birth in 1983 and was hailed as the first mom, or "astromom," in space on her 1984 mission. Both shared their stories in interviews, and

Seddon included hers in her memoir.[52] Their accounts reflect some of the considerations affecting the first generation of women astronauts.

Seddon recounted planning three pregnancies between her missions. As a result of her flight schedule, her children were spaced farther apart than she might otherwise have chosen, with births in 1982, 1989, and 1995. She recalled that at first NASA management seemed caught by surprise and had not yet considered policies for pregnancy. Decisions about what she could and couldn't do were made on the fly until a more deliberate process developed. Although they were congratulated and encouraged, she and Fisher typically were not consulted about those early rules but found them generally reasonable. Both Seddon and Fisher became pregnant before their first spaceflights, and both waited some time to announce their pregnancies, cautious about possible impacts on their careers. They made it clear that they intended to stay in the corps and do their jobs, and they did not face the consequences for becoming pregnant—demotion or firing—that women working in aerospace had suffered ten to twenty years earlier. Later astronauts did not feel a need to be discreet after the precedent of pregnancy was established.

Fisher received her mission assignment when she was eight and a half months pregnant and began training shortly after giving birth. She was thirty-three years old, had been at NASA five years, and did not want to miss this chance to fly. She intentionally returned to the office for the Monday morning staff meeting after giving birth on Friday to leave no doubt that she was not debilitated by pregnancy and childbirth and that she would continue to fulfill her astronaut responsibilities. After her second child was born in 1989, Fisher took successive leaves of absence and part-time work to be home with her young children until they reached school age.

Eileen Collins, from the second generation of women astronauts, became pregnant twice in the 1990s. She commented in her memoir on the trade-offs of trying to plan a pregnancy between missions, often a two- to three-year gap.[53] Mission scheduling can be somewhat erratic with technical problems causing delays or a shuffling of the flight sequence, so it is difficult to predict when you may have an open window and how long it will last. It is also difficult to predict how long it will take to conceive. A woman who is trying to become pregnant is required to alert the Astronaut Office so she will be put on nonflight status and not be assigned to an upcoming mission. For how long should she take herself out of consideration for flight, or should she delay pregnancy in hopes of being assigned to a mission? That decision is one of the balancing acts that only female astronauts face. One astronaut confided that she kept herself on nonflight status for three years without becoming pregnant before returning to active status to pursue a chance for another mission.

Most female astronauts who want to have children are successful.[54] Including Seddon, who bore three children, and Fisher, who had two, more than twenty women (about one third of the female astronauts to date) have had pregnancies and childbirths before, during, or after their time in the astronaut corps, yielding a total of forty children. All

At a baby shower: (*front*) Eileen Collins, Kalpana Chawla, Lisa Nowak, Stephanie Wilson; (*rear*) Kay Hire, Shannon Lucid, Ellen Ochoa, Tammy Jernigan, Anna Fisher, Linda Godwin, Bonnie Dunbar, Janice Voss (*standing*), Cady Coleman. *Ellen Ochoa*

of these women have flown at least once, and most have flown more often. At least three women astronauts successfully adopted children to start their families, and several others gained stepchildren by marriage.

Several women interviewed for this book commented on their experiences in becoming mothers while they were astronauts. They noted that the Astronaut Office was supportive during pregnancy and maternity leave, and they were readily assigned to flights when they returned to work. All reported being able to take the amount of leave they wanted to spend with their infants. Some negotiated part-time work assignments after childbirth to avoid falling behind in their responsibilities. One who adopted a child was able to negotiate a flexible work schedule for several weeks, allowing her to spend time at home bonding with the baby but come in for training sessions so she could remain on her assigned mission.

Some told anecdotes that probably have no counterpart among male astronauts.[55] One woman resumed her regular work schedule, which included underwater EVA training for six to eight hours a day, while she was still lactating, an experience that she would not care to repeat. Others who were working as CAPCOM on nine-hour shifts had to excuse themselves occasionally to pump milk in a restroom near Mission Control.

At least once a pregnant astronaut was summarily removed from a crew already training for a mission, although she would have delivered her baby well before flight. She was later reassigned to another mission but felt the situation could have been handled better by consulting with her. Except for that instance, women astronauts felt that their needs were accommodated, and becoming mothers while on the job had no adverse effect on their careers.

Fisher had a unique experience after having her first child, flying her first mission, having her second child, and deciding to stay home for a period longer than typical maternity leave. When she returned to full-time duty in the Astronaut Office in 1996 after an absence of seven years, her return proved more challenging than she expected because the office environment and technology had changed so much. During that time, many of her astronaut colleagues from the 1980s had moved on, and she didn't yet know the newcomers. Typewriters had disappeared in favor of desktop computers, paper correspondence was replaced by email, and overhead projectors and viewgraphs had yielded to PowerPoint presentations. The pace of work had quickened with these technologies, and she had to catch up quickly with new ways of doing things. She still knew the astronaut job, but technology had changed many of the ways of doing it. Fisher has cited that time as the most challenging in her career.[56]

Looking ahead to long-duration missions beyond Earth lasting a year or more, medical specialists assert that "it will be imperative to prevent pregnancies" in space until the effects of microgravity and radiation on pregnancy and human development are understood.[57] Nor is it yet clear how to deal with common complications, such as miscarriage, ectopic pregnancy, preeclampsia, breech birth, Caesarean delivery, or postpartum hemorrhage at such a distance from emergency medical care on Earth and with only limited available resources on the spacecraft. Younger astronauts in today's astronaut corps may face different issues and have somewhat different needs related to pregnancy and childbearing as they become eligible for significantly longer missions.

Fertility

The 1999 workshop report on gender differences discussed earlier included a fairly radical idea at the time: that assisted reproductive technologies should be included in the medical care of female astronauts who postponed pregnancy until completing their training and first missions and then sought to become pregnant at ages of declining fertility. Jennings, a workshop participant, wrote an internal white paper on the topic, and Jennings and Baker's review of gynecological issues for spaceflight, published the next year, recognized "a considerable need for infertility services and assisted reproductive technology (ART)," especially for older female astronauts who had delayed pregnancy.[58] After examining the data then available and caring for women astronauts, he noted that the average maternal age after spaceflight was forty-one to forty-two years, an age at which fertility was diminished and the chances of miscarriage and genetic defects were increased. At that time, he reported that the miscarriage rate in women astronauts after spaceflight exceeded 40 percent, most likely due to their age, because it was not yet (and still is not) known whether spaceflight affects fertility.

Jennings's opinion that NASA should support ART services for its astronauts arose from certain realities.[59] He noted that most women who entered the astronaut corps had not yet had children, and those who wanted children were waiting until they had served NASA in space, in essence sacrificing their most fertile years to their duties. It would be reasonable for the agency to repay their service by enabling the freezing of younger astronauts' ova and embryos for future fertilization/transfer and supporting fertility assistance such as in vitro fertilization (IVF) for older astronauts. He made the case that military astronauts received monetary support for fertility services while civilians bore their own expenses, and the disparity was unfair. Thereafter, almost everything NASA published on reproductive health included discussion of delayed pregnancy among female astronauts and the advisability of ART. Jennings continued to champion this cause until NASA adopted in 2012, after more than ten years' consideration, a policy of supporting infertility and reproductive services, to include freezing eggs and embryos, in vitro fertilization, and other assistive procedures for conception.

Before this happened, Rhea Seddon was among the first to seek such services to enable another pregnancy after her final mission, when she was well into her forties. She already had borne two children but felt that her family was not yet complete. When pregnancy did not happen quickly, she persuaded her husband, astronaut Robert L. "Hoot" Gibson, to "let me give all our money to the fertility experts."[60] She was thrilled to give birth again when she was forty-seven years old. Several women astronauts since have tried implantations and IVF with low and mixed success; the instances of successful pregnancies and miscarriages were about equal in astronauts aged forty and older, as also for the non-astronaut population.[61] It is not yet possible to distinguish effects of microgravity or radiation in spaceflight from the effects of normal aging on infertility.

Women astronauts, like countless women, have been able to exercise autonomy, choose among options, and make reproductive decisions that are compatible with their careers. NASA has supported their decisions without detriment to their professional success. However, public policies and laws that restrict or criminalize reproductive health management and care are a growing concern.

Behavioral Health and Performance

Some physicians think that behavioral health is as important as physical health for individuals and crews working in extreme environments, such as submarines, Antarctica, and space. Isolation and confinement are major psychological stressors that may affect mental state and behavior. People in those environments must depend on one another for safety and survival, so any behavior or mental condition that can cause a person to jeopardize safety, crew cohesion, or mission success is of concern for spaceflight.[62] That concern is addressed initially when astronaut applicants undergo psychological and psychiatric assessments during the semifinalist round of selection. Between 1 and 5 percent of applicants are disqualified by psychiatric illness, ensuring that no one with a known history or propensity for behavioral disorders or mental dysfunction is admitted into the astronaut corps.[63]

No matter how behaviorally and psychologically stable high-achieving astronauts may be naturally, as humans they may be at risk for stress from living and working in space, striving for excellence, and interacting with diverse personalities. Before the International Space Station existed for long-duration spaceflight, reports from cosmonauts, Skylab crews, and research expeditions indicated psychological issues that can arise on missions and the value of preparation and crew support.[64] Individual behavior and interpersonal relations, self-control, clear communication, and teamwork are crucial to mission success. At some point, though, friction may flare, patience may fail, and morale may dip. Poor sleep is often a trigger for irritability, which may then trigger other inappropriate behaviors. It is important to recognize and address such warning signals before they develop into anger, alienation, anxiety, or worse.

To that end, astronauts are trained in various techniques to manage their own, and support their crew's, behavioral health. Such training may include personality typing, personal counseling, participating in an analog mission like NEEMO or CAVES or another remote research expedition, taking a wilderness survival course together to improve teamwork and communication skills, developing leadership and followership skills, practicing problem solving and decision making, and learning more about the culture of their international partners. There have been no publicly reported behavioral incidents on space shuttle and International Space Station missions, and there is no scientific indication that men and women differ in reactions to stress or in behavioral health in space.[65] However, interest in research into the conditions that affect behavioral health grew during the shuttle and space station eras as mixed crews spent longer and longer periods in space.

Research and flight experience since the 1970s have settled on some of the triggers and remedies for behavioral health issues, with a view toward supporting astronauts individually and supporting the productivity of crews as a whole. Physician Patricia Santy took an early look at the relevance of psychiatry to aerospace medicine in 1983.[66] She described the traits of "psychological fitness" for successful performance as an astronaut: emotional stability, high motivation and energy level, emotional control, confidence, good interpersonal relationships, and commitment to the mission. She also discussed, based on studies of submariners and isolated Antarctic researchers, some of the symptoms that could negatively affect performance in an isolated environment: irritability, disorientation, deficits in cognition, and more. Without remedy, group compatibility and performance reportedly declined on prolonged expeditions. Santy postulated that training in group dynamics, problem solving, and conflict resolution would be helpful. She also noted the importance of comfort in the living-space environment, food, personal hygiene, adequate sleep, communication, time off, music, and the option of being alone for astronauts' well-being and morale. Santy's review was followed shortly by NASA volumes that addressed the stresses and behavioral implications of spaceflight.[67]

One of the workgroups tasked by the National Academy of Sciences to review the effects of sex and gender on adaptation to space focused on behavioral health. Their report, published in 2014, noted that there was some evidence, in the general population on Earth, that women had a heightened sensitivity to stress evidenced in higher cortisol and adrenaline levels, and that women were more subject to anxiety, panic, and depressive episodes than men.[68] It was unknown whether similar tendencies occurred in space because no such research existed, but it was considered unlikely that there would be significant differences among astronauts given their psychological screening. It was also noted how alike in personality traits male and female astronauts were. In another study of a mixed group of space station residents asked to keep journals on specific topics, there was no difference between men and women in self-rated parameters of fatigue, stress, workload, or sleep quality, or in accounts of their moods, but an interesting finding was that morale began to decline in the third quarter of their expeditions.[69] By that point, possible reasons included that having mastered their work, it was becoming routine, and that the long absence from home was taking a toll.

Astronauts train for and simulate most of what can go wrong on a mission, from balky equipment that must be repaired to life-threatening emergencies like a leak in cabin air pressure or a spacesuit. That training equips them to act in problem-solving mode rather than frustration or panic and thus prepares them to cope calmly with stress. Among the stressful situations handled by women astronauts were repair of an inconveniently nonfunctioning toilet (Mary Cleave), which really affected the quality of onboard life; a loss of thrust in a shuttle main engine during ascent (Eileen Collins); a jammed airlock hatch precluding her spacewalk (Tammy Jernigan); and an ill-fitting spacesuit that would compromise her performance and safety (Anne McClain). Surely there were many other stressful circumstances handled quietly that were not reported in the news media. Both women and men have dealt capably with stressors, and there have been no reported breakdowns in NASA astronauts' behavioral health that put a spaceflight in jeopardy.

Early astronauts had an aversion to psychologists, psychiatrists, and flight surgeons for fear of being grounded for some perceived weakness. They didn't want to be analyzed. That attitude gradually changed, in part because the culture changed within the aerospace medical community and society at large. The perceived stigma of counseling for behavioral and emotional issues has virtually disappeared, and the medical approach has become more supportive than negative. It is also likely that the contemporary term "behavioral health" is friendlier and draws less wariness than "psychiatry."[70] NASA has always had flight surgeons on staff to monitor the astronauts' physical heath, but in preparation for longer duration flights on the International Space Station, the agency drafted a team of psychologists to determine how best to prepare and support crews who would be living and working in space for months at a time. The Flight Medical Clinic at JSC established a Behavioral Health and Performance Center staffed with psychologists and psychiatrists whose goal is to help ensure mission success.[71] They have been closely involved with the astronauts from candidacy onward, conducting the initial screenings and consulting with astronaut selection boards. They don't have a vote in astronaut selection, but their assessments are taken seriously. Thereafter, their mission is to help their clients be as successful as possible in their job performance and management of stress or other difficulties. Until about 2005, they did not regularly follow individual astronauts but assisted them upon request. Now every crewmember is assigned a psychiatrist and a psychologist throughout their mission and evaluated for behavioral fitness to fly at least three times before launch.[72] They have private check-in conferences during flight and have relationships based on trust. Staff report that astronauts and their families, who also bear stress related to the spouse's or parent's job, now welcome this support and avail themselves of the counseling services offered there.[73]

Among behavioral health priorities is development of self-care habits to ameliorate the stresses of spaceflight, frequent travel, prolonged absences and disruptions to family life, and any worry, discord, or crisis in their lives. Self-care involves knowing one's own needs and making time to satisfy them.[74] These needs may be as ordinary as getting enough sleep and exercise or as unique as indulging in a hobby or a creative pursuit. Some are personal and some are social. An astronaut may need time alone to reflect and write, read, listen to music, or talk to family or friends. Social self-care in orbit includes celebrating birthdays and holidays, having occasional pizza nights or movie nights, watching a sports event, inventing weightless games and humorous pranks in off-duty time, expressing interest and empathy for others, laughing together, and especially fostering good conversation and camaraderie over dinner. Many astronauts keep a diary or journal to track their activities, record their feelings, and vent any frustrations, or they send daily email narratives home. One of the most satisfying self-care habits is regular communication with others not on the space station—family members, friends, and colleagues who are normally present in one's daily life—to stay connected despite isolation. These means of self-care are crucial for maintaining emotional balance, motivation, and the ability to perform well under stressful conditions.

Astronauts have remarked that a shuttle mission lasting at most about two weeks was a sprint; it was possible to put up with a nuisance or inconvenience for that short time. In contrast, a space station mission is a marathon lasting three to six to twelve months, and one has to know how to cope so little problems don't become big problems. Just as a runner controls his or her pace during a marathon and knows when to take water, an astronaut also paces for a long spaceflight and attends to self-care.

Shannon Lucid, the first woman to spend six months in space (188 days on *Mir* mission with two Russian men who didn't speak English), seemed to know instinctively how to practice self-care and also support the crew's behavioral health.[75] *Mir* was a stark and cluttered assembly of modules with few of the comforts of home, and the experiments that she was supposed to do had not yet been delivered when she arrived. She was there, but her work wasn't—a perfect setup for annoyance, restlessness, or melancholy. Instead, Lucid made *Mir* not just habitable but more livable by her attitude and actions. Perhaps being a wife and mother of three attuned her to the role of making *Mir* a home. She brought books and photos for company during her off-duty time, organized the food pantry, started some in-space traditions with the cosmonauts such as Sunday night Jello dessert, held a contest to find a missing item and awarded a prize for success, joked with them in her often unintelligible mix of technical and colloquial Russian, helped with the onboard chores, treated dinner as social time for halting conversations and hearty laughter, and maintained a positive attitude. Her family resupplied her with books and favorite snacks in care packages delivered by cargo ships, and she stayed in touch with them by daily email and occasional telephone or video calls, which she found essential to her well-being. Lucid emerged from her long stay in high spirits and beloved by her crewmates.

Lessons learned from Lucid and the other astronauts who stayed on *Mir* factored into International Space Station habitability design and behavioral health practices. With more privacy and generally a less hectic schedule than on the shuttle, everyone realized the importance of time for astronauts to simply be themselves and do something they enjoy beyond their work. Individuals have private sleep compartments like small closets that they personalize with photos and mementos from home. They can retreat there for solitude if they wish, or if it is not in use, they can escape to the panoramic cupola to watch scenery or read. Some bring musical instruments to space—flutes for Ellen Ochoa and Cady Coleman, a small keyboard for Susan Helms, her piccolo for Jessica Meir—for relaxation, or they load their devices with music and books. Nicole Stott brought a watercolor kit, quilter Karen Nyberg brought fabric and needlework supplies, and Sandy Magnus sent up seasonings so she could experiment with making new cuisine from the packaged foods. Suni Williams ran the Boston marathon and participated in a triathlon on the space station's exercise equipment. Others honed their photography skills or followed other interests. As the workload rhythm varies between very intense and less so, astronauts pay attention to sleep and self-care to avoid the behavioral hazards that may arise from being tired. No one wants to be the cranky, complaining, frustrated crewmate who could spoil the mission.

Families and support teams on the ground also help with self-care by including treats in the resupply cargo—fresh fruit and vegetables, snacks, surprise gifts, and other items to boost morale and well-being. Whatever the astronauts choose to do to keep themselves refreshed, motivated, and behaviorally stable affects the well-being of the

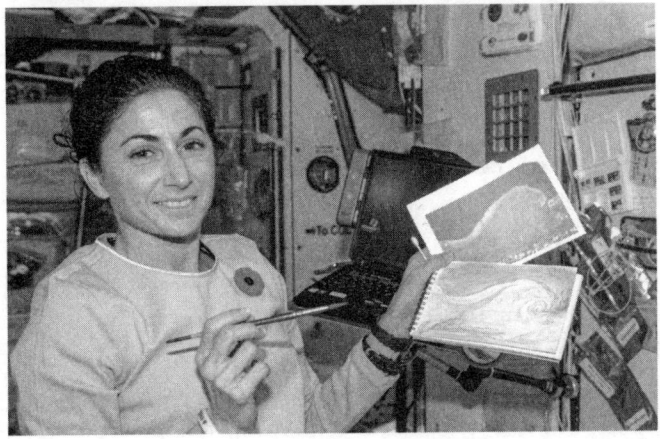

Personal time spent quilting (Karen Nyberg) and painting (Nicole Stott). *NASA*

team and the success of the mission. They have learned that coping skills and self-care practices are necessary to recharge their energy and stay balanced. Keeping themselves in balance mentally and emotionally reduces the risk of adverse reactions to stress, isolation, and confinement and increases the rewards of a mission well executed.

Astronauts have also learned that self-care is important after a mission, when they have to readapt physically and psychologically to being back on Earth. They often return elated from their time in space to find that aspects of "normal life" at home or the office have changed, and they must adjust. They may face uncertainty about their next assignment and prospects for another flight. After achieving a pinnacle, they may experience a letdown and a sense of "what do I do next?" For those who experience being an astronaut as their way of life more than their job, postflight can be a precarious time. Coping skills to meet the stresses of spaceflight are important after flight, too, to maintain the behavioral health that will keep them ready for whatever their future holds.

Family Life: A Balancing Act for Everyone

What is often called "work-life balance" can be especially challenging for astronauts, whose work is extremely demanding with long hours at NASA, days and nights in simulations, weeks of travel around the world, months of deployment in Russia, weeks to months in space, all repeated again and again. This isn't an issue unique to astronauts; spouses in military service, law enforcement, medicine, and other highly demanding professions also face danger, long absences, and similar stresses on work-life balance. Prolonged family separations may be the hardest condition of employment to cope with, because they affect all members of the family and prompt changes in their normal roles, especially in child-rearing and management of the household.

As within any family, roles and responsibilities can compete with the demands of the parents' careers and the children's needs, leading to trade-offs and shifting balances. Spouses left behind assume greater responsibilities for the home and children, often needing help from other family members or hired aids. Vacations are rare, holidays and birthdays are missed, and parents cannot attend every sports event and recital, attend every parent-teacher conference, or be present for every bedtime, meltdown, sickness, or gleeful moment. These stresses increase when both parents have comparably intense and irregular work, as when both are astronauts, or the spouse is a pilot, active-duty military officer, in-demand physician, or first responder.

Gendered social expectations typically assume that family responsibilities fall unequally on the woman as the primary parent for childcare and domestic life.[76] She essentially holds two jobs, one at work and the other at home. A woman astronaut's long absence from home usually shifts that imbalance to the spouse. As an example—even in a marriage of equally well-educated and committed professional peers—Laurel Clark's widowed husband admitted that after her death on *Columbia* he was clueless upon finding himself the sole parent and homemaker. "She was 100% the caregiver for [their son's] . . . formative eight years," he said, "and then when she left, I'm like, wow, I have no idea what I'm doing."[77]

So far, ten married couples have served in the NASA astronaut corps.[78] All but the Fishers, who married shortly after submitting their applications, met at NASA. She was selected in 1978 and he in 1980. The only couple to fly together on a mission were Jan Davis and Mark Lee. Some of these marriages ended in divorce or widowhood.

Coordinating work schedules and providing full childcare coverage are the biggest challenges for astronaut families, as they are for almost all working parents. Astronauts must become creative in finding solutions that fit their lifestyle, whether they engage full-time live-in nannies or babysitters with flexible schedules or grandparents, relatives, friends, and neighbors, to provide continuity of care while both parents work and travel. Sometimes daycare centers can be part of the solution, but most have set hours that align better with parents who work regular eight- to ten-hour days than with the schedules of astronauts. Again, this challenge is not unique to astronaut families, and they earn enough to afford help with childcare, but the demands for their time away from home are heavy.

Women astronauts have used various strategies to solve their childcare needs. Shannon Lucid arrived with three school-age children and a self-sufficient husband who had a regular work schedule and little travel; he was able to cover their childcare. Her children grew increasingly independent as she progressed through her astronaut career, and her husband readily took care of everything when she was away. Rhea Seddon and Anna Fisher led the way as the first to need infant care and the first whose husbands also were astronauts. Both settled on nannies who became part of their families. Seddon had a reliable babysitter and an au pair to help with her first child but welcomed a live-in nanny as her family grew. The nanny tended to all three children over a period of twenty-five years as a beloved member of the family. Fisher found a beloved nanny for her first child and then chose to stay home with her two youngsters until they were school age. Ellen Ochoa with two children and a husband who traveled frequently found it imperative to have a nanny. They credit these caregivers with making it possible for them to do their astronaut jobs while mothering.[79]

Two of the women had unusual domestic arrangements with a hybrid mix of familial care and babysitters.[80] Kathy Thornton married a professor at the University of Virginia who had two sons, and they had a daughter together. When Thornton became an astronaut, she and her two-year-old daughter moved to Texas, while the boys and their father stayed in Virginia. Thornton and her husband maintained two homes and traveled back and forth frequently. Two more daughters were born while she was at NASA, and they also lived with her in Texas. Her husband cared for the girls when she was in space, and she had help from live-in college students or an overnight babysitter as needed when she was otherwise traveling or working. The family reunited twelve years later when she left NASA, but the two-household arrangement worked well for them while she was an astronaut. Cady Coleman, who was already an astronaut living in Texas, married an artist whose homestead, studio, business, and son were in Massachusetts. The couple regularly traveled back and forth to be together. When they added another son to their family, he lived with Coleman in Texas but stayed with his dad while she was traveling or in space. Their long-distance marriage and dual-home solution was complicated, but it worked for their family—and earned them many frequent flyer miles.

Astronauts' normal work schedules on the ground vary from fast-paced to "insane." Depending on their technical assignments, training, and flight assignments, every hour is scheduled, often around the clock. Each crew has a dedicated scheduler before flight to ensure that they make all their training sessions, keep up with their T-38 flight hours, get their time in the gym and the simulators, participate in practice simulations with the mission control team, fly to Florida or anywhere else they are needed, complete training at the international partner sites including months in Russia, attend required meetings, do public appearances, and keep up with studying and reporting on their work. Then they spend weeks or months in space flying their missions. When the couple are both astronauts, these complex schedules are interleaved. Having a normal life around all the coming and going is a challenge for astronaut families. It takes exceptional effort for a couple to stay in sync with each other when their work schedules are so regimented yet changeable. This stress can wreak havoc on family life. Some have remarked on the need for more family-friendly schedules, but it is hard to see how that would be possible while keeping up the pace of spaceflight. Although it was a death blow to an astronaut's career in the 1960s, divorce is no longer uncommon in the astronaut corps; high-intensity work schedules and prolonged absences are among the factors in family breakups.

Participation in a Shuttle-*Mir* mission or an International Space Station expedition forced another decision-making balancing act. Astronauts had to look at the trade-offs of a much longer absence from home while studying and training in Russia—learning a new language and technologies, living a more spartan life without many of the conveniences Americans take for granted, and rare visits with family given the travel expenses and rigid schedules—versus the appeal of spending months in space, working with international crews, helping to assemble the space station, and doing more complex scientific research in space. Some, like Shannon Lucid, welcomed the adventure. By the time she spent most of a year training in Russia and residing on *Mir*, her kids were young adults who could fend for themselves. They took her absence in stride, bolstered by regular email and occasional telephone contact. Others, like Rhea Seddon, did not see time in Russia squaring with their family's life or their own goals. Each woman made her own decision about the balance of values in her work and family life. Before getting an ISS assignment, Cady Coleman had to counter her managers' assumption that she would not want to be away from her child that long. Despite military women being deployed abroad without their families, old ideas died hard: the first four women assigned to ISS expeditions through 2008 were either single or childless. Nicole Stott (2009) and Cady Coleman (2010) were the first moms assigned to the ISS, and since then women who are or are not parents (just as it has always been for men) have served on the space station.

Most of the women interviewed for this book were emphatic that the key to making the astronaut career work for women, especially women with children, is a supportive, involved husband or partner. Their mates need to understand and accept that the astronaut job is demanding and stressful for everyone, and the long workdays, irregular schedule, and extensive travel are inherent to the job, not optional. They also have to be or become comfortable with others' perspective that the woman is the more important person or has the more important job and be comfortable with her being in the spotlight. At the same time, women astronauts recognize the need to reciprocate by supporting husbands as equal partners with dreams and ambitions of their own. As examples, Susan Still Kilrain resigned from NASA when her military husband was reassigned to another country because she prioritized his advancement and wanted to keep her family together. Another, Janet Kavandi, who married a commercial airline pilot, knew, "He wanted to be a pilot as much as I wanted to be an astronaut. We struggled and did that thing where you don't let things stop you." It was a challenge to make it work.[81]

Eileen Collins also married a pilot and had to figure out childcare with both parents traveling. She wrote eloquently about how her life changed upon having her first child. She was so accustomed to her fast-paced life and having control over her work that at first she had trouble adjusting. Before the baby, "I worked all day, came home, and worked all evening, seven days a week. I never let up."[82] On maternity leave with a newborn she had a much different schedule in response to the baby's needs. She intentionally made adjustments to slow down and reduce her stress but stay productive. One of her cleverest ways of coping was to read technical material aloud to her newborn daughter. Collins later found the right woman who stayed with her children for nine years. She also found a daycare center for supplemental care during regular hours so the nanny could handle the odd schedules.

Both sex and gender factors may influence decisions to have children, how many, and when. Male astronauts can add children to their families at any point in their career without the temporary disqualifications that women astronauts face. It was reported in 2014 that male astronauts have a higher average number of children (2.5) than female astronauts (1.9), with a range of 1–8 children among the men and 1–4 children among the women.[83] Those disparities may be slightly different now, but the contributing factors remain: more male than female astronauts, shorter fertility of females, possibly second families after divorce, possibly non-astronaut mothers not delaying pregnancy as long as astronaut mothers or having children at younger ages, and possibly differences in career demands of non-astronaut mothers.

Seddon did something unheard of in the astronaut corps: she declined a mission assignment. No one turns down a flight for fear of never being asked again. She was already assigned to a future flight that had been repeatedly delayed and was trying to become pregnant with her second child. The new offer for an interim flight would have meant delaying pregnancy and possibly never having another child. She knew that at the end of her career, she would rather have more children than more flights. "I was sure no one had ever said, 'No, thanks. I'd rather have a baby," she wrote, but "it was the balance that I would be happiest with."[84]

Even without marriage or parenting, a woman astronaut's life may be affected by gender expectations. One astronaut remarked that dating partners struggled with her job's time demands, her public stature and fame, and the priority she gave to professional success.[85] Whether such issues factored into any of the other women's personal lives is unknown; a number of them remained single or divorced while they were astronauts without revealing their reasons.

The inherent risks of spaceflight can be hard on families, and astronauts find themselves balancing their family's unease with their own commitment to the job. Collins admitted that astronauts "put on brave faces when talking to our friends and the media: but there is always an unspoken fear that everyone thinks about."[86] Seddon noted that "it was always more difficult being the one staying at home,"[87] and Sandra Magnus observed "the difficulty for loved ones to stand to the side and watch what we do and the fear that comes with it. It is much harder to watch than to do."[88] The astronaut is confident in her decisions and acceptance of risk, but family members live with the stress of "what ifs" and consequences if things don't go well. Children often vacillate between excitement and anxiety about their parents going into space, and the children of the *Challenger* and *Columbia* crews suffered the heartbreak and lasting effects of losing a parent in a spaceflight catastrophe.

Astronauts use a variety of coping strategies to prepare children for their absence during spaceflight and to stay in touch with them while in space.[89] For example, with young children they may continue bedtime rituals by recording stories to be played each night or doing a live video story time, or they may leave notes and mementos for the child to find each day until mom returns. For older children, daily email messages and occasional phone calls help bridge the distance, and sometimes the astronauts arrange a video call with their child's class at school. Karen Nyberg made a stuffed dinosaur for her young son while she was on the space station and showed it to him during a video call. Cady Coleman kept one of her young son's stuffed animals on the space station and used it to enact what she was doing and seeing. Women astronauts are intentional and creative in being present for their families while they are physically absent. From

Shannon Lucid's long stay in Russia and on *Mir* to today's occupants on the International Space Station, all the astronauts rely on email, telephone calls, and video visits to maintain continuity in family life. They can now connect with their families conveniently at home without them having to go to JSC to make the link. The importance of these communication tools in reducing the strain of separation cannot be overstated. Staying connected to their families is sustenance, as crucial as food to everyone's health.

Two documentaries featuring women astronauts that appeared in 2024 gave intimate insights into family separation.[90] In *Space: The Longest Goodbye*, Kayla Barron shared how she and her husband were thinking about the possibility of her absence for one to three years at the Moon or Mars, how that might affect their marriage and plans for a family, and how that would rank with other values they hold. She wasn't yet sure what they would decide if the opportunity arose for her to go. In the same film, Cady Coleman shared clips from her video visits with her son and husband that revealed some of the strains of absence. Sometimes the calls didn't connect or the connection was lost, frustrating all of them. Once, she had to find the right tone of gentle disapproval to probe a misbehavior incident, and other times she and her son read together or cheerfully chatted. As she neared the end of her stay on the space station and was feeling blue about what more she might have accomplished, her husband tried to soothe her with praise and encouragement. Both women missed their families and wanted to return to them, but they also found it hard to leave space. Barron noted that going back to normal life might be tough after this special life in space. In *Spacewoman*, a film based on her memoir, Eileen Collins talked about how she was able to compartmentalize work and family life, but her family could not readily separate them. She and her daughter revealed that their relationship suffered for a while over Collins' decision to command the return-to-flight mission after the *Columbia* tragedy. Her daughter explained that at age ten she couldn't understand why her mother would choose that risk over their family. Her confessed anger was a striking contrast to the images of astronaut family members stoically, even happily, waving goodbye. Both films poignantly captured the emotional strain of spousal/parental absence on family life.

When shuttle missions resumed after the 1986 *Challenger* tragedy, NASA instituted a family support plan to help relieve stress on astronaut families—a plan originated in the Astronaut Office by those who knew the stress firsthand. Instead of families making their own arrangements, spouses and children traveled together on a NASA plane to Florida for launches and landings, and they were housed in the same condominium complex. Being together rather than scattered made it easier to communicate and coordinate between the families and support personnel, and the camaraderie was beneficial for kids and adults alike. It relieved so much tension at an already stressful time that Seddon declared it "one of the finest things NASA had ever done for the astronauts."[91]

Watching mom Nicole Stott in training. *NASA*

Going home embraced by mom. *Dorothy Metcalf-Lindenburger*

Astronaut Nicole Mann may have put it best by doubting the concept of "work-life balance." She has observed, "There's no such thing as finding a 'perfect balance' because certain things need to be prioritized at different times." She constantly tweaks balance based on what is happening and keeps modifying as she goes.[92] That may be the best-kept secret to success as an astronaut.

Extravehicular Activity: Being Small in a Big-Suit World

Going outside in the life-sustaining spacesuit is a quintessential astronaut experience with an endless panoramic view.[93] Spacewalking is a solitary and intimate exposure to the void and hazards of space. Despite being in audio contact with the crew, an astronaut is enclosed in a private, personal spacecraft; the crew can see *it*, but they usually can't see *you* behind the helmet visor. As thrilling as it is to launch, be weightless, and stare with awe at the fabulous views through the windows of a space shuttle or space station, nothing compares to extravehicular activity (EVA). It is the most dangerous planned mission event other than launch and entry. All astronauts may be trained and evaluated for EVA, but not all are fortunate enough to be tapped to do even one spacewalk. Almost all astronauts covet an assignment to perform EVA. It is mentally and physically challenging; it adds more risk to the job; it requires refined technical skills and extraordinary concentration; it is exhilarating but exhausting; and it results in dramatic images and thus more visibility to the public than what happens inside a spacecraft. For some astronauts, spacewalking is a career pinnacle.

In interviews for this book, women astronauts made two main points about extravehicular activity. First, it is actually a team effort. The two or four designated spacewalkers per mission are most visible, but the entire crew supports them. Most EVAs rely on use of a robotic arm operated by two crewmates—the prime and backup remote manipulator system operators—who guide the arm to maneuver large hardware and spacewalkers into position. A designated intravehicular activity (IVA) crewmate assists the crew in "pre- and post-activities" for the spacewalk—inspecting the suits, preparing the tools, suiting up, depressurizing, and then repressurizing and doffing the suit and stowing everything upon return. That in-cabin crewmate also serves as the EVA choreographer and monitors the EVA checklists and timeline to keep the spacewalkers on task and on schedule. For shuttle-based EVAs, the commander and pilot often needed to maneuver the orbiter or hold it in a certain position while the EVA was in progress. Another crewmate or two are tasked with video and photographic documentation of the EVA. Without these supporting roles, spacewalks wouldn't happen.

Although fewer than twenty US women to date have actually gone outside on EVA, most women support spacewalks as members of the EVA teams working on the inside. Whether or not they ever do a spacewalk themselves, they are well experienced in extravehicular activity procedures and often are trained to do contingency EVAs that don't become necessary. They are sensitive to perceptions that the ones who go outside are a separate or more elite group of astronauts, because others are also certified for EVA but don't get the chance. Women find these support roles satisfying, but some still want to be chosen to go outside. They may have to be proactive to get into the EVA lineup, not by currying favor or playing political games but by asking outright or by earning the best scores while training. For example, Kathy Sullivan asked for extra EVA training, which led to her assignment to work on tool design and EVA procedures for servicing the Hubble Space Telescope, which then positioned her well for EVA assignment on the 1990 deployment mission.[94]

The second point is that spacewalking is a balancing act of the astronaut's body size and strength, mental and physical skills, and spacesuit technology, and some women are disadvantaged by their size and the size of available spacesuits. Although weightlessness negates much of the difference in size and strength between males and females, weightlessness in underwater training differs enough from weightlessness in spaceflight that it can affect a woman's performance if the spacesuit does not fit her well. Anthropometrics—body size, shape, posture, reach, weight, center of gravity, muscle mass, and upper and lower body strength—affect both suit design and performance.[95] These parameters generally differ in male and female bodies. Fitness training and genetics can reduce some differences, but the fact remains that bodies are different. In general, women are not simply shorter than men; they are proportioned differently.

Because most space equipment was designed with men in mind, some of it was not optimal for women's performance in training and in space. The 1999 workshop on gender-related issues in spaceflight (mentioned earlier) cited this area of concern for female astronauts: spacesuit, equipment, and task design must enable both men and women to do their jobs effectively.[96] The report argued that poorly fitting spacesuits, for example, "should not be used as an excuse to exclude women from certain jobs." Human factors design and engineering needed to better account for female anthropometrics so equipment could be used effectively by all astronauts. The argument included this prescient statement: "A space 'glass ceiling' should not exist based on size or gender." In fairness, the spacesuits for the shuttle and space station were meant to fit a broad range of bodies, male and female, from the fifth to the ninety-fifth percentile in height. Suits were not designed specifically for men only or for women, and most of the parts were adjustable for length and girth so almost anyone could wear them. In reality, though, almost anyone was not everyone.

EVA is exhausting, whether in underwater training or in space. The sustained muscular exertion of working within a stiff, pressurized suit increases heart rate, blood pressure, and energy expenditure during sessions often lasting for six to eight hours without nourishment or rest. Because the suits are not custom-fit, most astronauts emerge from them chafed and bruised. Spacewalking looks graceful, but it is hard and cumbersome, even with a great deal of practice. While training, astronauts are constantly rated by instructors on their proficiency in certain tasks and criteria and observed by EVA-experienced astronauts. An EVA selection board then weighs their competencies in making crew assignments, deciding who can best carry out EVAs, who needs more training, and who has washed out.[97] The best are then certified for in-orbit EVAs and reevaluated for proficiency after their flight.

EVA is a sensitive topic among some women astronauts because fewer have been certified and chosen for spacewalks. Since Kathy Sullivan became the first US woman to perform an EVA in 1984, only seventeen other American women have had the opportunity, as compared to more than a hundred American men. During the forty years after Sullivan's spacewalk, multiple years passed without any woman doing EVA. Even accounting for the lower number of women in the astronaut corps at any given time, barely one-third of all women astronauts have done EVAs compared to about one-half of the men.

Only four women were assigned as prime EVA crewmembers during the first two decades of the shuttle era. Initially, there were not many EVAs, and the number of women in the astronaut corps then was small but increasing slowly. By 1994 when the twenty-fifth woman was selected for the astronaut corps, twenty-three men but only two women were experienced spacewalkers. Some women were simply too petite to wear the smallest spacesuit (Anna Fisher, Rhea Seddon, Mary Cleave, Marsha Ivins, and Nancy Currie, for example, were shorter than sixty-four inches and weighed close to a hundred pounds), and they were assigned to other EVA-related roles. In fact, most of the smaller women became expert operators of the remote manipulator arm so crucial for EVA success. Some women were trained, certified, and assigned to be contingency spacewalkers, but when a contingency EVA was not necessary on their missions, they didn't do a spacewalk (for example, Ellen Baker, Cady Coleman, Bonnie Dunbar, and Mary Ellen Weber). Sometimes an expected EVA didn't work out as planned. On her second mission as a prime EVA crewmember, Kathy Sullivan was completely prepared to assist in deploying the Hubble Space Telescope if any number of problems arose; however, when a glitch occurred in the activation of a solar array, the problem was resolved by computer commands, and she was not sent out on an EVA after all. Tammy Jernigan missed her first EVA when the airlock hatch jammed and would not open for the two spacewalkers to exit, prompting cancellation of the EVA and disappointment for everyone involved in the mission.

The number and percentage of women doing EVA has increased since space station assembly began in 1998, with Tammy Jernigan and Susan Helms being the first women to do EVAs at the ISS. The end of the shuttle era in 2011 also made more opportunities for women, because the astronaut corps shrank in number and everyone serving on the International Space Station or eventually going on an Artemis mission to the Moon had to be certified for EVA. If an astronaut doesn't pass advanced EVA training or can't wear an EVA suit well enough to work effectively, he or she probably will not be assigned to ISS or Artemis missions. Most of NASA's women astronauts have done their EVAs since 2010, with Peggy Whitson performing the most spacewalks (ten) thus far by a woman.

Numbers of women astronauts who have been assigned to conduct EVAs during the past twenty years don't necessarily tell the whole story.[98] Some look at the numbers and see a simple case of supply and demand: more men were available to be assigned to EVA than women, so it is logical that men outnumber women spacewalkers. Or someone might suggest that in the small group of women, most were already training for other missions to which they had been assigned, or they might not have had the particular skills needed for certain EVA missions. However, only one woman, Kathy Thornton, was assigned to do EVA on a Hubble Space Telescope Servicing mission, whereas fifteen men, some more than once, were assigned to those five spacewalk-heavy missions. One can wonder whether there were no other qualified, available women between 1993 and 2009 when the Hubble servicing missions were flown. Might a woman have been chosen to go outside on an EVA instead of operating the robotic arm or choreographing the spacewalk?

Some women astronauts of the 1990s thought that disparities in spacewalking might have been intentional—perhaps unconsciously so—and they perceived objective and subjective barriers that made it difficult to impossible for them to qualify and be assigned to EVA. It was frustrating to train, be evaluated after every training session by trainers and experienced spacewalkers, have high ratings, and not be assigned to an EVA mission or for a contingency EVA. Many astronauts were competing for EVA slots, and it was hard for women to break into a role that generally went to larger men. Women who trained well for EVA but were not assigned felt the stigma of rejection. They saw two main problems that limited their ability to earn EVA assignments: spacesuits that didn't fit properly and possible unconscious bias or disrespect among the EVA planners. These perceptions may be explored in greater depth with access to more information and discussion, but they are the lived experience of those women who felt passed over.

For space shuttle and eventual space station missions, NASA abandoned the earlier practice of manufacturing EVA suits custom-fitted to each astronaut. That was not practical or economical for a large astronaut corps, so the philosophy shifted to standard-size "off the rack" suits that would be reworn. Each Extravehicular Mobility Unit (EMU) suit would be assembled from interchangeable components—torso, upper arm segments, lower arms, thighs, lower legs, and soft boots that could be adjusted slightly for better fit. Gloves originally came in standard sizes as well, but the wide variety of hand shapes and sizes precluded good fits. The only custom-fitted component was the gloves, made to fit individual molds of each astronaut's hands because fit is critical for performance. The inner structure of the suit was a vestlike, fiberglass Hard Upper Torso (HUT), a hard shell that bore the weight of the suit, its chest-mounted control unit, and its life support systems backpack. To dress for EVA, after donning undergarments, the astronaut stepped into and pulled up the lower torso pants with boots already attached, squatted below the waist of the upper torso, wiggled up into it, threaded both arms into the attached sleeves, stood up to push the head through the neck ring, and then the top and bottom of the suit were locked together in a ring around the waist that sealed the suit airtight. The upper torsos came in four sizes: small, medium, large, and extra-large. Therein was a problem.

Although the HUT and other components were designed to fit a wide range of standard body sizes, some women astronauts were so petite in height and frame that even the size small was too large. A few of them worked on a task team to explore the feasibility of an extra-small size, but technical and financial hurdles impeded consideration. Women taller than sixty-four inches with a larger frame had better luck wearing a small or medium suit. Even when the size worked, the suits generally didn't fit women as well as men. In particular, the torso did not fit women's shoulders and curves or the suit's waist was below the women's waists, affecting their ability to bend. Even with sizing adjustments, the suit's knees and elbow parts didn't always align well with their bodies. Some women wearing a suit in underwater training drifted around in the interior volume with too much empty space around their torso and legs. They would fall back in the suit, pulling their arms in and their hands out of the gloves, or they would sink down in the suit, compromising their view through the helmet. The extra air inside shifted like a big bubble as women moved and interfered with their position in the suit. That loose fit allowed the body to move around in the suit and thereby alter the center of mass, which made it difficult to maintain neutral buoyancy and also difficult to do a task properly. They had limited ability to reach the controls on the chest pack if the HUT held their arms out too much like a "T," or they strained against the hard frame jutting into their shoulders and underarms as they tried to reach a task. Men and women alike often emerged from underwater training with bruises from working against the inflated suit and the drag of surrounding water, but movement in the suit was much easier in space. Kathy Sullivan and several other women wore a medium torso comfortably enough for their EVA training and orbital EVA. Kathy Thornton found that the small size suited her well; she felt as if she could have worn a ski vest inside the medium suit. Cady Coleman naturally fit into a small suit but was advised to use a medium because another problem was looming.

During the 1990s one of the biggest responsibilities of the Astronaut Office was to determine how many EVAs would be required for space station construction and to ensure that enough astronauts were trained and ready for the job. They predicted the need for more than two hundred ISS assembly EVAs increasing in frequency to a "wall of EVA" by 2003 that would plateau at that peak for years. More astronauts needed to be trained and ready for that workload. At the same time, the cost of the space station kept increasing, and pressure mounted to trim costs wherever possible. As a result, some large components of the ISS were sacrificed—a dormitory-like habitation module, for example—as well as some scientific equipment and other features. That pressure extended even to the number of spacesuits supplied and maintained for use on the shuttle and ISS. Instead of developing a new and improved EVA suit for the space station years, NASA decided to enhance the baseline shuttle EMU suit, maintaining the same design but upgrading some of its systems.[99]

For the ISS era, the small and extra-large HUTs were to be discontinued for economy's sake. The reasoning behind this 1994 decision was not fully transparent, and some of the women astronauts were concerned that it was an attempt to edge them out of EVA. They suspected that an "EVA mafia" of larger men might be trying to hoard EVA assignments for themselves and supporting elimination of the small HUT to reduce their competition, or worse, were convinced that women couldn't do the job and didn't belong in EVA. It soon became evident that some of the men who wore the extra-large torso couldn't fit in the size large. When the decision was finalized, the small size was discontinued but the extra-large was kept to accommodate the greater number of astronauts needing that size. Some women think that abandoning the small torso on the basis of cost, even though not explicitly on the basis of gender, actually was a gendered decision for its disparate effect on women, limiting their opportunities and flaunting inequity. Up to one third of the twenty-four women astronauts then active were affected by that decision.[100] It has not been reported whether any smaller men were similarly affected.

Coleman was warned that this change was coming and that if she really wanted an EVA assignment she should wear the medium size torso and make it work. Thornton offered to help her.[101] For underwater training she had to stuff the suit full of foam padding to keep her position in the too-large volume, and even so she emerged from training sessions as battered and bruised as everyone else. Once she was faulted for poor performance when in fact the suit fit had not been adjusted properly for her; she thought it was almost a reflex to fault the woman than to consider other causes for the problem. When the suit fit was adjusted, her performance soared. She never complained, but after she flew her mission she used her new credibility to explain to the suit techs how much harder it was to work in a poorly fitted suit and how they could better help astronauts solve that problem. Mary Ellen Weber also wore a medium suit stuffed with padding although a small suit fit her better. Although it was inconvenient to do, and generally unnecessary for men, use of padding reflected the women's resourcefulness and problem-solving ability. Both Coleman and Weber were assigned for unlikely contingency EVA duty, but neither had to do one in space.

A complicating factor in the torso size and fit problem was that NASA tried to improve the HUT design during the 1990s after shoulder joint reliability issues arose. This introduced a new set of fit problems. The original torso was called the "pivoted" HUT, because the shoulder joint was hinged. Astronauts found it to be comfortable, flexible, with good reach across the front, but failure of the shoulder joint had endangered an astronaut—an unacceptable safety issue. In 1997, an enhanced HUT having a different shoulder joint design, the "planar" HUT, was introduced, and it soon replaced the pivoted HUT in EVA suits used on the ISS. This design changed the geometry but decreased the range of motion at the shoulder and made the suit harder to don and doff, but it solved the reliability problem of the pivoted HUT. Kathy Thornton had worn a small pivoted HUT on her three spacewalks but later tried a medium planar HUT and found it far too large. Some men who had worn a size large pivoted torso could not wear a large planar torso and had to move up into an extra-large size to accommodate their broader shoulders and longer arms. A few other men went in the opposite direction, downsizing from extra-large pivoted to large planar torso to get the best fit. Astronaut Jerry Ross, one of the most experienced spacewalkers and a former chief of the EVA branch in the Astronaut Office, noted that the extra-large planar HUT worn by many of the "musclemen" was absolutely critical to completion of the ISS.[102]

For the last few years of the 1990s, astronauts could choose between the two HUT versions until the first fully enhanced EMU was used on an ISS assembly mission in 1998. The last time the baseline suit with pivoted HUT was used in space was in 2002. Thereafter, every spacewalker had to wear a suit with the new torso. Women were not the only ones affected by torso size and design issues. A number of men reported shoulder and rotator cuff injuries, some requiring surgery, caused by repetitive motion, limited range of motion, and other factors while wearing the planar HUT during their frequent underwater training sessions. The injuries did not occur in space. Based on this history, a 2003 study found the extra-large planar HUT unacceptable for long-term use in training because of its greater risk of injury due to restricted arm movement in its fit.[103]

Discontinuing the small torso affected women disproportionately, and several cited it as having a significant effect on their careers. It didn't help their morale when in 1997 NASA named spacewalkers for ISS assembly flights—a cadre of fourteen men, including two rookies who had not flown at all and a Canadian astronaut.[104] A spacewalking male astronaut later told a reporter, "Our spacesuits only come in medium, large, and extra-large. Anybody who is on the smaller side . . . will not be able to have a chance to go outside," as if that privilege were the natural order of things, not a deliberate decision.[105] In the early 2000s, Nancy Currie and then Wendy Lawrence led an effort to develop a redesigned small torso, but funding was denied and the effort ended.

The reason these suit size issues mattered so much was that EVA was often a step toward greater stature and leadership within the astronaut corps. Qualification (or not) for EVA could profoundly affect an astronaut's career opportunities. Small women who might have been effective spacewalkers but could not work safely in the medium torso suit were effectively barred from EVA, which barred them from a long-duration stay on the ISS. During the 1990s, the Astronaut Office decided that given the small number of an ISS crew, in the event of an emergency and for workload management, everyone had to be qualified to do everything, including EVA. So, if a woman couldn't qualify for EVA because no suit fit her properly for safety and work efficiency, she also couldn't be assigned to a long-duration ISS expedition. She might be able to visit the ISS on a shuttle mission, but she would never be an ISS resident. Justifying the decisions to discontinue or redesign the small size torso as a cost-cutting measure signaled to some of the women that smaller astronauts were not valued enough to warrant the expense.

One instance pushed the suit fit issue out into the public. In 2019, astronaut Anne McClain wore a medium torso suit for her first EVA at the ISS, and it fit perfectly. She had also trained in a large torso and planned to wear a large suit on her second EVA while her intended partner Christina Koch wore the medium suit. (Just these two suits were configured and available for use during their ISS expedition.) However, McClain bowed out of that EVA upon realizing that the large suit did not fit her properly in space and would affect her ability to work safely. Suits fit differently and are stiffer in space than on the ground; they aren't as "broken in" as the training suits. Also, the body changes in length and girth in space, another reason for suits fitting differently. McClain stepped back in favor of her first partner astronaut, a man who wore the large suit, and he did the EVA with Koch instead. To McClain and the crew, this was a logical solution to a problem. McClain wore the medium suit on her second EVA, when she was again paired with a man wearing the large one.[106] Having been alerted to an imminent, historic all-woman EVA, the news media made much of the missed opportunity and framed the lack of enough suits to fit women as sexism. Although this interpretation missed the details of the situation, reporters did flag the problem that women astronauts had long recognized.[107]

Weightlessness offsets to a great extent the male advantages of size and strength.[108] There are instances when a smaller-size astronaut can do a task as well as or more easily than a larger one. Mary Cleave, a very small woman, was able to reach through tight spaces and fit behind a wall of storage lockers in the shuttle crew cabin to do repair tasks that stymied her larger crewmates. Some ISS assembly tasks could be done by a smaller astronaut—fine motor work in installing or relocating components, for example, or running cables and making electrical connections. Other tasks may have logically required a larger astronaut with longer reach, such as mating massive modules or boom-like truss segments or working at the outer limits of a robotic arm's length. Because some tasks require inordinate upper body strength, both men and women do rigorous weight training to build up their stamina using those muscles, but sometimes men may have a natural biomechanical advantage. Astronauts also work on hand and forearm exercises to prepare for the exhaustion of repetitive hand tasks, like tool use.

In reality, anyone doing EVA must be able to perform *all* the tasks if things don't go as planned. In addition, everyone must be able to rescue his or her EVA partner in an emergency and quickly bring the partner back to safety. Cady Coleman included in her memoir such an urgent rescue scenario.[109] No one would argue that a woman should be sent on EVA simply because she is a woman; EVA is too important and hazardous to send anyone but the best-trained, most capable person for all that the job entails. However, some of the women eager to be involved in ISS assembly, if not long-duration stays, perceived that the idea of tall, broad-shouldered, long-armed musclemen as space-walkers was so ingrained in EVA planners' thinking that their own potential as smaller human beings was simply not seen or considered. They felt invisible as candidates for EVA assignments.

There were two solutions to that problem: design an EVA suit that any astronaut can wear comfortably and safely, or select more larger-framed women into the astronaut corps. The latter option seems facetious, but about half of the women selected since 2000 have performed EVAs without needing a small suit. The redesign option, called for and studied since 1998, is achievable with commitment despite the expense. Resolving the suit size and fit problem will indicate that women are fully and equally valued as participants in extravehicular activity, and that cost is no longer an excuse for not providing them equally enabling spacesuit technology as men receive. As NASA moves into the Artemis era, it has committed to have a next-generation spacesuit available to fit everyone and enable everyone who qualifies to move and work productively in EVA. That suit has not yet been unveiled, but several companies are working hard on designs to meet the requirement. It is expected to have a much-improved shoulder design and improved mobility for all. NASA has also committed that the next landing on the Moon will include a woman, which implies that qualified women (and smaller men) will have equal standing for future EVAs in a suit that fits them well.

Over time, the number of women on a crew increased from one to two, then three, but never a full crew. In 2007 it happened by coincidence that two NASA women were simultaneously in command of both a shuttle mission (Pamela Melroy) and an ISS expedition (Peggy Whitson). It also happened that four women were aboard the ISS at once in 2010. And in 2019 it finally happened naturally that two women, Anne McClain and Christina Koch, were scheduled to do an EVA together for the first time. McClain chose to bow out because the spacesuit she had planned to wear didn't fit properly. Instead, Koch and Jessica Meir became the first pair of women to do a spacewalk about seven months later and then did two more. In 2023, Jasmin Moghbeli and Laurel O'Hara also did a spacewalk together. These spacewalks were not arranged to be historic; given the composition of the astronaut corps, it was inevitable that two women eventually would work outside together. That happened without contrivance, no differently than two men had always been paired, on the basis of their competence. Now that the precedent has been established, two women simultaneously in EVA is just another day at work.

The eighteen US women spacewalkers to date have performed a variety of EVA tasks, typically working with a man as their partner. EVAs in the 1980s focused primarily on satellite servicing tasks. Kathy Sullivan's main EVA task was to demonstrate a technique for transferring fluids through valves from one container to another, a function that would be useful for refueling satellites in orbit. In the 1990s, EVA tasks often focused on demonstrating techniques for space station assembly and maintenance. Kathy

Thornton's first EVA tested different methods of assembling a structure in the payload bay of the shuttle. Her second and third spacewalks occurred during the first Hubble Space Telescope servicing mission, where she and her EVA partner removed a large instrument and installed a new optics system to correct the telescope's faulty vision. They also replaced two large solar arrays. Three years later, Linda Godwin and her EVA partner were the first Americans to do an EVA at the Russian space station *Mir* during a shuttle docking mission; they installed canisters of environmental effects experiments on the *Mir* exterior and tested new tethers and foot restraints for future use on *Mir* and the ISS. From 1999 on, all NASA EVAs except those on Hubble servicing missions were carried out on shuttle missions to the ISS or by resident ISS crewmembers. ISS EVAs involved installing large and small components, linking modules together, working with electrical cables and fluid lines, replacing or repositioning items, doing inspections, cleaning and lubricating equipment, making repairs, and other tasks. As of 2024, fifteen more NASA women had logged their EVA hours on the ISS.[110]

Of the eighteen American women to date who have done EVAs, Peggy Whitson has completed a total of ten, all at the ISS, and Sunita Williams closely follows with nine EVAs to date. Besides the US women spacewalkers, only five other women astronauts have participated in EVA. Svetlana Savitskaya of the Soviet Union did a spacewalk in July 1984 ahead of Kathy Sullivan's EVA in October, thus claiming the title of first woman spacewalker. No other Soviet or Russian woman has done an EVA since. Almost four decades later, taikonaut Wang Yaping did an EVA at China's Tiangong space station in late 2021 as the first Chinese woman spacewalker. Liu Yang followed with an EVA in 2022. Samantha Cristoforetti of Italy and the European Space Agency did an EVA at the ISS during her time as an Expedition 67/68 crewmember in 2022, the first woman spacewalker from one of the ISS partner nations. American commercial astronaut Sara Gillis did a stand-up EVA in the open hatch of a SpaceX vehicle on the private Polaris Dawn mission in 2024, but she did not fully exit the spacecraft.

While America's women astronauts of the 1980s and 1990s had some doubts about equity in EVA assignments, EVA history in this century demonstrates a more complete integration of NASA women into spacewalking than in any other space program in the world. Moreover, the frequency of women in EVA has noticeably improved during the space station era, and it is likely to be required that every woman who becomes a NASA astronaut also become EVA-qualified and properly suited. Thus far, every woman except one from the 2013 and 2017 astronaut classes has been assigned to do EVAs, a record that no longer looks like disparity. However, as recently as 2019, the two women on the first all-woman spacewalk, commented on the factors that still affect women in EVA despite efforts to support and accommodate them.[111] It is too soon to see EVA assignments for the 2021 class that graduated from initial training in early 2024. Thus far, Europe, Canada, and Japan have chosen only men for EVA opportunities except

Cristoforetti. Some among the commercial purveyors of spaceflight for private space-farers are considering eventual EVA training for their clientele. For now, if a woman aspires to be a spacewalker, her best bet is to be a NASA astronaut.

In the Spotlight: Privacy and Celebrity

NASA receives thousands of requests annually for public appearances by astronauts; they are sought to appear at schools, ceremonies, civic events, groundbreakings, festivals, conferences and conventions, and in media interviews. They may be asked to testify before Congress or make remarks at a White House or diplomatic event. Most of their appearances occur on the ground, but they also appear on the in-orbit schedule as media availabilities and chats with students in their classrooms. Some astronauts confessed to initial hesitance about this requirement, citing discomfort or stage fright or claiming that they had always been shy and stayed out of the limelight, but they soon learned that being in the spotlight comes with the job and they had to get comfortable with it. With experience, most embraced public appearances as both a responsibility and a privilege, a way to speak to the public that supports the nation's space program and inspires the next generation of explorers. The topics they typically addressed—NASA programs and missions, STEM education, teamwork/leadership, their own mission narratives, and achieving big dreams—were topics they knew well and felt passionate about. Some came to enjoy it so much that in retirement they signed on with speakers' bureaus and spent much of their time traveling for public appearances around the world.

Astronauts were not unfamiliar with public recognition; most received honors and awards as students and early in their careers.[112] But celebrity is a more expansive kind of recognition that can bring problems if not handled well. Being selected as astronaut candidates and being assigned to first missions brought a degree of celebrity to every astronaut. They were introduced to the media in press conferences, made available for group and individual interviews, reported on in the national media as well as their hometown press and alma mater alumni news, and followed during their missions. Some were featured in popular national magazines and in the magazines and websites of professional science and engineering societies to which they belonged. Some were invited to appear on the morning television talk shows or assigned to provide launch commentary for the national news networks. Several appeared in IMAX films edited from footage shot during their missions. Over time, they were elected to numerous halls of fame and received awards from civic and professional organizations. A few were elected to prestigious national academies, and Shannon Lucid was awarded the highest civilian honor, the Congressional Space Medal of Honor.[113]

Despite the celebrity that prompts requests for public appearances, though, most of the astronauts do their work without fanfare and become almost anonymous. Not many members of the public can name more than a few NASA astronauts from the shuttle and space station eras. Audiences discover them when they appear in public. It is telling that someone in their audiences almost always asks, "What was it like to go to the Moon?"

The first women astronauts were advised to think carefully about what they wished to reveal and set their own boundaries for interviews. Many reporters at that time seemed flummoxed about how to interview professional women, and they asked sexist questions (tagged by *People* magazine as "those dumb chauvinist questions") about their personal lives, dating, hopes for marriage, or interests in cooking and other so-called feminine activities. Reporters seemed to feel entitled to personal information from the women that they would never ask of male astronauts, such as their weight or how they maintained their figure.[114] Sally Ride and Judy Resnik, zealous about guarding their privacy, were well known to be reluctant about interviews but did their duty. Ride deflected such questions as, "Do you weep when things go wrong?" without deigning to answer, and Resnik stuck to objective information about her work when probed about her personal life. Rhea Seddon and Anna Fisher were targeted for media attention even before they became pregnant and had babies, a novelty in the astronaut corps but a hardly newsworthy fact of life. Nevertheless, they agreed to interviews on the topic if only to indicate how normal they were. Shannon Lucid decided not to speak about her husband and children or allow reporters access to her family to maintain their privacy, a policy that she maintained throughout her career. Lucid declined many public appearances and interviews if they required traveling away from her family or addressing organizations that were discriminatory; she chose not to speak to groups limited only to boys or only to girls.[115] It was possible to decline some requests, but everyone was expected to engage in at least one public appearance a month.

Dealing with gender-related insensitivity in public was another balancing act. As ambassadors for NASA, the women had to remain polite and composed, no matter how insulting or inappropriate the overtures. Eventually most astronauts went to "charm school" for training in media relations and learned how to pivot away from questions they preferred not to answer and make their own points in interviews. As time went by and women in space became more common, some later astronauts loosened up about mentioning their families and permitting them to be photographed, and with the advent of social media astronauts were able to engage with the public on their own terms beyond making formal public appearances. Privacy may be less a consideration today when people readily share their activities and feelings online for everyone to see.

Celebrity brought a degree of wariness to some of the women astronauts. Most seriously, Eileen Collins reported that NASA discovered a serious threat against her life and suggested that she avoid a public celebration honoring her in her hometown, to which she was already en route.[116] She returned to Houston immediately and, at her request, all of her remaining public events for the year were canceled. Some of the women felt cautious about people wanting to befriend or date them, unsure whether they were genuine friends or wannabe groupies. They wondered whether some people wanted to know them for who they were as a person or only because of their celebrity as an astronaut. In the days before cell phones, some of the women protected a zone of privacy with unlisted telephone numbers and unpublished addresses, and even today they are careful to guard personal email addresses.

Most were cautious about how much personal information they revealed in interviews, because reporters assumed that women were approachable and trusting and would respond candidly. But the astronauts realized that some personal information might be misconstrued or used to present them in an unfavorable light or simply expose more than they wished about their private lives. Society considers astronauts role models of achievement who inspire young people, and in that role they are expected to reflect well on NASA and their work. Role models must be careful to present themselves well in public view, especially when they are sharing their experiences and advising youngsters. Women astronauts have deemed it prudent to maintain some privacy and not to seek celebrity but to use it to share their experiences of spaceflight.

Rewards of Being on a Mission

Beyond the rarity and privilege of flying in space, experiencing weightlessness, and seeing the whole Earth from above, each astronaut found her own rewards in spaceflight.[117] For some, the reward was in certain favorite jobs. Several cited being a CAPCOM in Mission Control, living a mission minute-by-minute with an orbital crew and the ground support teams, as the next best thing to being in space oneself. Others found leadership positions such as payload commander, NEEMO analog mission commander, or ISS commander especially satisfying. A few mentioned contributing to the return-to-flight efforts after the *Challenger* and *Columbia* tragedies, solving problems and instituting practices to reduce the risk of future accidents. Others found great reward in laying the foundation for International Space Station operations by participating in the Shuttle-*Mir* missions and the ISS assembly missions, working to prove that former space rivals and international partners could cooperate successfully in this massive engineering and scientific program. One found her joy in a specific ongoing task— reviewing all in-flight photography and helping to select images for public release—a vivid way to share with the public the myriad activities and amazing scenery of spaceflight.

Many of the women astronauts cited existential rewards arising not from aspects of their job but from the totality of their experiences: being part of something bigger than oneself, doing something challenging and meaningful that benefits humanity, constant learning and new challenges, discovering one's own full potential, being a valuable and essential crewmate, and gaining a heightened sense of the interconnectedness and fragility of life on our home planet. They found it rewarding to work in an international environment, connecting with people around the world through missions and public appearances. It was deeply rewarding to be a trailblazer for others to follow and to prove that women can do anything in space. More than one has remarked that the spaceship doesn't care if you are male or female. Both can do the job.

Each woman had her own reasons for leaving spaceflight behind, and each knew when the time had come. For some, the decision was to retire from NASA after serving the requisite twenty or thirty or more years as a federal employee and reaching retirement age. Others left earlier for a variety of reasons: uncertainty that she would fly again while many others were in the queue for mission assignments, knowing that she would not be assigned to an ISS expedition or not wanting to spend so much time away from home while training abroad for an ISS expedition, seeing lucrative opportunities in other organizations or wanting to launch her own enterprise, deciding to prioritize family life without the stresses of travel and separation or to support her husband's career as he had supported hers, feeling she had contributed and accomplished all she could as an astronaut, or even having had enough of Houston's heat and humidity. A few moved up into senior management positions at NASA, but some left because they didn't see a path to leadership or felt their full potential wasn't recognized or developed. A few left dissatisfied with the influence of politics and bureaucracy on decisions about the space program and the near-constant uncertainties about funding; it was demoralizing to work hard on a new project only to have it delayed or canceled. Those women wanted to choose their own future freely, to do something they couldn't do while at NASA, and to do it without constraints that could hem them in if they stayed. Regardless of their reasons for ending a career as an astronaut, each woman was positive about the experience. They expressed no regrets and would do it all again, implying that the rewards of being an astronaut far surpassed any stresses or disappointments.

A NASA administrator introducing a new class of astronaut candidates remarked, "Astronauts represent the best of humanity and our most fervent hopes for the future. No pressure."[118] There is no longer any question about women standing out or fitting in or being competent in all the necessary skills, whether technical, interpersonal, or leadership. These women have proven that a woman's place is in space just like a man's. As the many limits on women in aviation, science, engineering, and spaceflight gradually fell in the 1970s and 1980s, sixty-one women met the same highest standards to take their place in space. More will follow them, and the next ones may establish women's place soon on the Moon or someday on Mars.

When asked what they most want people to know about women who are astronauts, the women presented in this history have an emphatic answer: "We are astronauts. Period."

ACKNOWLEDGMENTS
|||||||||||||||||||||||||||||||||||||||||||||||||||

It has been my good fortune to communicate with almost all of the women astronauts who are no longer on active duty at NASA. Their interest and generosity as I worked on this book are acknowledged with my sincere appreciation. Special gratitude goes to Sandra Magnus, Ellen Ochoa, and Kathryn Sullivan for their initial assistance in contacting some of the women and vouching for me. For their answers to my many questions and their patient explanations when I needed clarity, it is an honor to thank Ellen Baker, Mary Cleave, Catherine "Cady" Coleman, Eileen Collins, Nancy Currie-Gregg, Jan Davis, Bonnie Dunbar, Anna Fisher, Linda Godwin, Susan Helms, Joan Higginbotham, Kathryn "Kay" Hire, Marsha Ivins, Tamara Jernigan, Janet Kavandi, Susan Still Kilrain, Wendy Lawrence, Shannon Lucid, Sandra Magnus, Pamela Melroy, Dorothy Metcalf-Lindenburger, Barbara Morgan, Lisa Nowak, Karen Nyberg, Ellen Ochoa, Rhea Seddon, Heide Stefanyshyn-Piper, Nicole Stott, Kathryn Sullivan, Kathryn Thornton, and Mary Ellen Weber. In addition to the information and insights they offered, their personal warmth heightened my pleasure in our conversations.

Likewise, I am grateful to Tam O'Shaughnessy and Karen "Bear" Ride for conversations and insights about their beloved Sally. I appreciate that Bear and her son Whit Scott granted permission to use Whit's photograph taken at the reunion reported in the introduction, and that Tam gave me a copy of Whit's video from that special gathering of astronauts.

Andy Turnage of the Association of Space Explorers graciously posted a notice about this book project to the membership, which led some of the women astronauts to contact me. NASA Johnson Space Center public affairs officers circulated a notice to former and present women astronauts on my behalf as well.

I am indebted to several staff members of the NASA Johnson Space Center for interviews: Duane Ross, longtime manager of the astronaut selection process, now retired; Anne Roemer, current director of the JSC Human Resources office and former manager of astronaut selections; and Dr. Albert Holland and Dr. James Picano of the JSC Behavioral Health and Performance Operations Group in the Space Medicine Division. Retired astronauts Jerry L. Ross and Michael López-Alegría, two of the most accomplished spacewalkers, shared their experiences with the space shuttle extravehicular activity spacesuit and perspectives on the suit-fit problems that plagued many of the women astronauts. Sharon McDougle, former spacesuit technician who suited up astronauts for training and flight, provided an elusive photo that I had been seeking. Dr. Richard T. Jennings, former flight surgeon in the JSC Flight Medicine Clinic, helped me understand NASA's responses to the reproductive health concerns of female astronauts.

Some of my colleagues who share interests in the history and topics I have explored in this book have been generous with suggestions, corrections, and resources. Thank you especially to NASA Chief Historian Brian C. Odom; NASA Johnson Space Center Historian Jennifer M. Ross-Nazzal; Margaret A. Weitekamp and Emily A. Margolis, space history curators at the Smithsonian's National Air and Space Museum (NASM); NASM archivists Melissa Keiser and Kate Igoe; NASM registrar Erik Satrum; independent radio and documentary producer Richard Paul; and Robert Pearlman of CollectSPACE.

For detective work, documents, and memories related to the Tektite II underwater mission discussed in chapter 1, NASA Marshall Space Flight Center retirees Jack W. Stokes and Kenny Mitchell and the center's archivist Jordan E. Whetstone were invaluable. Kathryn Sullivan kindly introduced me to Tektite II aquanaut Sylvia A. Earle, who then helped me contact her mission crewmate Peggy Lucas Bond. Both scientist-aquanauts provided details that I could not find in published materials. Tektite 2020 conference participants Tierney Thys and James Merle Thomas pointed me to resources from that event. Deborah Shapiro in the Smithsonian Institution Archives gave me insight into its Tektite II records.

It would have been impossible to do the research for this book without ready access to excellent libraries. The Smithsonian Libraries and Archives and its Interlibrary Loan Service never fail to locate obscure articles and deliver them promptly. Katrina M. Brown, supervisory librarian in the Smithsonian Libraries and Archives, provided special assistance. The Library of Congress is indispensable to researchers, and I continue to mine it remotely since leaving Washington. The Boulder Public Library's network with academic libraries throughout the region gave me ready access to books not in its local collection. I am grateful that the University of Colorado Library grants privileges to researchers who are not affiliated with the university. The University of Alabama in Huntsville Library archivist, Charlie Gibbons, went to great lengths to satisfy my remote request, as did Troy Eller English, reference librarian for the Society of Women Engineers Collection at the Wayne State University Libraries, and Nicholas Schiller, reference librarian for the Ms. magazine collection at the Drew University Library.

I appreciate several people who graciously stepped away from their primary duties to answer a stranger's inquiry. Erica Beade kept me in touch with traveler Cady Coleman. Amy Jaramillo tried her best to locate an elusive document and field other questions, and Kelly Gerald scoured the Phi Beta Kappa Society membership database to verify astronauts awarded the honor society's key. Elizabeth Howell of Space.com generously compiled a list with links to her interviews and profiles of women astronauts to aid my research. Marcia Franklin, a producer at Idaho Public Television, located for me Barbara Morgan's Teacher-in-Space application in her long-ago interview file. Ido Mizrahy granted me advance viewing of his film The Longest Goodbye. Often my 24/7 tech support son Bryan Guido Hassin rescued me when my laptop misbehaved or when I thought I had lost an entire draft chapter.

For their expertise in the art of transforming a manuscript into a better book, Carolyn Gleason, Julie Huggins, and Gregory McNamee of Smithsonian Books deserve credit and praise, as do Sarah Fannon and Matt Litts for their marketing. Cover designer Pete Garceau did a wonderful job crafting a striking cover for this book. I'm also appreciative of David Griffith lending his expertise to design such a readable interior. Editors and designers are unsung heroes in refining an author's text and crafting a package of words and images into a work of art. It has been a pleasure to work together and see the book benefit from their many talents.

Personal acknowledgments are as well-earned as professional ones. I have been blessed by curious family members and friends who ask about my work in progress, offer ideas and encouragement, patiently read samples so I can test potential reader reactions, ask questions so I know where to clarify, and generally grace my life with their affection and good cheer. Among them I count for this project Lee Crane Wood, Sam Wood, Kitty Williamson, Susan Quick, Susan Nilsson-Weiskott, Rebecca Stokes, Laura Saba, Patrice Neal, Janet Neal Fotioo, Katherine Barrett, and Bryan Guido Hassin. Although he isn't reading at this level yet, young Oberon Hassin is impressed that I have written a book of more than 100,000 words, and his sister Minerva Hassin loves learning words as she "reads" her little board books. It is a joy to share with them the adventures of reading and exploration.

APPENDIX A

|||||||||||||||||||||||||||||||

Order	Name	Born	Selected	Class	Field	Title	Flights	Status
1	Ride, S	1951	1978	8	Astrophysics	PhD	1983, 1984	1987 left d. 2012
2	Resnik, J	1949	1978	8	Engineer	PhD	1984, 1986	d. 1986
3	Sullivan, K	1951	1978	8	Geology	PhD	1984, 1990, 1992	1993 left
4	Fisher, A	1949	1978	8	Medicine	MD	1984	2017 ret.
5	Seddon, R	1947	1978	8	Surgery	MD	1985, 1991, 1993	1997 ret.
6	Lucid, S	1943	1978	8	Chemical Engineer	PhD	1985 1989, 1991 1993, 1996	2012 ret.
7	Dunbar, B	1949	1980	9	Engineer	PhD	1985 1990, 1992 1995, 1998	2005 ret.
8	Cleave, M	1947	1980	9	Engineer	PhD	1985, 1989	2007 ret. d. 2023
9	Baker, E	1953	1984	10	Medicine	MD	1989, 1992, 1993	2011 ret.
10	Thornton, K	1952	1984	10	Physics	PhD	1989, 1992, 1993, 1995	1996 left
11	Ivins, M	1951	1984	10	Engineer		1990, 1992, 1994, 1997, 2001	2010 ret.
12	Godwin, L	1952	1985	11	Physics	PhD	1991, 1994, 1996, 2001	2010 ret.
13	Jernigan, T	1959	1985	11	Astrophysics	PhD	1991, 1992, 1995, 1996, 1999	2001 ret.
14	Davis, J.	1953	1987	12	Engineer	PhD	1992, 1994, 1997	2005 ret.
15	Jemison, M	1956	1987	12	Medicine	MD	1992	1993 left
16	Helms, S	1958	1990	13	Flight Test Engineer	LTGEN, USAF	1993, 1994, 1996, 2000, 2001	2002 back to USAF; 2014 USAF ret.
17	Ochoa, E	1958	1990	13	Engineer	PhD	1993, 1994, 1999, 2002	2018 ret.
18	Voss, J	1956	1990	13	Engineer	PhD	1993, 1995, 1997, 1997, 2000	d. 2012
19	Currie, N (Gregg)	1958	1990	13	Engineer	COL, USA	1993, 1995, 1998, 2002	2017 ret.
20	Collins, E	1956	1990	13	Mathematics, Test Pilot	COL, USAF	1995, 1997, 1999, 2005	2006 ret.
21	Lawrence, W	1959	1992	14	Engineer, Pilot	CAPT, USN	1995, 1997, 1998, 2005	2006 ret.
22	Weber, ME	1962	1992	14	Chemistry	PhD	1995, 2000	2002 left
23	Coleman, C	1960	1992	14	Chemistry, Engineer	Ph.D, COL USAF	1995, 1999, 2010	2016 ret.
24	Still, S (Kilrain)	1961	1994	15	Engineer, Test Pilot	CDR	1997, 1997	2002 ret.
25	Chawla, K	1961	1994	15	Engineer	PhD	1997, 2003	d. 2003
26	Hire, K	1959	1994	15	Engineer	CAPT, USN	1998, 2010	2019 ret
27	Kavandi, J	1959	1994	15	Chemistry	PhD	1998, 2000, 2001	2019 ret.
28	Melroy, P	1961	1994	15	Physics, Astronomy, Planetary Science, Test Pilot	COL, USAF	2000, 2002, 2007	2009 left; 2021 returned; 2025 ret.

Order	Name	Born	Selected	Class	Field	Title	Flights	Status
29	Whitson, P	1960	1996	16	Biochemistry	PhD	2002, 2007, 2016, 2023/ Axiom2	2018 ret.
30	Magnus, S	1964	1996	16	Engineer	PhD	2002, 2008, 2011	2012 left
31	Clark, L	1961	1996	16	Medicine	MD, CAPT, USN	2003	d. 2003
32	Wilson, S	1966	1996	16	Engineer		2006, 2007, 2010	Active
33	Nowak, L	1963	1996	16	Engineer, Test Pilot	CDR, USN	2006	2007 left
34	Stefanyshyn-Piper, H	1963	1996	16	Engineer	CAPT, USN	2006, 2008	2009 ret.
35	Williams, S	1965	1996	16	Engineer, Test Pilot	CAPT, USN	2006, 2012, 2024	Active
36	Higginbotham, J	1964	1996	16	Engineer		2006	2007 ret.
xx	Cagle, Y	1959	1996	16	Medicine	MD	no flights	Management
37	Caldwell, T (Dyson)	1969	1998	17	Chemistry	PhD	2007, 2010, 2024	Active
38	Morgan, B	1951	1998	17	Education		2007	2008 ret.
39	Nyberg, K	1969	2000	18	Engineer	PhD	2008, 2013	2020 ret.
40	McArthur, M	1971	2000	18	Oceanography	PhD	2009, 2021	Active
41	Stott, N	1962	2000	18	Engineer		2009, 2011	2015 ret.
xx	Robertson, P	1962	2000	18	Medicine	MD	no flights	d. 2001
42	Metcalf-Lindenburger, D	1975	2004	19	Earth Sci, Astronomy		2010	2014 ret.
43	Walker, S	1965	2004	19	Physics	PhD	2010, 2020	Active
44	Rubins, K	1978	2009	20	Microbiology	PhD	2016, 2020	Active
45	Auñón, S (Chancellor)	1976	2009	20	Medicine	MD	2018	Management
46	McClain, A	1979	2013	21	Engineer, Test Pilot	COL, USA	2018, 2025	Active
47	Koch, C	1979	2013	21	Engineer		2019, 2026?	assigned to Artemis II
48	Meir, J	1977	2013	21	Marine Biology	PhD	2019	Active
49	Barron, K	1987	2017	22	Engineer	LCDR, USN	2021	Active
50	Watkins, J	1988	2017	22	Geology	PhD	2022	Active
51	Mann, N	1977	2013	21	Engineer, Test Pilot	COL, USMC	2022	Active
52	Moghbeli, J	1983	2017	22	Engineer, Test Pilot	LTCOL, USMC	2023	Active
53	O'Hara, L	1983	2017	22	Engineer		2023	Active
54	Epps, J	1970	2009	20	Engineer	PhD	2024	Active
55	Ayers, N	1989	2021	23	Engineer, Fighter Pilot	MAJ, USAF	2025	Active
	Cardman, Z	1987	2017	22	Geobiology		Assigned for 2025	Space X Crew 11
	Birch, C	1986	2021	23	Bioengineering	PhD	Not yet assigned	Active
	Burnham, D	1985	2021	23	Engineer	LT, USNR	Not yet assigned	Active
	Wittner, J	1983	2021	23	Engineer, Test Pilot	CDR, USN	Not yet assigned	Active

APPENDIX B

||||||||||||||||||||||||||||

US Women Astronaut Spacewalkers

Ranked by extravehicular activity (EVA) time
As of January 2025

Rank	Name	EVA Year (s)	Number of EVAs	Total Duration in EVA
1	Sunita L. Williams	2006, 2007, 2012, 2025	9	62:06
2	Peggy A. Whitson	2002, 2007, 2008, 2017	10	60:21
3	Christina H. Koch	2019, 2020	6	42:15
4	Heidemarie M. Stefanyshyn-Piper	2006, 2008	5	33:42
5	Kathleen H. Rubins	2016, 2021	4	26:46
6	Tracy Caldwell Dyson	2010, 2024	4	23:20
7	Jessica U. Meir	2019, 2020	3	21:44
8	Kathryn C. Thornton	1992, 1993	3	21:11
9	Nicole A. Mann	2023	2	14:02
10	Kayla S. Barron	2021, 2022	2	13:26
11	Anne C. McClain	2019	2	13:08
12	Linda M. Godwin	1996, 2001	2	10:14
13	Susan J. Helms	2001	1	08:56
14	Tamara E. Jernigan	1999	1	07:55
15	Jasmin Moghbeli	2023	1	06:42
15	Loral A. O'Hara	2023	1	06:42
17	Nicole P. Stott	2009	1	06:39
18	Kathryn D. Sullivan	1984	1	03:29

In twenty-one of the more than forty years since the first US woman spacewalked, at least one US woman astronaut has performed at least one EVA.
Only Kathy Sullivan went on EVA in the 1980s.
Three women spacewalked in the 1990s.
The number and frequency increased in the 2000–2009 decade, with six women spacewalking.
In the 2010–2019 decade, seven women went on EVAs.
The 2020 decade started with nine women on EVAs in the first four years.
Altogether from 1984 through January 2025, eighteen women have performed fifty-eight EVAs. Several women did spacewalks in more than one year.

ACRONYMS AND ABBREVIATIONS
||||||||||||||||||||||||||||||||||||||||||||||||||||||||||||||||||||||||||||

ART — Assisted Reproductive Technologies, services for infertility

ATLAS — Atmospheric Laboratory for Applications and Science, a series of three Spacelab missions

CACO — Casualty Assistance Calls Officer

CAPCOM — Capsule Communicator on the Mission Control Center team

CAVES — Cooperative Adventure for Valuing and Exercising Human Behaviour and Performance Skills, a training program of the European Space Agency

CVT — Concept Verification Test

EEOA — Equal Employment Opportunity Act of 1972

EMU — Extravehicular Mobility Unit, EVA suit used during Space Shuttle and ISS eras

ESA — European Space Agency

EVA — Extravehicular Activity, or spacewalk, outside the spacecraft

FCOD — Flight Crew Operations Directorate at NASA JSC

HUT — Hard Upper Torso of the EMU

ISS — International Space Station

IVA — Intravehicular Activity, inside the spacecraft

JPL — Jet Propulsion Laboratory

JSC — NASA's Johnson Space Center in Houston, Texas

KSC — NASA's Kennedy Space Center at Cape Canaveral, Florida

MCC — Mission Control Center at NASA JSC

Mir — Soviet-Russian space station, 1986–2001

MSE — Manned Spaceflight Engineers, US Air Force

MSFC — NASA's Marshall Space Flight Center in Huntsville, Alabama

NACA — National Advisory Committee on Aeronautics

NASA — National Aeronautics and Space Administration

NBL, NBS — Neutral Buoyancy Laboratory and Neutral Buoyancy Simulator, at JSC and MSFC, respectively; used for underwater EVA training

NEEMO — NASA Extreme Environment Mission Operations, conducted at the Aquarius underwater habitat and laboratory off Key Largo in the Florida Keys National Marine Sanctuary, about 62 feet beneath the surface

NOAA — National Oceanographic and Atmospheric Administration

RMS — Remote Manipulator System robotic arm

SAIL — Shuttle Avionics and Integration Laboratory

STEAM — Science, Technology, Engineering, Arts, and Mathematics

STEM — Science, Technology, Engineering, and Mathematics

NOTES
||||||||||||||

Introduction

1 This account is based on correspondence during March, April, and August 2023 with Tam O'Shaughnessy, Bear Ride, and attending astronauts Anna Fisher, Cady Coleman, Marsha Ivins, and Ellen Ochoa, as well as a personal video provided by Tam O'Shaughnessy.

2 Louis Menand, "Generation Overload: Can We Retire the Concept?," *New Yorker*, October 18, 2021, 63–67.

3 NASA Press Release 23-040, "NASA Names Astronauts to Next Moon Mission, First Crew Under Artemis," April 3, 2023, https://www.nasa.gov/news-release/nasa-names-astronauts-to-next-moon-mission-first-crew-under-artemis.

4 Among the most informative, albeit not comprehensive, books about women astronauts are Meredith E. Bagby, *The New Guys: The Historic Class of Astronauts That Broke Barriers and Changed the Face of Space Travel* (New York: William Morrow, 2023); Umberto Cavallaro, *To the Stars: Women Spacefarers' Legacy* (Chichester, UK: Springer-Praxis, 2023); Umberto Cavallaro, *Women Spacefarers: Sixty Different Paths to Space* (Chichester, UK: Springer-Praxis, 2017); David J. Shayler and Ian Moule, *Women in Space: Following Valentina* (Chichester, UK: Springer, 2005); David J. Shayler, ed., *Walking in Space: Development of Space Walking Techniques* (Chichester, UK: Springer, 2004); and Loren Grush, *The Six: The Untold Story of America's First Women Astronauts* (New York: Scribner, 2023).

5 As of 2024 the corps of astronauts for the shuttle-ISS-Artemis period selected since 1978 comprised 286 new recruits plus 28 carryovers from earlier groups for a total of 314. This count does not include the full number of Mercury, Gemini, Apollo, and Skylab astronauts, but only those who, along with the Air Force Manned Orbiting Lab astronauts, stayed on for the space shuttle program. Because they were excluded until 1978, it is irrelevant to calculate percentage of women in the total astronaut corps since 1958.

6 https://www.nasa.gov/astronauts.

7 Margaret A. Weitekamp, *Right Stuff, Wrong Sex: America's First Women in Space Program* (Baltimore: Johns Hopkins University Press, 2004).

8 J. W. Miller, J. G. VanDerwalker, and R. A. Waller, eds., *Tektite 2: Scientists in the Sea* (Washington, DC: US Department of the Interior, 1971).

9 Sylvia A. Earle, correspondence with author, April 30, 2023.

10 C. S. Griner et al., "The Concept Verification Testing of Materials Science Payloads," NASA Technical Memorandum X-73320 (Huntsville, AL: NASA Marshall Space Flight Center, June 1976), https://ntrs.nasa.gov/search?q=TM-X-73320.

Chapter 1

1 John Uri, "60 Years Ago: NASA Introduces Mercury 7 Astronauts," April 9, 2019, https://www.nasa.gov/history/60-years-ago-nasa-introduces-mercury-7-astronauts/.

2 Joseph D. Atkinson and Jay M. Shafritz, *The Real Stuff: A History of NASA's Astronaut Recruitment Program* (New York: Praeger, 1985).

3 Matthew H. Hersch, *Inventing the American Astronaut* (New York: Palgrave Macmillan, 2012).

4 Margaret W. Rossiter, *Women Scientists in America: Before Affirmative Action, 1940–1972* (Baltimore: Johns Hopkins University Press, 1995).

5 Robert L. Rosholt, *An Administrative History of NASA, 1958–1963* (Washington, DC: Scientific and Technical Information Division, NASA, 1966), 56. Sylvia Doughty Fries, *NASA Engineers in the Age of Apollo* (Washington, DC: NASA, 1992), 198–200, 205. Tellingly, the index to Fries's book does not include "woman" or "women."

6 Atkinson and Shafritz, *The Real Stuff*, 28–53.

7   Hersch, *Inventing the American Astronaut*; Deborah G. Douglas and Amy E. Foster, *American Women and Flight since 1940* (Lexington: University Press of Kentucky, 2004); Jeanne Holm, *Women in the Military: An Unfinished Revolution*, rev. ed. (Novato, CA: Presidio Press, 1992); Yvonne C. Pateman, *Women Who Dared: American Female Test Pilots, Flight-Test Engineers, and Astronauts, 1912–1996* (Laguna Hills, CA: Norstahr Publishing, 1997).

8   Fred Erisman, *Boys' Books, Boys' Dreams, and the Mystique of Flight* (Fort Worth: Texas Christian University Press, 2006); Fred Erisman, *In Their Own Words: Forgotten Women Pilots of Early Aviation* (West Lafayette, IN: Purdue University Press, 2021).

9   Margaret A. Weitekamp, *Space Craze: America's Enduring Fascination with Real and Imagined Spaceflight* (Washington, DC: Smithsonian Books, 2022), ch. 1.

10 NASA News Release, "NASA Will Recruit 10 to 20 Scientist-Astronauts," October 19, 1964. nasa.gov/centers/johnson/pdf/83118main_1964.pdf/

11 NASA News Release MSC 66-70, "Scientists Invited to Become Astronauts, Do Research in Space," September 22, 1966, nasa.gov/centers/johnson/pdf/83118main_1966.pdf.

12 No women applicants are mentioned in Atkinson and Shafritz, *The Real Stuff*, except for a passing remark that any minority or women applicants were given "little consideration" (83). In *NASA's Scientist-Astronauts* (Chichester, UK: Springer-Praxis, 2007), David Shayler and Colin Burgess state that four women applied (37), but there is no citation to support that claim, and the nearest citation to that statement does not mention women applicants. An article published in *Ms.* magazine in September 1973, "The 13 Left Behind," claimed that seventeen women had applied, without citing evidence. These claims are mysterious and unverified.

13 Margaret A. Weitekamp, *Right Stuff, Wrong Sex: America's First Women in Space Program* (Baltimore: Johns Hopkins University Press, 2004), 150–51.

14 Louis Lasagna, "Why Not 'Astronauttes' Also?: Comparing the Sexes in All Relevant Particulars," *New York Times Magazine*, October 21, 1962; Richard Witkin, "Training for Space: Soviet and U.S. Differ in Assessing the Qualifications of an Astronaut," *New York Times*, June 18, 1963.

15 AP, "Good Health All That's Needed to Ride Space Shuttle," *Palm Beach Post*, July 29, 1970.

16 Clare Booth Luce, "But Some People Simply Never Get the Message," *Life*, June 28, 1963.

17 Margaret A. Weitekamp, "NASA's Early Stand on Women Astronauts: 'No Present Plans to Include Women on Space Flights,'" March 17, 2016, https://airandspace.si.edu/stories/editorial/nasas-early-stand-women-astronauts.

18 Nur Ibrahim, "Did NASA Reject Hillary Clinton's Childhood Dream of Becoming an Astronaut?" Snopes, February 25, 2021, https://www.snopes.com/fact-check/nasa-hillary-clinton-astronaut/. Marsha Ivins was the engineering student who became an astronaut from 1984 to 2011.

19 "Women's Place In Outer Space," *Washington Post*, March 17, 1968.

20 Isaac Asimov, "No Space For Women?," *Ladies' Home Journal*, March 1971.

21 This narrative is primarily informed by Douglas and Foster, *American Women and Flight since 1940*, and Pateman, *Women Who Dared*.

22 Dorothy Cochrane and P. Ramirez, "Meet Jacqueline Cochran," October 28, 2021, https://airandspace.si.edu/stories/editorial/meet-jacqueline-cochran.

23 For perspectives on these two women, see Margaret Weitekamp and Dorothy Cochrane, "Remembering Geraldyn 'Jerrie' Cobb, Pioneering Woman Aviator," April 18, 2019, https://airandspace.si.edu/stories/editorial/remembering-geraldyn-jerrie-cobb-pioneering-woman-aviator; Dorothy Cochrane, "Celebrating Jerrie Mock, the First Solo World Flight by a Woman," April 16, 2024, https://airandspace.si.edu/stories/editorial/celebrating-jerrie-mock.

24 Weitekamp, *Right Stuff, Wrong Sex*. Unless noted otherwise for particular facts, this source informs the narrative.

25 Ben Kocivar, "The Lady Wants to Orbit," *Look*, February 2, 1960. The writer also estimated the traits that a woman astronaut should have, some serious and some odd—flat chest, boyish figure, and soothing personality.

26 Rebecca Rissman Siegel, *To Fly among the Stars: The Hidden Story of the Fight for Women Astronauts* (New York: Scholastic Focus, 2020), 135–36.

27 "Science: From Aviatrix to Astronautrix," *Time*, August 29, 1960; "A Lady Proves She's Fit for Space Flight," *Life*, August 29, 1960; "A Woman Passes Tests Given to 7 Astronauts," *New York Times*, August 19, 1960; "Woman Qualifies for Space Training," *Washington Post*, August 19, 1960; Jack Smith, "The Craftier Sex Is Cleared for Space," *Los Angeles Times*, August 28, 1960; "Science: Damp Prelude to Space: A Potential Lady Orbiter Excels in Lonesome Test," *Life*, October 24, 1960.

28 "12 Women to Take Astronaut Tests," *New York Times*, January 26, 1961.

29 Tanya Lee Stone, *Almost Astronauts: 13 Women Who Dared to Dream* (Somerville, MA: Candlewick Press, 2009), 62–63. According to Cobb in a 2007 interview, Vice President Lyndon Johnson was polite but not supportive. He remarked that "if we let you or other women into the space program, we'd have to let blacks in. . . . We'd have to let every minority in, and we just can't do it."

30 Ison Wong, "Wally Funk Becomes Oldest Person to Fly to Space 60 Years after She Was Denied the Opportunity," *NBC News*, July 20, 2021, https://www.nbcnews.com/science/space/wally-funk-becomes-oldest-person-fly-space-60-years-after-n1274486; Margaret A. Weitekamp, "Never Say Never: Wally Funk's Space Flight Dream Comes True," July 16, 2021, https://airandspace.si.edu/stories/editorial/never-say-never-wally-funks-space-flight-dream-comes-true.

31 "13 Women Triumphing Vicariously: Shuttle Pilot Lives Their Early Dream," *New York Times*, February 5, 1995.

32 Weitekamp, *Right Stuff, Wrong Sex,* 44.

33 This narrative is informed primarily by Douglas and Foster, *American Women and Flight since 1940*; Pateman, *Women Who Dared*; Holm, *Women in the Military*.

34 Holm, *Women in the Military*, 175–85.

35 Douglas and Foster, *American Women and Flight since 1940*, 170, mentions the article by Bruce Callander, "Why Can't a Woman Be a Military Pilot?"

36 Pateman, *Women Who Dared*, uses the phrase "Decade of Change" in a chapter title and discussion, 63–70. Holm, *Women in the Military*, similarly calls this period "A New Beginning and "The Seventies: A Decade of Expansion."

37 Douglas and Foster, *American Women and Flight since 1940*, 186–87.

38 Holm, *Women in the Military*, ch. 21, "The Service Academies," details the difficulties of integrating women into these all-male institutions.

39 Helms and two other 1990 NASA classmates—Air Force Major Eileen Collins and Army Captain Nancy Sherlock—were the first military women to become astronauts.

40 To look ahead in the story of women astronauts, later in the 1990s the agency selected three Navy women who were test pilot graduates and also the third woman to graduate from the Air Force test pilot school. By 2023, four other women graduates of the services' military test pilot schools had become astronauts, for a total of ten: Eileen Collins, Susan Helms, and Pamela Melroy, all from the USAF test pilot school; and from the Navy test pilot school, Susan Still (Kilrain), Sunita Williams, Lisa Nowak, Nicole Mann, Jasmin Moghbeli, Anne McClain, and Jessica Wittner. Only Collins, Melroy, and Still were selected as pilot astronauts, the only women so designated. They went on to pilot or command a total of nine space shuttle missions.

41 Holm, *Women in the Military*. Even as women gained access to coveted positions and aircraft, egregious sexism remained. Holm (428) refers to a woman reconnaissance pilot who had to sign a contract to have biweekly pregnancy tests and not get pregnant for a year. Holm also notes (429) how galling it was that Canadian forces, trained by Americans, produced the first women jet fighter pilots.

42 Holm, *Women in the Military*; Eileen A. Bjorkman, *The Fly Girls Revolt: The Story*

of the Women Who Kicked Open the Door to Fly in Combat (New York: Knox Press, 2023), esp. 220–40 on the battle over combat exclusion.

43 Bjorkman, The Fly Girls Revolt, x.

44 Elisabeth Bumiller, "Pentagon Allows Women Closer to Combat, but Not Close Enough for Some," New York Times, February 10, 2012.

45 As it happened, Air Force pilot Eileen Collins, who had flown combat support missions during US action in Grenada, became in 1990 the first woman military pilot astronaut recruited by NASA. The first woman assigned to a combat aircrew in 1993, Navy maritime patrol navigator/communicator Kathleen "Kay" Hire, joined NASA as an astronaut candidate in 1994. Pilot astronaut Pamela Melroy and other women with impressive combat records followed them into the astronaut corps.

46 Janice Delaney, Mary Jane Lupton, and Emily Toth, The Curse: A Cultural History of Menstruation, rev. ed. (Urbana: University of Illinois Press, 1988).

47 Neil Amdur, "Boston Marathon to Recognize Women's Figures at Last Today," New York Times, April 17, 1972; Jere Longman, "How the Women Won," New York Times, June 23, 1996; Charlie Lovett, "The Fight To Establish The Women's Marathon Race," 1995, https://www.marathonguide.com/history/olympicmarathons/chapter25.cfm.

48 Johnnie R. Betson Jr. and Robert R. Secrest, "Prospective Women Astronauts Selection Program: Rationale and Comments," American Journal of Obstetrics and Gynecology 88, no. 3 (February 1964): 421–23.

49 Delaney, Lupton, and Toth, The Curse, ch. 6.

50 "The Tampax Report" (Research & Forecasts, Inc., 1981); Delaney, Lupton, and Toth, The Curse, 62–63.

51 Boston Women's Health Book Collective, Our Bodies, Ourselves (New York: Simon & Schuster, 1973); "The History & Legacy of Our Bodies, Ourselves," Our Bodies Ourselves Today, https://www.ourbodiesourselves.org/about-us/our-history/.

52 Rossiter, Women Scientists in America, 1995.

53 Amy Sue Bix, Girls Coming to Tech! A History of American Engineering Education for Women (Cambridge, MA: MIT Press, 2013). Accounts of such discrimination are found in the Society of Women Engineers archives. See Troy Eller, "Publicity, Recruitment, and History: Society of Women Engineers," Centaurus: An International Journal of the History of Science and Its Cultural Aspects 54, no. 4 (2012): 299–304.

54 Bix; Amy Sue Bix, "From 'Engineeresses' to 'Girl Engineers' to 'Good Engineers': A History of Women's U.S. Engineering Education," NWSA Journal 16, no. 1 (2004): 27–49.

55 Rossiter, Women Scientists in America, xvii–xviii.

56 Margaret Rossiter's two volumes on Women Scientists in America and Amy Bix, Girls Coming to Tech! inform this and the following paragraphs.

57 Rossiter, Women Scientists in America, xvii.

58 Sylvia Doughty Fries, NASA Engineers and the Age of Apollo, NASA SP 4104 (Washington, DC: NASA, Scientific and Technical Information Program, 1992).

59 UPI, "Women Play Big Role in Space Plan: But There Are None to Ride Rockets," Chicago Daily Tribune, December 25, 1960; UPI, "Women Needed in Space Work," June 28, 1962; Christine K. Roberts, "Petite Engineer Likes Math, Music," in NASA and the Long Civil Rights Movement (Gainesville: University Press of Florida, 2019), 219–32.

60 This section is informed by Nathalia Holt, Rise of the Rocket Girls: The Women Who Propelled Us, from Missiles to the Moon to Mars (New York: Little, Brown, 2016); George D. Morgan, Rocket Girl: The Story of Mary Sherman Morgan, America's First Female Rocket Scientist (Amherst, NY: Prometheus Books, 2013); Margot Lee Shetterly, Hidden Figures: The American Dream and the Untold Story of the Black Women Mathematicians Who Helped Win the Space Race (New York: Morrow, 2016); Jennifer M. Ross-Nazzal, ed., Making Space for Women: Stories from Trailblazing Women of NASA's Johnson

Space Center (College Station: Texas A&M University Press, 2022).

61 Mary Finch Hoyt, *American Women of the Space Age* (New York: Atheneum, 1966).

62 Holt, *Rise of the Rocket Girls*, 236.

63 This claim is stated but not supported by Peggy Orenstein, "Champion of the Deep," *New York Times*, June 23, 1991; James Merle Thomas and Meghan O'Hara, "Tektite Revisited," *Triple Canopy*, July 26, 2011; Amy Crawford, "The Forgotten Women Aquanauts of the 1970s," *Atlas Obscura*, March 24, 2023. Additional evidence presented here strengthens the claim.

64 J. W. Miller, G. VanDerwalker, and R. A. Waller, *Tektite 2: Scientists in the Sea* (Washington, DC: US Department of the Interior, 1971); Jack W. Stokes, "Tektite II: An Analog, Habitability and Behavioral Studies," briefing charts, personal papers courtesy of Jack W. Stokes, August 4, 2011.

65 Robert Helmreich, "The TEKTITE II Human Behavior Program" (Austin: University of Texas at Austin, Dept. of Psychology, March 15, 1971), https://apps.dtic.mil/sti/citations/AD0721364.

66 H. H. Watters and J. W. Miller, "Tektite: Experience with an Underwater Analog of Future Space Operations," *AIAA Paper No. 71-828*, July 1971, 5; Stokes, "Tektite II."

67 "Sylvia Earle Video Time—How She Became the First Woman to Become Chief Scientist of U.S. NOAA," YouTube, October 31, 2017, https://www.youtube.com/watch?v=FCHX6Ei940g&t=5s.

68 Judy Klemesrud, "The Women Aquanauts Did the Job, Despite 'Male Chauvinism,'" *New York Times*, July 28, 1970.

69 "The Tektite Underwater Habitat Documentary," https://tektitedocumentary.wordpress.com/; "Sylvia Earle Video Time."

70 "5 Women Scientists Start 2-Week Stay Under Sea," *New York Times*, July 7, 1970.

71 Earle, married to scientist Giles W. Mead at the time, was also known then as Sylvia Mead.

72 Nate Haseltine, "5 Women to Spend 14 Days Under Sea," *Washington Post*, March 3, 1970.

73 Multiple articles with this line appeared in the *New York Times*, *Boston Globe*, *Chicago Tribune*, and elsewhere in July 1970 and reappeared as recently as 2023 in Amy Crawford's *Atlas Obscura* article.

74 Earle correspondence with author, April 30, 2023.

75 Stokes, "Tektite II"; D. P. Nowlis, E. C. Wortz, and H. Watters, "Tektite II Habitability Research Program" (Los Angeles: Airesearch, January 14, 1972), Bruce B. Collette and Sylvia A. Earle, *Results of the Tektite Program: Ecology of Coral Reef Fishes* (Los Angeles: Natural History Museum, 1972), 1–9.

76 Earle, "All-Girl Team Tests the Habitat," *National Geographic*, August 1971.

77 Earle correspondence with author, April 30, 2023.

78 Collette and Earle, *Results of the Tektite Program*; E. L. Beckman and E. M. Smith, "Tektite II: Medical Supervision of the Scientists in the Sea. IX. Mission-by-Mission Experiences," *Texas Reports on Biology and Medicine* 30, no. 3 (1972): 112–17.

79 Earle, "All-Girl Team Tests the Habitat"; Katie Couric Media, "Sylvia Earle: First Woman to Walk the Ocean Floor," January 21, 2020.

80 Earle correspondence with author, April 30, 2023; Dorothy Townsend, "Times Woman of the Year: Dr. Mead: Sea-Probing Pioneer," *Los Angeles Times*, December 23, 1970.

81 "Women of Sea and Space," July 17, 2020, https://tektite2020.com/; "The Tektite Underwater Habitat Documentary."

82 Richard D. Lyons, "President Hails Aquanauts' Feat," New York Times, April 16, 1969; D. C. Pauli and H. A. Cole, "Summary Report on Project Tektite I, ONR Report DR-153-S" (Office of Naval Research, January 16, 1970), Foreword.

83 Watters and Miller, "Tektite," 1.

84 "People Under Stress, and Under Water Too, in Tektite Studies," *Smithsonian Magazine*, October 1970.

85 Jim Clash, "Deep Thoughts with Oceans Legend Sylvia Earle," *Forbes*, March 29,

2023, https://www.forbes.com/sites
/jimclash/2023/03/29/deep-thoughts
-with-oceans-legend-sylvia-earle/.

86 Lee Edson, "Will Man Ever Live in Space?
(If So, Woman Will Live There Too)," *New
York Times,* December 31, 1972.

87 Earle correspondence with author,
April 30, 2023.

88 "Tektite II Research Peggy Lucas,"
Marshall Star, June 24, 1970, NASA
Marshall Space Flight Center Archives;
Lucas correspondence with author,
June 12, 2023. She was testing a
closed-cycle "rebreather" unit that
cleansed and recirculated air for the diver
much like the astronauts' life support
backpack. This alternative to regular
scuba tanks was used on the Tektite
missions to give the divers more time on
their excursions. It had the added
advantage of producing little to no
bubbles, a boon for observing sea life
without disturbance.

89 "Tektite II: Men Undersea: A Two-Day
Symposium" (NASA Marshall Space Flight
Center, May 4, 1971), University of Alabama
in Huntsville Library and Archives, Richard
Heckman Papers, Box 7, Folder 28.

90 Clash, "Deep Thoughts with Oceans
Legend Sylvia Earle"; Earle correspondence with author, April 30, 2023; Lucas
correspondence with author, June 12,
2023.

91 "Four Marshall Women Begin Five Day
Laboratory Test," *Marshall Star,* December 18, 1974.

92 C. S. Griner, M. H. Johnston, and A.
Whitaker, "The Concept Verification
Testing of Materials Science Payloads,
NASA Technical Memorandum TM-X-
73320" (Huntsville, AL: NASA Marshall
Space Flight Center, June 1976),
https://ntrs.nasa.gov/search?q=TM-X
-73320.

93 "Four Marshall Women Begin Five Day
Laboratory Test."

94 David Shayler and Ian A. Moule, *Women
in Space: Following Valentina* (Chichester, UK: Springer, 2005), 156–60. Over
time, photos of the women simulating
weightlessness in the neutral buoyancy
pool have appeared during March
(Women's History Month) on NASA and

NASA Marshall Center websites. Shayler
and Moule include a photo of them
wearing Apollo-era suits for a publicity
shot and states that they underwent
spacesuit training.

95 "Six, Including Woman Named Finalists
for Seat on Spacelab," *New York Times,*
December 23, 1977.

96 Amy E. Foster, *Integrating Women
into the Astronaut Corps: Politics
and Logistics at NASA, 1972–2004*
(Baltimore: Johns Hopkins University
Press, 2011), 85.

97 Robert Helmreich et al., "NASA Technical
Paper 1364, A Critical Review of the Life
Sciences Project Management at Ames
Research Center for the Spacelab Mission
Development Test III," January 1979.

98 "Patricia S. Cowings, Ph.D.—The
Banneker Institute for Science &
Technology," April 12, 2016, https://web
.archive.org/web/20160412053014
/http://www.thebannekerinstitute.org
/resources/featured-scientist/patricia-s
-cowings-ph-d/.

99 Shayler and Moule, *Women in Space,*
148–49; Foster, *Integrating Women into
the Astronaut Corps,* 80–81; Harold
Sandler and David L. Winter, "Physiologi-
cal Responses of Women to Simulated
Weightlessness: A Review of the
Significant Findings of the First Female
Bed-Rest Study" (Washington, DC:
National Aeronautics and Space
Administration, 1978).

100 This section is largely informed by Ruth
Rosen, *The World Split Open: How the
Modern Women's Movement Changed
America* (New York: Viking, 2000), which
includes a timeline of major events and
legislation.

101 Martha Weinman Lear, "The Second
Feminist Wave," *New York Times,*
March 10, 1968; Betty Friedan, "Feminism
Takes a New Turn," *New York Times
Magazine,* November 18, 1979.

102 As Weitekamp notes in *Right Stuff, Wrong
Sex* (185), this kind of support system did
not exist for the Lovelace program
participants as they sought women's
access to space.

103 Friedan, "Feminism Takes a New Turn";
Rosen, *The World Split Open,* 295–330.

104 "The Statement by President Nixon, 5 January 1972," https://history.nasa.gov/stsnixon.htm.

105 "Racism, Sexism and Space Ventures," *Washington Post*, November 24, 1973.

106 Atkinson and Shafritz, *The Real Stuff*, 133–38; Kim McQuaid, "Race, Gender, and Space Exploration: A Chapter in the Social History of the Space Age," *Journal of American Studies* 41, no. 2 (2007): 405–34. The Harris story and NASA's resistance to affirmative action is informed by McQuaid, who researched it in great detail.

107 McQuaid, "Race, Gender, and Space Exploration."

108 Atkinson and Shafritz, *The Real Stuff*, 135.

109 Ibid., 137–38.

110 Valerie Neal, *Spaceflight in the Shuttle Era and Beyond: Redefining Humanity's Purpose in Space* (New Haven, CT: Yale University Press, 2017).

111 NASA News Release 76-44, "NASA to Recruit Space Shuttle Astronauts," July 8, 1976; Atkinson and Shafritz, *The Real Stuff*; David J. Shayler and Colin Burgess, *NASA's First Space Shuttle Astronaut Selection: Redefining the Right Stuff* (Cham, Switzerland: Springer International, 2020). The announcement also referred to payload specialists responsible for certain sponsored payloads, a function that mission specialists might serve; this concept was not yet fully formed. Shafritz informs most of the discussion of recruitment and selection, except as other sources are noted.

112 Nichelle Nichols, *Beyond Uhura: Star Trek and Other Memories* (New York: Putnam's, 1994), 217–32.

113 Todd Thompson Director, "Woman in Motion: Nichelle Nichols, Star Trek and the Remaking of NASA," https://tv.apple.com/us/movie/woman-in-motion-nichelle-nichols-star-trek-and-the-remaking-of-nasa/umc.cmc.3xq6dlh-qb32xjjx59bzea6nh; Elizabeth Howell, "New Documentary Explores 'Star Trek,' Nichelle Nichols and NASA's 1970s Astronaut Search," Space.com May 26, 2021, https://www.space.com/star-trek-nichelle-nichols-woman-in-motion-space-documentary.

114 NASA News Release 77-39, "Over 8,000 Apply for Space Shuttle Astronaut Program at JSC," July 15, 1977. Shayler and Burgess's *NASA's First Space Shuttle Astronaut Selection* presents a somewhat different tally of applications but without an explanation for the different numbers.

115 NASA New Release 77-75, "Tenth Group of Astronaut Applicants Report to JSC November 14," November 11, 1977.

116 NASA News Release 78-03, "NASA Selects 35 Astronaut Candidates," January 16, 1978.

117 Shayler and Burgess, *NASA's First Space Shuttle Astronaut Selection*, 16–31.

Chapter 2

1 NASA Press Conference, "Space Shuttle Astronauts, Transcript," January 16, 1978.

2 Kathryn D. Sullivan, NASA Johnson Space Center (JSC) Oral History Project interview, May 10, 2007.

3 Ibid.

4 Jennifer M. Ross-Nazzal, "'You've Come a Long Way, Maybe': The First Six Women Astronauts and the Media," in *Spacefarers: Images of Astronauts and Cosmonauts in the Heroic Era of Spaceflight*, ed. Michael J. Neufeld (Washington, DC: Smithsonian Institution Press, 2013), 175–201.

5 Interviews conducted with author, February–September 2023.

6 Margaret Rhea Seddon, NASA Johnson Space Center (JSC) Oral History Project interview, May 20, 2010.

7 Rhea Seddon, *Go for Orbit: One of America's First Women Astronauts Finds Her Space* (Murfreesboro, TN: Your Space Press, 2015), 21.

8 Anna Fisher interview with author, August 8, 2023.

9 Joseph D. Atkinson and Jay M. Shafritz, *The Real Stuff: A History of NASA's Astronaut Recruitment Program* (New York: Praeger, 1985), 163.

10 Amy E. Foster, *Integrating Women into the Astronaut Corps: Politics and Logistics at NASA, 1972–2004* (Baltimore: Johns Hopkins University Press, 2011), 100–102; Carolyn Leach Huntoon, NASA

Johnson Space Center (JSC) Oral History Project interview, June 5, 2002, and April 21, 2008.

11 Jessica Contrera, "She Was Pregnant When NASA Offered to Send Her to Space. Anna Fisher Didn't Hesitate," *Washington Post*, May 13, 2019.

12 Seddon, JSC interview.

13 Ibid.

14 R. Mike Mullane, *Riding Rockets: The Outrageous Tales of a Space Shuttle Astronaut* (New York: Scribner, 2006), 29–30, 36–43.

15 Seddon, JSC interview; Sullivan, JSC interview; Anna L. Fisher, NASA Johnson Space Center (JSC) Oral History Project interview, February 17, 2009.

16 Carolyn Leach Huntoon, NASA Johnson Space Center (JSC) Oral History Project interview, May 10, 2002.

17 Interviews with author, February–September 2023.

18 Astronaut Selection chart, Sally K. Ride Papers, National Air and Space Museum Archives, NASM 2014.0025-bx59-fd001_001.

19 Interviews with author, February–September 2023, and correspondence with author at various times in 2023 and 2024.

20 Ellen Keerdoja and Louis Alexander, "The Astronauts: Ready for Launch," *Newsweek*, January 5, 1981.

21 Seddon, JSC interview.

22 In the collection of the Smithsonian National Air and Space Museum, Catalog Number A20140359000.

23 Profiles of the astronauts are informed by NASA information sources in the public domain and these well-researched books: Meredith E. Bagby, *The New Guys: The Historic Class of Astronauts That Broke Barriers and Changed the Face of Space Travel* (New York: William Morrow, 2023); Umberto Cavallaro, *To the Stars: Women Spacefarers' Legacy* (Chichester, UK: Springer-Praxis, 2023); David Shayler and Colin Burgess, *NASA'S Scientist-Astronauts*, Springer-Praxis Books in Space Exploration (Chichester, UK: Springer-Praxis, 2007); David J. Shayler

and Ian Moule, *Women in Space: Following Valentina* (Chichester, UK: Springer-Praxis, 2005).

24 Judy Klemesrud, "A Marriage That Was Made for the Heavens," *New York Times*, June 3, 1980.

25 These jobs are described in more detail in chapter 5.

26 Contrera, "She Was Pregnant When NASA Offered to Send Her to Space."

27 Interview with author, August 8, 2023.

28 "No Room for Failure," *The Arrow of Pi Beta Phi*, Winter 2023, 10–14.

29 Shannon Lucid, *No Sugar Added: One Family's Saga of Dementia and Caretaking* (Vista, CA: MkEk Publishing, 2019), 61–64.

30 Correspondence with author, January 1, 2024.

31 "Space Shuttle Astronauts, Transcript."

32 Kathy Sawyer, "Extended Space Travel Has Advantages, Lucid Says," *Washington Post*, October 25, 1996; Shannon W. Lucid, "Six Months on Mir," *Scientific American*, May 1998, 46–55.

33 Peggy Whitson, Sunita Williams, Kate Rubins, Christina Koch, Shannon Walker, Samantha Cristoforetti, and many men have since logged more total time in space.

34 Shannon Lucid, "Mission to Mir," 1997, https://www.imax.com/movie/mission-mir.

35 Shannon Lucid, interview with author, April 11, 2023.

36 "The Shannon Lucid Story," a public affairs compilation of interview and appearance requests for her consideration, author's collection.

37 Interview with author.

38 The crews of the final *Challenger* and *Columbia* missions, including Judith Resnik, Christa McAuliffe, Kalpana Chawla, and Laurel Clark, received this medal posthumously.

39 Her other favorite role was flying T-38s. Interview with author, April 11, 2023, and correspondence with author, January 1, 2024.

40 "Astronaut Judy Resnik Interview 9th April 1981," YouTube, July 25, 2016, https://www.youtube.com/watch?v=EaafRyuwA8w.

41 "Judith Resnik Becomes First Jewish American Astronaut and Second Woman in Space," Jewish Women's Archive, August 30, 1984, https://jwa.org/thisweek/aug/30/1984/this-week-in-history-judith-resnik-first-american-jewish-astronaut-and-second; Kate Bigam, "Remembering Judith Resnik, the First Jewish American Woman in Space," Jewish Women's Archive, January 26, 2012, https://jwa.org/blog/remembering-judith-resnik-first-jewish-american-woman-in-space; "Science Honorees, Judith Resnik & Jeffrey Hoffman, The Jewish-American Hall of Fame," n.d., https://amuseum.org/index.php/science-honoree-resnik-hoffman/.

42 Umberto Cavallaro, *Women Spacefarers: Sixty Different Paths to Space* (Chambersburg, PA: Springer International, 2017), 32; Elizabeth Kolbert, "Two Paths to the Stars: Turnings and Triumphs, Judith Resnik," *New York Times*, February 9, 1986.

43 Fisher, JSC interview.

44 Lynn Sherr, "Remembering Judy," *Ms. Magazine*, June 1986. In this article published after the Challenger tragedy in 1986, the five remaining women of the TFNG class share their impressions of Resnik. Rhea Seddon, Mike Mullane, and other astronauts of that era also have described Resnik in their memoirs.

45 Luther Young, "Judith Resnik: Astronaut Is Eager for Orbit but Not the Sex-Role Analysis," *Baltimore Sun*, September 26, 1983.

46 Thomas O'Toole, "Judith Resnik To Be Second U.S. Woman To Orbit Earth," *Washington Post*, May 23, 1984.

47 John Glenn, "Remarks of Senator John Glenn at the Memorial Service for Judith Resnik," February 3, 1986, https://www.nasa.gov/missions/space-shuttle/sts-51l/remarks-of-senator-john-glenn-at-the-memorial-service-for-judith-resnik/.

48 Fisher, JSC interview.

49 The most thorough accounts of Sally Ride's life are Lynn Sherr, *Sally Ride: America's First Woman in Space* (New York: Simon & Schuster, 2014) and Tam O'Shaughnessy, *Sally Ride: A Photobiography of America's Pioneering Woman in Space* (New York: Roaring Brook Press, 2015).

50 Thomas O'Toole, "Sally Ride Soars at Her First News Session," *Washington Post*, May 25, 1983.

51 Sara Sanborn, "Sally Ride, Astronaut: The World Is Watching," *Ms. Magazine*, January 1983; Ann Friedman, "Astronaut Sally Ride and the Burden of Being The First," *The American Prospect*, June 19, 2014, https://prospect.org/api/content/ab495f3a-b46d-5a9a-9c49-d6f9a78df33a/.

52 Sally Ride, *Leadership and America's Future in Space: A Report to the Administrator* (Washington, DC: NASA, 1987).

53 Original announcement written by Tam O'Shaughnessy, www.sallyridescience.com/sallyride/bio, July 24, 2012. Denise Grady, "Sally Ride, Trailblazing Astronaut, Dies at 61," *New York Times,* July 23, 2012; T. Rees Shapiro and Brian Vastag, "Sally Ride Dies at 61; Was First American Woman Sent into Space," *Washington Post*, July 23, 2012.

54 Boyle, Alan, "Why Sally Ride Waited until Her Death to Tell the World She Was Gay," *NBC News*, July 25, 2012; Denise Grady, "A Deadline Call on Posthumous Privacy," *New York Times*, July 22, 2017, https://www.nytimes.com/2017/07/22/insider/sally-ride-obituary-posthumous-privacy.html; Francis French, "Gay Astronauts: A Final Frontier," *The Vintage Space* (blog), December 19, 2020, https://medium.com/the-vintage-space/gay-astronauts-a-final-frontier-9892d0987fa0.

55 Lynn Sherr, *Sally Ride: America's First Woman in Space* (New York: Simon & Schuster, 2014), 175.

56 Seddon, *Go For Orbit* and Seddon, JSC interview.

57 Skylab crewmember Joe Kerwin was the first NASA physician-astronaut to fly ten years earlier in 1973.

58 Seddon, *Go for Orbit*, 332, and JSC interview.

59 Kathy Sullivan, *Handprints on Hubble: An Astronaut's Story of Invention* (Cambridge, MA: MIT Press, 2019).

60 Ibid.

61 This feat set three world records with a precise recorded distance of 386.546 miles (622.085 km). Elizabeth Montoya, "Record-Breaking Highs and Lows: Meet the NASA Astronaut That Travelled to the Deepest Point on Earth," Guinness World Records, November 24, 2020, https://www.guinnessworldrecords.com/news/2020/11/record-breaking-highs-and-lows-meet-the-nasa-astronaut-that-traveled-to-the-deep-638666.html.

62 Chelsea Gohd, "Kathy Sullivan, 1st American Woman to Walk in Space, on the All-Female Spacewalk That Could Have Been," Space.com, March 26, 2019, https://www.space.com/kathy-sullivan-on-all-female-spacewalk.html. See also Sullivan, JSC interview.

63 NASA News Release 79-50, "NASA to Recruit Space Shuttle Astronauts," August 1, 1979; NASA News Release 79-56, "Hispanics Are Encouraged to Apply for Astronaut Program," September 12, 1979.

64 NASA News Release 79-74, "NASA Receives 3278 Applications for Astronaut Program," December 11, 1979.

65 NASA News Release 80-038, "NASA Selects 19 Astronaut Candidates," May 29, 1980; Cavallaro, *To the Stars*, 124–30.

66 Mary L. Cleave, NASA Johnson Space Center (JSC) Oral History Project interview, March 5, 2002.

67 Mary Cleave, interview with author, March 4, 2023.

68 Cleave, JSC interview.

69 Primary sources for this profile are Dunbar's five NASA Johnson Space Center (JSC) Oral History Project interviews in 1998, 2004, and 2005. Other sources are cited separately.

70 Bonnie J. Dunbar, NASA Johnson Space Center (JSC) Oral History Project interview, December 22, 2004.

71 As of 2024, only four other women astronauts had flown in space five times: Susan Helms, Marsha Ivins, Tamara Jernigan, and Janice Voss.

72 NASA News Release 83-015, "Astronaut Recruitment," May 16, 1983.

73 NASA News Release 84-028, "NASA Selects 17 Astronaut Candidates," May 23, 1984.

74 "Ellen Baker, MD: Astronaut—Women Shaping History Education Update," March/April 2014, http://www.educationupdate.com/archives/2014/MAR/HTML/cov-EllenBaker.html.

75 Cavallaro, *To the Stars*, 131–34.

76 Victoria Zunitch, "The Atmosphere Is No Glass Ceiling," *Queens Chronicle*, October 17, 2019, https://www.qchron.com/qboro/stories/the-atmosphere-is-no-glass-ceiling/article_04fccb7f-b617-5241-90aa-43c02e1604fc.html.

77 Marsha Ivins, interview with author, March 18, 2023.

78 Ibid.

79 Kathryn Thornton, interview with author, March 23, 2023; Kathy Sawyer, "The Astronaut's Two Orbits Kathryn Thornton, Shuttling Between Her Role as Mom and Space Pioneer," *Washington Post*, February 8, 1994.

80 Ben Evans, "Top Ten US EVA Missions of All Time: No. 3—'Repackage Our Margin,'" AmericaSpace, June 8, 2015, https://www.americaspace.com/2015/06/08/top-ten-u-s-eva-missions-of-all-time-no-3-everyones-gonna-hate-us-now/; Ben Evans, "Top Ten US EVA Missions of All Time: No. 7—'We've Got a Satellite!'" AmericaSpace, June 4, 2015, https://www.americaspace.com/2015/06/04/top-ten-u-s-eva-missions-of-all-time-no-7-weve-got-a-satellite/.

81 Sawyer, "The Astronaut's Two Orbits."

82 Kathryn Thornton, interview with author, March 23, 2023.

83 Sawyer, "The Astronaut's Two Orbits."

84 Ibid.

85 Thornton, interview with author.

86 NASA News Release 85-024, "NASA Alters Astronaut Selection Process," June 7, 1985.

87 Shayler and Moule, *Women in Space*, 176; NASA News Release 85-023, "NASA Selects 13 Astronaut Candidates," June 4, 1985.

88 Linda Godwin, interview with author, April 2, 2023.

89 Godwin, interview with author.

90 Those women were Eileen Collins, Linda Godwin, Marsha Ivins, Janet Kavandi, and Wendy Lawrence.

91 University of Missouri News Bureau, "MU's 'Intro to Astronomy' Taught by Former Astronaut," September 12, 2011, https://munewsarchives.missouri.edu /news-releases/2011/0912-mu%E2%80% 99s-%E2%80%9Cintro-to-astronomy%E2 %80%9D-taught-by-former-astronaut/.

92 Tamara Jernigan, interview with author, April 15, 2023.

93 Ibid.

94 John Dreyfuss, "Two Adventures in Outer Space Are Heading for the Launching Pad: Life Style All Systems Go for Tamara Jernigan, Nation's Youngest Astronaut Candidate," *Los Angeles Times*, July 1, 1985.

95 Jernigan, interview with author.

96 NASA News Release 87-028, "NASA Selects 15 New Astronaut Candidates," June 5, 1987; Shayler and Moule, *Women in Space*, 177; UPI, "New Astronauts Report for Duty," UPI, August 17, 1987, https://www.upi.com/Archives/1987/08/17 /New-astronauts-report-for-duty /9769556171200/.

97 Philip M. Boffey, "Despite Numbers of Applicants, Few Civilians Are Selected as Astronauts," *New York Times*, August 17, 1987.

98 Jan Davis, interview with author, May 29, 2023.

99 Jan Davis interview.

100 Elizabeth Howell, "NASA Astronaut Jan Davis Recounts Career, Father's POW Experience in New Book," Space.com, September 28, 2023, https://www.space .com/nasa-astronaut-jan-davis-book-air -born.

101 Jan Davis, correspondence with author, January 8 and 12, 2024. The couple divorced in 1998.

102 "Married Astronauts Can Fly Together, NASA Says," UPI, https://www.upi.com /Archives/1991/03/06/Married-astronauts -can-fly-together-NASA-says /6048668235600/.

103 Howell, "NASA Astronaut Jan Davis Recounts Career."

104 Ibid.

105 Jan Davis, *Air Born: Two Generations in Flight* (West Palm Beach, FL: Ballast Books, 2023).

106 Mae Jemison, *Find Where the Wind Goes: Moments from My Life* (New York: Scholastic, 2001). After her flight she appeared with Nichols in "Second Chances," a 1993 episode of *Star Trek, The Next Generation*.

107 Jemison, 138–48; Ben Evans, "Of Marriage, Medaka Fish and Multicultural-ism: The Legacy of STS-47," Ameri-caSpace, October 28, 2012, https://www .americaspace.com/2012/10/28/of -marriage-medaka-fish-and -multiculturalism-the-legacy-of-sts-47/; J. Alfred Phelps, *They Had a Dream: The Story of African-American Astronauts* (Novato, CA: Presidio, 1994), 215–24.

108 NASA News Release 93-043, "Astronaut Jemison to Leave NASA," March 5, 1993; Phelps, *They Had a Dream*, 223–25.

109 Jemison, *Find Where the Wind Goes*, 227–29; "Astronaut Jemison to Leave NASA."

110 Jemison, *Find Where the Wind Goes*, 105–6, 137–56.

111 NASA News Release 88-018, "NASA Announces 2-Year Astronaut Selection Cycle," May 11, 1988.

112 NASA News Release 89-046, "Astronaut Class of 1990 Selection Underway," September 8, 1989.

113 NASA News Release 90-7, "1990 Astronaut Candidates Selected," January 17, 1990.

114 The first two Hispanic men selected before Ochoa (1990) were Franklin Chang-Diaz (1980) and Sidney Gutierrez (1984).

115 Eileen Collins and Jonathan H. Ward, *Through the Glass Ceiling to the Stars: The Story of the First American Woman to*

Command a Space Mission (New York: Arcade Publishing, 2021). Unless noted otherwise, her memoir is the main source for this profile.

116 NASA News Release H98-37, "Collins Named First Female Shuttle Commander," March 5, 1998, https://www3.nasa.gov/home/hqnews/1998/98-037.txt; "First Lady Hillary Clinton Names Eileen Collins First Female Space Shuttle Commander, March 5, 1998," YouTube, October 26, 2016, https://www.youtube.com/watch?v=MyNgd2QmNZM.

117 NASA News Release 06-208, "NASA Astronaut Eileen Collins Completes Career of Space Firsts," May 1, 2006, https://www3.nasa.gov/home/hqnews/2006/may/HQ_06208_Collins_retires.html.

118 Hannah Berryman Director, "Spacewoman," spacewoman.film, 2024, https://spacewoman.film/.

119 Cavallaro, *To the Stars*, 388.

120 Ibid., 437–43.

121 Nancy Currie-Gregg, interview with author, June 1, 2023.

122 Ibid.

123 Laura S. Woodmansee, *Women of Space: Cool Careers on the Final Frontier* (Burlington, ON: Apogee Books, 2003), 95–98; "Lessons Learned from Great Colorado Women—Susan Helms, Former NASA Astronaut and Retired Lieutenant General, US Air Force," *Denver Post*, February 4, 2019, https://yourhub.denverpost.com/blog/2019/02/lessons-learned-from-great-colorado-women-susan-helms-former-nasa-astronaut-and-retired-lieutenant-general-us-air-force/235244/.

124 Lauren Ferrara and Josh Copitch, "#girlcrush: Lt. Gen. Susan Helms Is a Trailblazer in Space," *FOX21 News Colorado* (blog), August 15, 2018, https://www.fox21news.com/news/local/girlcrush-lt-gen-susan-helms-is-a-trailblazer-in-space/.

125 Cavallaro, *To the Stars*, 431.

126 Podcast, "A Home Out of This World with Susan Helms," Kathy Sullivan Explores, August 11, 2022, https://www.kathysullivanexplores.com/podcast/a-home-out-of-this-world-with-susan-helms; Knott, "Female Astronaut Pioneers Last Frontier," News Article, Air Force, March 28, 2003, https://www.af.mil/News/Article-Display/Article/139670/female-astronaut-pioneers-last-frontier/https%3A%2F%2Fwww.af.mil%2FNews%2FArticle-Display%2FArticle%2F139670%2Ffemale-astronaut-pioneers-last-frontier%2F.

127 NASA JSC Press Kit, "Expedition Two: Open for Business," November 8, 2001. https://web.archive.org/web/20011108080503/http://spaceflight.nasa.gov/station/crew/exp2/exp2_presskit.pdf.

128 "A Home Out of This World with Susan Helms."

129 Ibid.; Rose Gudex, "Retired General Talks Issues Facing Females in Military," News article, Air Force, November 4, 2015, https://www.af.mil/News/Article-Display/Article/627626/retired-general-talks-issues-facing-females-in-military/https%3A%2F%2Fwww.af.mil%2FNews%2FArticle-Display%2FArticle%2F627626%2Fretired-general-talks-issues-facing-females-in-military%2F.

130 Ellen Ochoa, correspondence with author, January 22, 2024; Charley Locke, "'What We Keep' Is the Story of the Objects We Treasure," *Texas Monthly*, September 25, 2018.

131 Jade Scipioni, "How Ellen Ochoa, the First Hispanic Woman in Space, Dealt with 'People Who Didn't Think I Should Be There,'" CNBC, September 21, 2021, https://www.cnbc.com/2021/09/21/ellen-ochoa-first-hispanic-woman-in-space-on-challenges-leadership.html.

132 Quotations in this paragraph are from Scipioni, "How Ellen Ochoa."

133 Ellen Ochoa, interview, March 15, 2023, and correspondence with author, January 22, 2024.

134 NASA, "Ellen Ochoa—NASA," undated, https://www.nasa.gov/people/ellen-ochoa/.

135 Ochoa, interview with author; Alex Stuckey, "First Latina in Space Retires from Johnson Space Center, Leaves Legacy of Inclusion, Equality in Wake," *Houston Chronicle*, May 25, 2018.

136 "STS 63: Post Flight Presentation," YouTube, August 2, 2011, https://www.youtube.com/watch?v=fz1NjPhEMn8.

137 Dennis Hevesi, "Janice Voss, 55, Shuttle Astronaut and Scientist," February 10, 2012; "L'Engle's Fiction Inspired Real Science," *NPR All Things Considered*, September 8, 2007, https://www.npr.org/templates/story/story.php?storyId=14266537.

138 Purdue College of Engineering, "Dream Realized—Scholarship Established in Memory of Astronaut Alum Janice Voss," College of Engineering, Purdue University, https://engineering.purdue.edu/Engr/Giving/Imprints/2012/dream-realized.

139 Hevesi, "Janice Voss"; Nicole Cloutier-Lemasters, "Astronaut Janice Voss Dies—NASA," June 5, 2012, https://www.nasa.gov/news-release/astronaut-janice-voss-dies/.

140 Robert Pearlman, "Astronaut Janice Voss (1956–2012)," February 7, 2012, http://www.collectspace.com/ubb/Forum38/HTML/001463.html.

141 Melvin Croft and John Youskauskas, *Come Fly with Us: NASA's Payload Specialist Program* (Lincoln: University of Nebraska Press, 2019).

142 Curtis Peebles, *High Frontier: The U.S. Air Force and the Military Space Program* (Washington, DC: Air Force History and Museums Program, 1997), 27; Aeryn Avilla, "Blue Shuttle: The Manned Spaceflight Engineer Program," SpaceflightHistories, January 24, 2022, https://www.spaceflighthistories.com/post/manned-spaceflight-engineer; Croft and Youskauskas, *Come Fly with Us*, 195–202.

143 Cavallaro, *To the Stars*, 296.

Chapter 3

1 Elizabeth Adell Cook, Sue Thomas, and Clyde Wilcox, eds., *The Year of the Woman: Myths and Realities* (Boulder, CO: Westview Press, 1994).

2 John Noble Wilford, "The Shuttle's Future: NASA Looks Toward Space Stations," *New York Times*, March 16, 1982; Valerie Neal, *Spaceflight in the Shuttle Era and Beyond: Redefining Humanity's Purpose in Space* (New Haven, CT: Yale University Press, 2017), ch. 5. NASA issued several publications titled *Space Station: The Next Logical Step* by James M. Beggs (1982), Walter Froelich (1984), and Andrew J. Stofan (1986).

3 NASA, "Space Station—Key to the Future," EP-75, 1970, https://ntrs.nasa.gov/citations/19700022268.

4 Anna L. Fisher, NASA Johnson Space Center (JSC) Oral History Project interview, May 3, 2011.

5 NASA News Release 91-67, "New Astronaut Candidates Sought by NASA," May 2, 1991.

6 NASA News Release 92-018, "1992 Astronaut Candidates Selected," March 31, 1992.

7 Profiles of the astronauts are informed by NASA information sources in the public domain and these well-researched books: Umberto Cavallaro, *To the Stars: Women Spacefarers' Legacy* (Chichester, UK: Springer-Praxis Books, 2023); David Shayler and Colin Burgess, *NASA'S Scientist-Astronauts* (Chichester, UK: Springer-Praxis, 2007); David J. Shayler and Ian Moule, *Women in Space: Following Valentina* (Chichester, UK: Springer-Praxis, 2005).

8 "Preflight Interview: Catherine Coleman," October 28, 2010, https://web.archive.org/web/20121102121507/http://www.nasa.gov/mission_pages/station/expeditions/expedition26/coleman_interview.html; Cady Coleman, *Sharing Space: An Astronaut's Guide to Mission, Wonder, and Making Change* (New York: Penguin Random House, 2024), 6–12.

9 "Preflight Interview: Catherine Coleman."

10 The spacesuit size issue, which affected several of the women astronauts, is discussed in chapter 6. See Coleman's account in her book, Chapter 7, "If the Spacesuit Doesn't Fit, Wear it Anyway (and Wear it Well)."

11 Cady Coleman, "Flutes in Space: Astronaut Plays aboard Space Station," PBS, 2011, https://www.npr.org/2011/02/15/133780067/flutes-in-space-astronaut-plays-aboard-space-station; "Ian Anderson + Cady Coleman Flute Duet in Space," YouTube, April 8, 2011, https://www.youtube.com/watch?v=XeC4nqBB5BM.

12 Robert Pearlman, "An Post (Ireland) 'Space Exploration' Stamps," CollectSpace, July 23, 2019, http://www.collectspace.com/ubb/Forum20/HTML/001421.html.

13 Robert Pearlman, "Astronaut Gave 'Gravity' Advice to Sandra Bullock from Space," CollectSpace, September 13, 2013, http://www.collectspace.com/news/news-091313a.html; Clara Moskowitz, "Gravity's Astronaut Describes the Trials of Space," *Scientific American*, https://www.scientificamerican.com/article/gravity-astronaut-cady-coleman-interview/.

14 Sean Mobley, "Ad Astra with Astronaut Wendy Lawrence," podcast, Museum of Flight, December 19, 2023, https://blog.museumofflight.org/flightdeck/astronaut-wendy-lawrence-1.

15 Ibid.

16 Ibid.

17 Wendy Lawrence, correspondence with author, April 4, 2024.

18 NASA News Release J98-40, "Lawrence Moves to Rotational Assignment at National Reconnaissance Office," August 25, 1998.

19 "2019 Distinguished Graduate Awards: Captain Wendy B. Lawrence '81, USN (Ret.)," YouTube, May 6, 2019, https://www.youtube.com/watch?v=LZabqdx_ly8.

20 "LGBTQI+ Pride Month—Wendy Lawrence, U.S. Navy Officer and NASA Astronaut," *Transportation History*, June 30, 2023, https://transportationhistory.org/2023/06/30/lgbtqi-pride-month-wendy-lawrence-u-s-navy-officer-and-nasa-astronaut/; Wendy Lawrence, correspondence with author, May 5, 2024.

21 Mobley, "Ad Astra with Astronaut Wendy Lawrence."

22 Ibid.; "LGBTQI+ Pride Month."

23 Mary Ellen Weber, correspondence with author, June 10, 2024.

24 Ibid.

25 "Dean's Lecture Series: Mary Ellen Weber," YouTube, September 24, 2019, https://www.youtube.com/watch?v=ppwpOolz7pl.

26 Mary Ellen Weber, interviews with author, June 26, 2023, and July 10, 2024.

27 NASA News Release 93-101, "Biennial Astronaut Recruiting Begins," June 1, 1993.

28 NASA News Release 94-205, "1995 Astronaut Candidates Selected," December 8, 1994.

29 Jean-Pierre Harrison, *The Edge of Time: The Authoritative Biography of Kalpana Chawla* (Los Gatos, CA: Harrison Publishing, 2011). This is the reference for personal information in Chawla's profile unless otherwise noted.

30 Cavallaro, *To the Stars*, 50–57.

31 Darryl Fears, "Sky-High Ambition and a Love of Flight," *Washington Post*, February 6, 2003.

32 Kay Hire, correspondence with author, July 15 and 20, 2024.

33 Kay Hire, interview with author, August 13, 2023; Cavallaro, *To the Stars*, 461–67.

34 Captain Kathryn P. Hire, United States Navy Biography, 2016, provided to the author by Hire.

35 Bill the Goat, "USNA Hero Highlight: Kathryn P. Hire," October 2021, https://go.navyonline.com/blog/usna-hero-highlight-kathryn-p.-hire.

36 NASA News Release J19-004, "Astronaut Kathryn (Kay) Hire Retires from NASA," March 21, 2019.

37 Hire, interview with author.

38 Ibid.

39 Cavallaro, *To the Stars*, 313–19.

40 Elizabeth Howell, "Retired NASA Astronaut Janet Kavandi Brings Inclusion to Sierra Space Missions," Space.com, March 19, 2023, https://www.space.com/womens-history-month-astronaut-janet-kavandi-sierra-space; "Paths to the New Age of Spaceflight with Janet Kavandi (Episode 086, Kathy Sullivan Explores)," Kathy Sullivan, November 24, 2022, https://www.kathysullivanexplores.com/podcast/paths-to-the-new-age-of-spaceflight-with-janet-kavandi.

41 Jillian Kramer, "An Astronaut Who Built Paths to Space for Other Women," *New York Times*, October 2, 2019.

42 "Pamela Melroy—Her STEM Story," undated, https://www.samuseum.sa.gov.au/visit/her-story-online/pamela-melroy.

43 Jennifer Levasseur, "In Her Own Orbit," March 22, 2022, https://airandspace.si.edu/air-and-space-quarterly/spring-2022/her-own-orbit.

44 Cavallaro, *To the Stars*, 394–402.

45 Pamela Melroy, conversations with author over time, 2003–2019.

46 Torsten Kriening, "Space Café Podcast Episode 016 Featuring Pamela Melroy Is Now Available," SpaceWatch.Global, November 24, 2020, https://spacewatch.global/2020/11/space-cafe-podcast-episode-016-featuring-pamela-melroy-is-now-available/.

47 https://www.nasa.gov/podcasts/small-steps-giant-leaps/small-steps-giant-leaps-episode-145-pam-melroy-behind-the-strategy/.

48 Juliet Homes, "From Star Trek Posters to NASA: The Journey of Pamela Melroy '83," Wellesley, May 26, 2023, https://www.wellesley.edu/news/from-star-trek-posters-to-nasa-the-journey-of-pamela-melroy-83.

49 Cavallaro, *To the Stars*, 389–94.

50 Susan Still Kilrain, interview with author, March 1, 2023.

51 Ibid.

52 N2K Networks, "Forging Your Journey with NASA Astronaut Susan Kilrain," Deep Space, https://space.n2k.com/podcasts/t-minus/ds55

53 Stephen Luntz, "Astronaut Susan Kilrain Tells Her Story Of Piloting Space Shuttle In Controversial Place," IFLScience, March 8, 2023, https://www.iflscience.com/astronaut-susan-kilrain-tells-her-story-of-piloting-space-shuttle-in-controversial-place-67867.

54 Maria Elizabeth Kallukaren, "'You Can Be Girlie, and Still Do STEM,' Says Former Astronaut," *Gulf News*, November 11, 2022, https://gulfnews.com/friday/art-people/you-can-be-girlie-and-still-do-stem-says-former-astronaut-1.91540039.

55 Kilrain interview.

56 N2K Networks, *Forging Your Journey*.

57 NASA News Release 95-045, "Biennial Astronaut Selection Process Begins," June 15, 1995; NASA News Release 96-84, "NASA Selects Astronaut Class of 1996," May 1, 1996.

58 Several astronauts commented on this situation in interviews with the author in 2023.

59 "Yvonne Cagle, UW Medical School Grad Turns Astronaut," *Columns, The University of Washington Alumni Magazine*, September 1996, www.washington.edu/alumni/columns/sept96/cagle.html; Betty Kaplan Gubert, Miriam Sawyer, and Caroline M. Fannin, *Distinguished African Americans in Aviation and Space Science* (Westport, CT: Oryx Press, 2002), 63–65.

60 "GSS Exclusive: 'Hidden Figure' Katherine Johnson Speaks to NASA's Dr. Yvonne Cagle," Fordham GSS, February 17, 2017, https://gss.news.fordham.edu/events/gss-exclusive-hidden-figure-katherine-johnson-speaks-nasas-dr-yvonne-cagle/.

61 Kevin Hymel, "From Flight Surgeon to Astronaut: Air Force Colonel (Dr.) Yvonne Cagle," February 22, 2018, https://www.airforcemedicine.af.mil/DesktopModules/ArticleCS/Print.aspx?PortalId=1&ModuleId=5872&Article=1448142; Yvonne Cagle, "Poetry of Space on Earth," TED Talk, November 27, 2018, https://www.ted.com/talks/yvonne_cagle_poetry_of_space_on_earth.

62 "Mission: 'Space for All' Yvonne Cagle," MIT News, May 21, 2015, https://news.mit.edu/2015/mission-space-for-all-yvonne-cagle-0521.

63 "HSF—STS-107 Crew Interviews—Laurel Clark," June 10, 2002, https://web.archive.org/web/20041113141500/http://spaceflight.nasa.gov/shuttle/archives/sts-107/crew/intclark.html.

64 "STS 107 Shuttle Press Kit: Providing 24/7 Space Science Research," December 16, 2002, https://ntrs.nasa.gov/citations/20030011376.

65 Chris Kidler, "Columbia's Astronauts Find Small Miracles of Life and Light," Space.com, January 29, 2013, https://www.space.com/19468-shuttle-columbia-final-flight-sts107-small-miracles.html.

66 "HSF—STS-107 Crew Interviews—Laurel Clark."

67 Kathy Sawyer, "Shuttle Astronaut's Videotape Recovered: Recording Ends 4 Minutes Before Data Suggest Reentry Trouble Began," *Washington Post*, February 26, 2003; Associated Press, "Columbia Crew Cheerful on Tape before Disaster, *Tucson Citizen*, March 1, 2003.

68 Laurel Clark, "Laurel Clark Letter Home," January 31, 2003, https://web.archive.org /web/20070313102022/http://racine.wi .net/clarkletter.html."Laurel Clark Obituary," Text, Legacy.com, February 1, 2003, https://www.legacy.com/link.asp?i =ls000000764391; Timothy Dwyer, "Celebrating an Explorer's Life: Astronaut's Passion Recalled at Funeral," *Washington Post*, March 11, 2003; Alan Cooperman, "Pushing Past the Barriers With a Smile: Voyage Into Space Was the Fruit of Navy Commander's Labors and Dreams," *Washington Post*, February 7, 2003.

69 Patrick Paolantonio, "Shuttle Columbia Tragedy 20 Years Later, Husband Talks Loss and Discovery," WISN, February 3, 2023, https://www.wisn.com/article /shuttle-columbia-tragedy-20-years-later -husband-talks-loss-and-discovery /42750278; Ashley Strickland, "Families of the Columbia Crew Members Keep Their Memories Alive," CNN, April 12, 2024, https://www.cnn.com/2024/04/12 /world/nasa-columbia-shuttle-crew -families-scn/index.html.

70 Joan Higginbotham, "An Astronaut's Inspiring and Winding Road to Space," TED Talk, October 2019, https://www.ted .com/talks/joan_higginbotham_an _astronaut_s_inspiring_and_winding _road_to_space.

71 Adam Lowenstein, "Alumna Joan Higginbotham Builds Successful, Varied Post-NASA Career," *Florida Tech News*, May 20, 2024, https://news.fit.edu/alumni /alumna-joan-higginbotham-builds -successful-varied-post-nasa-career/.

72 Higginbotham, "An Astronaut's Road."

73 Todd Halvorson, "From KSC Worker to Astronaut: Higginbotham," *Florida Today*, October 17, 2007; Stephen Withers, "iTWire—Shuttle Astronaut Quits," November 22, 2007, https://itwire.com /science-news/space/shuttle-astronaut -quits.html.

74 Cavallaro, *To the Stars*, 339–45.

75 Higginbotham, "An Astronaut's Road."

76 Gordon J. Bernhardt, "Sandra H. Magnus: A Space of One's Own," Profiles in Success, March 2014, https://www .profilesinsuccess.com/profile/sandra -magnus; NASA, "Sandra Magnus ISS Expedition 18 Preflight Interview," September 24, 2008, nasa.gov/mission _pages/station/expeditions/expedition18 /magnus_interview.html.

77 Sandra Magnus, interview with author, February 22, 2023.

78 Ibid.

79 Sandra Magnus, "Sandra Magnus' Expedition 18 Journal," 2009; Sandra Magnus, "ISS 18: Sandy Magnus on Cooking in Space," http://www .collectspace.com/ubb/Forum30/HTML /000767.html; Tariq Malik, "Astronaut Chef Redefines Cooking on High," Space .com, March 27, 2009, https://www.space .com/6501-astronaut-chef-redefines -cooking-high.html.

80 NASA, "Sandra Magnus STS-135 Preflight Interview," 2011.

81 USNA, "USNA Notable Graduates, Lisa Nowak, NASA Astronaut," 2017, https:// www.usna.edu/Notables/astronauts /1985nowak.php.

82 "Preflight Interview: Lisa Nowak, STS-121," August 11, 2005, https://web.archive.org /web/20210610022904/https://www.nasa .gov/vision/space/preparingtravel/sts121 _interview_nowak.html Wayback Machine.

83 Lisa Nowak, interview with author, July 2, 2023.

84 Anna Sale and Jonathan Clark, "An Astronaut's Husband, Left Behind," Slate Magazine, https://slate.com/transcripts/M WJXMFZPUnE2aVVaTG8zVnJacVIsNX- hzQIVJb0I5QIYzSUFiVk9rSmV2OD0=.

85 Heide Stefanyshyn-Piper, interview with author, April 2, 2023; milkywaykiwi, "Inspiring Tomorrow's Innovators: The Heidemarie Stefanyshyn-Piper Story," https://milky-way.kiwi/space-news /heidemarie-stefanyshyn-piper/.

86 Stefanyshyn-Piper interview.

87 Ibid.

88   Ann Parson, "Scientist at Work: Peggy Whitson Testing Limits, 220 Miles above Earth," *New York Times*, September 5, 2006.

89   Amy Held, "'American Space Ninja' Back on Earth after Record-Breaking Flight," *NPR*, September 3, 2017, https://www.npr.org/sections/thetwo-way/2017/09/03/548295156/-american-space-ninja-peggy-whitson-back-on-earth-after-record-breaking-flight.

90   Parson, "Scientist at Work"; Cavallaro, *To the Stars*, 402–10.

91   Ibid.

92   Bianca Brosh, "America's Most Experienced Astronaut Was Rejected 4 Times before Joining NASA," *MSNBC*, August 7, 2023, https://www.msnbc.com/know-your-value/career-growth/america-s-most-experienced-astronaut-was-rejected-4-times-joining-n1306953.

93   Cavallaro, *To the Stars*, 409.

94   NASA News Release 18-054, "Record-Setting NASA Astronaut Peggy Whitson Retires," June 15, 2008, https://www.nasa.gov/news-release/record-setting-nasa-astronaut-peggy-whitson-retires/.

95   Parson, "Scientist at Work."

96   Preflight Interview: Stephanie Wilson, STS-121 (2005), August 11, 2005, nasa.gov/vision/space/preparingtravel/sts121_interview_wilson.html.

97   Cavallaro, *To the Stars*, 333–39.

98   Ibid.; NASA, "Stephanie Wilson: Becoming an Astronaut Kicking and Swimming," November 20, 2006, nasa.gov/astronauts/s_wilson_profile.html.

99   Cavallaro, *To the Stars*, 333–39.

100 NASA News Release 20-128, "NASA Names Artemis Team of Astronauts Eligible for Early Moon Missions," December 9, 2020, https://www.nasa.gov/news-release/nasa-names-artemis-team-of-astronauts-eligible-for-early-moon-missions/; "The Artemis II Astronauts," podcast, April 7, 2023, https://www.nasa.gov/podcasts/houston-we-have-a-podcast/the-artemis-ii-astronauts/.

101 NASA News Release 21-035, "Acting NASA Administrator Statement on Agency FY 2022 Discretionary Request," April 9, 2021, https://www.nasa.gov/news-release/acting-nasa-administrator-statement-on-agency-fy-2022-discretionary-request/.

102 Cavallaro, *To the Stars*, 336–37.

103 Robert Stanton, "Astronaut Waits for Chance to Blast Off—Stephanie Wilson," *Houston Chronicle*, September 5, 2002.

104 NASA News Release 97-97, "Biennial Astronaut Selection Process Begins," May 14, 1997.

105 "Tracy Caldwell-Dyson—Personal Essay," undated, https://www.nasa.gov/people/tracy-caldwell-dyson/; Preflight Interview, Tracy Caldwell Dyson, ISS Expedition 23, March 23, 2010, https://www.nasa.gov/mission_pages/station/expeditions/expedition23/dyson_interview.html.

106 Cavallaro, *To the Stars*, 223–29.

107 Career Girls.org, "Career Advice from Tracy Caldwell Dyson, Astronaut," June 1, 2015, https://www.careergirls.org/role-models/astronaut-1/.

108 Bianca Bosker, "NASA Astronaut Tracy Caldwell Dyson on the Challenges Women Face in Space," *HuffPost*, July 19, 2011, https://www.huffpost.com/entry/nasa-astronaut-tracy-caldwell-dyson_n_904005; "Focusing on the Big Picture—from 250 Miles above the Earth," *Orange County Register* (blog), April 14, 2020, https://www.ocregister.com/2020/04/14/focusing-on-the-big-picture-from-250-miles-above-the-earth/.

109 "Astronaut Moments: Tracy Dyson," YouTube, March 19, 2024, https://www.youtube.com/watch?v=UeMB1BunMsA; Career Girls.org, "Career Advice."

110 Tracy Caldwell Dyson, "Living in Space," October 13, 2020, https://www.nasa.gov/podcasts/small-steps-giant-leaps/small-steps-giant-leaps-episode-46-iss-20-living-in-space/.

111 Kathy Sawyer, "12 Years After Blast, Perseverance Pays Off," *Washington Post*, January 21, 1998; William Harwood, "McAuliffe's Backup Will Go to Space in 2004," *Washington Post*, April 13, 2002.

112 Richard Luscombe, "Teacher's Shuttle Voyage Fulfils Dream," *The Guardian*, July 19, 2007, sec. World news, https://

www.theguardian.com/world/2007/jul/19
/spaceexploration.usa.

113 "Barbara Morgan: No Limits," Idaho Public
Television, 2008, https://video.idahoptv
.org/video/idaho-public-television
-specials-barbara-morgan-no-limits/ and
Barbara Morgan, conversation with
author, September 20, 2024.

114 Experience listed on her Teacher-in-
Space application, referenced in the
Idaho Public Television No Limits special
program cited above.

115 Cavallaro, To the Stars, 346–54.

116 John Schwartz, "Astronaut Teaches in
Space, and Lesson Is Bittersweet," New
York Times, August 15, 2007.

117 Special Collections and University
Archives et al., "Patricia Hilliard Robertson: A Tribute to IUP's Astronaut," Indiana
University of Pennsylvania, https://www
.iup.edu/library/departments/archives
/digital-projects-and-exhibits/patricia
-hilliard-robertson.html. Collection
includes a memory book made by her
colleagues at NASA Johnson Space
Center.

118 Cavallaro, To the Stars, 411–19.

119 Ibid., 413.

120 Jennifer M. Ross-Nazzal, "Sunita L.
Williams Oral History," September 8,
2015, https://historycollection.jsc.nasa
.gov/JSCHistoryPortal/history/oral
_histories/ISS/WilliamsSL/williamssl.htm.

121 Ibid.

122 NASA News Release 15-148, "NASA
Selects Astronauts for First U.S.
Commercial Spaceflights," July 9, 2015,
https://www.nasa.gov/news-release/nasa
-selects-astronauts-for-first-u-s
-commercial-spaceflights-2/.

123 NASA News Release 18-067, "NASA
Assigns Crews to First Test Flights,
Missions on Commercial Spacecraft,"
August 3, 2018, https://www.nasa.gov
/news-release/nasa-assigns-crews-to
-first-test-flights-missions-on-commercial
-spacecraft/.

124 Ross-Nazzal, "Sunita L. Williams Oral
History."

125 NASA News Release J99-17, "Application
Deadline Near for NASA Astronaut
Selection," May 17, 1999.

126 The four women were Kathy Sullivan,
Kathy Thornton, Nancy Currie, and
Megan McArthur.

127 Matthew Chin and Judy Lin, "10 Questions for Bruin Astronaut Megan
McArthur," May 6, 2009, https://samueli
.ucla.edu/10-questions-for-bruin
-astronaut-megan-mcarthur/; Cavallaro,
To the Stars, 361–68.

128 Megan McArthur, "A NASA Astronaut's
Lessons on Fear, Confidence and
Preparing for Spaceflight," TED Talk,
November 2020, https://www.ted.com
/talks/megan_mcarthur_a_nasa
_astronaut_s_lessons_on_fear
_confidence
_and_preparing_for_spaceflight.

129 "Robots in Space," Blaze and the Monster
Machines, Season 4, Episode 9,
October 4, 2018, https://www.imdb.com
/title/tt9083242/?ref_=ttep_ep_9.

130 Cavallaro, To the Stars, 354–61.

131 Elizabeth Howell, "Saving a Spaceman
from Drowning," NBC News, July 20,
2017, https://www.nbcnews.com/mach
/science/saving-spaceman-drowning
-ncna784431.

132 Robert Pearlman, "'Made in Space!'
Astronaut Sews Dinosaur Toy from Space
Station Scraps," CollectSpace.com,
September 27, 2013, http://www
.collectspace.com/news/news-092713c
-astronaut-sews-dinosaur-space.html;
Robert Pearlman, "Astronaut-Artist Karen
Nyberg Unveils Quilt Portraits of First
Women to Fly in Space," CollectSpace
.com, August 17, 2023, http://www
.collectspace.com/news/news-081723a
-astronaut-karen-nyberg-fabric-art
-women-in-space.html.

133 NASA News Release 13-320, "NASA
Astronaut Karen Nyberg Invites Quilters
to Contribute a Star Block," October 31,
2013, https://www.nasa.gov/news-release
/nasa-astronaut-karen-nyberg-invites
-quilters-to-contribute-a-star-block/;
Robert Pearlman, "Astronaut's Sewn-in-
Space Star Shines at Quilt Festival,"
CollectSpace.com, November 4, 2014,
http://www.collectspace.com/news/news
-110414c-astronaut-quilt-festival-stars.

html; Lisa Boone, "NASA Astronaut Karen Nyberg Hosts Quilting Bee from Space Station," *Los Angeles Times*, October 31, 2013.

134 Elizabeth Howell, "Astronaut Captures Space Station View in a Quilting Pattern," *Forbes*, April 25, 2022, https://www.forbes.com/sites/elizabethhowell1/2022/04/25/astronaut-captures-space-station-view-in-a-quilting-pattern.

135 Pearlman, "Quilt Portraits of First Women to Fly in Space."

136 Anna Moselein, "What Is It Like to Actually Parent from Space?," *Glamour*, September 4, 2020, https://www.glamour.com/story/away-netflix; "Out of This World: Retired Astronaut Karen Nyberg Reflects," *UND Today*, November 30, 2021, https://blogs.und.edu/und-today/2021/11/out-of-this-world-retired-astronaut-karen-nyberg-reflects/; *Astronaut*, 2021, https://www.youtube.com/watch?v=QUjFhIms2WM.

137 Moselein, "What Is It Like to Actually Parent from Space?"

138 Nicole Stott, correspondence with author, March 2024.

139 Cavallaro, *To the Stars*, 230–41.

140 From Stott's website, www.npsdiscovery.com.

Chapter 4

1 Elizabeth Adell Cook, Sue Thomas, and Clyde Wilcox, eds., *The Year of the Woman: Myths and Realities* (Boulder, CO: Westview Press, 1994).

2 Glen R. Asner and Stephen J. Garber, *Origins of 21st-Century Space Travel, 1994–2001* (Washington, DC: National Aeronautics and Space Administration, 2019); Stephen Garber and Glen Asner, "21st Century Space Travel: Space Policy and the Vision for Space Exploration, 1999–2004," NASA podcast, n.d., https://www.nasa.gov/podcasts/houston-we-have-a-podcast/21st-century-space-travel/.

3 "President Bush Announces New Vision for Space Exploration Program," January 14, 2004, https://georgewbush-whitehouse.archives.gov/news/releases/2004/01/20040114-3.html; NASA, "The Vision for Space Exploration," Febru-

ary 2004, https://www.nasa.gov/wp-content/uploads/2023/01/55583main_vision_space_exploration2.pdf; David E. Sanger and Richard W. Stevenson, "Bush Backs Goal of Flight to Moon to Establish Base," *New York Times*, January 15, 2004; Joseph N. Tatarewicz, "The 'Vision for Space Exploration' of President George W. Bush, Space Science, and U.S. Space Policy," *Futures* 41, no. 8 (October 1, 2009): 531–40; Guy Gugliotta, "Bush's 'Vision' for Space Clouded," *NBC News*, May 1, 2004, https://www.nbcnews.com/id/wbna4875321.

4 Jason Davis, "'Apollo on Steroids': The Rise and Fall of NASA's Constellation Moon," Planetary Society, August 1, 2016, https://www.planetary.org/articles/20160801-horizon-goal-part-2.

5 Augustine Committee, "Report of the Review of U.S. Human Spaceflight Plans Committee, Seeking a Spaceflight Program Worthy of a Great Nation," October 2009, https://www.nasa.gov/wp-content/uploads/2015/01/617036main_396093main_HSF_Cmte_FinalReport.pdf.

6 Mark Stencel, "NASA's Flight Plan Gets Small Course Corrections," *NPR*, April 15, 2010, https://www.npr.org/2010/04/15/126023150/nasas-flight-plan-gets-small-course-corrections; Jonathan Amos, "Obama Cancels Moon Return Project," *BBC News*, February 1, 2010, http://news.bbc.co.uk/2/hi/science/nature/8489097.stm; Joel Achenbach, "NASA's Mission Improbable (Asteroid Mission)," *Washington Post* (blog), August 17, 2013, http://www.washingtonpost.com/sf/national/2013/08/17/nasas-mission-improbable/.

7 Office of Management and the Budget, "Terminations, Reductions, and Savings, Budget of the U.S. Government, Fiscal Year 2011"; William Harwood, "Obama Kills Moon Program, Endorses Commercial Space," February 1, 2010, https://spaceflightnow.com/news/n1002/01nasabudget/; Jason Davis, "Space in Transition: How Obama's White House Charted a New Course," Planetary Society, August 22, 2016, https://www.planetary.org/articles/20160822-horizon-goal-part-3; Kenneth Chang, "Obama Vows Renewed Space Program," *New York Times*, April 15, 2010.

8   NASA, "NASA's Commercial Crew Program Essentials," no date, https://www.nasa.gov/humans-in-space/commercial-space/commercial-crew-program/commercial-crew-program-essentials/; W. H. Lambright, "NASA, Industry, and the Commercial Crew Development Program: The Politics of Partnership," in *NASA Spaceflight*, ed. Roger D. Launius and Howard McCurdy (Cham, Switzerland: Palgrave Macmillan, 2017), 349–78; NASA, "Commercial Crew Program Overview," n.d., https://www.nasa.gov/humans-in-space/commercial-space/commercial-crew-program/commercial-crew-program-overview/; NASA, "Commercial Crew Program Press Kit," no date, https://www.nasa.gov/commercial-crew-program-press-kit/.

9   William Harwood, "Trump Budget Blueprint Focuses on Deep Space Exploration, Commercial Partnerships," Space Flight Now, March 16, 2017, https://spaceflightnow.com/2017/03/16/trump-budget-blueprint-focuses-on-deep-space-exploration-commercial-partnerships/; Tariq Malik, "Trump's NASA Budget Eliminates Crewed Mission to Asteroid," *Scientific American*, March 16, 2017, https://www.scientificamerican.com/article/trump-rsquo-s-nasa-budget-eliminates-crewed-mission-to-asteroid/; Jeff Foust, "NASA Closing out Asteroid Redirect Mission," *SpaceNews* (blog), June 14, 2017, https://spacenews.com/nasa-closing-out-asteroid-redirect-mission/.

10   NASA News Release 17-097, "New Space Policy Directive Calls for Human Expansion across Solar System," 2017, 17–097, https://www.nasa.gov/news-release/new-space-policy-directive-calls-for-human-expansion-across-solar-system/.

11   "Artemis Plan: NASA's Lunar Exploration Program Overview" (NASA, September 2020), https://www.nasa.gov/wp-content/uploads/2020/12/artemis_plan-20200921.pdf; NASA News Release 20-092, "NASA Publishes Artemis Plan to Land First Woman, Next Man on Moon in 2024," September 21, 2020, https://www.nasa.gov/news-release/nasa-publishes-artemis-plan-to-land-first-woman-next-man-on-moon-in-2024/; Chelsea Gohd, "NASA to Land 1st Person of Color on the Moon with Artemis Program," April 9, 2021, https://www.space.com/nasa-sending-first-person-of-color-to-moon-artemis. "NASA's First [Artemis] Flight with Crew Important Step on Long-Term Return to the Moon," Missions to Mars, August 27, 2018, https://www.nasa.gov/missions/artemis/nasas-first-flight-with-crew-important-step-on-long-term-return-to-the-moon-missions-to-mars/.

12   Miriam Kramer, "Biden Supports NASA's Artemis Program to Send Humans Back to the Surface of the Moon," *Axios*, February 4, 2021, https://www.axios.com/2021/02/04/artemis-program-biden-moon; Danielle Haynes, "White House: Biden Supports NASA's Artemis Program," UPI, February 4, 2021, https://www.upi.com/Top_News/US/2021/02/04/White-House-Biden-supports-NASA-program-to-send-astronauts-to-moon/9021612463940/; NASA News Release 21-035, "Acting NASA Administrator Statement on Agency FY 2022 Discretionary Request," April 9, 2021, https://www.nasa.gov/news-release/acting-nasa-administrator-statement-on-agency-fy-2022-discretionary-request/; Gohd, "NASA to Land 1st Person of Color"; Ashley Strickland, "NASA's Artemis Program Will Land the First Person of Color on the Moon," *CNN*, April 9, 2021, https://www.cnn.com/2021/04/09/world/nasa-artemis-person-of-color-crew-scn/index.html; White House, "United States Space Priorities Framework" (White House, December 2021); Joey Roulette, "Trump's Moon Program Survived a Transfer of Power, So What's Next?," *The Verge*, March 12, 2021, https://www.theverge.com/2021/3/12/22323621/trump-moon-program-artemis-biden-nasa-timeline.

13   NASA News Release 18-067, "NASA Assigns Crews to First Test Flights, Missions on Commercial Spacecraft," August 3, 2018, https://www.nasa.gov/news-release/nasa-assigns-crews-to-first-test-flights-missions-on-commercial-spacecraft/.

14   Kenneth Chang, "NASA Delays Artemis Astronaut Moon Missions," *New York Times*, January 9, 2024; Kenneth Chang, "NASA Artemis Moon Missions Delayed Until 2026 and 2027," *New York Times*, December 5, 2024; NASA News Release

24-145, "NASA Shares Orion Heat Shield Findings, Updates Artemis Moon Missions," December 5, 2024, https://www.nasa.gov/news-release/nasa-shares-orion-heat-shield-findings-updates-artemis-moon-missions/.

15 "Artemis Plan: NASA's Lunar Exploration Program Overview," NASA, September 2020, https://www.nasa.gov/wp-content/uploads/2020/12/artemis_plan-20200921.pdf; "Artemis," no date, https://www.nasa.gov/humans-in-space/artemis/.

16 NASA News Release 23-040, "NASA Names Astronauts to Next Moon Mission, First Crew Under Artemis," April 3, 2023, https://www.nasa.gov/news-release/nasa-names-astronauts-to-next-moon-mission-first-crew-under-artemis/; "The Artemis II Astronauts," NASA podcast, April 7, 2023, https://www.nasa.gov/podcasts/houston-we-have-a-podcast/the-artemis-ii-astronauts/; NASA, "Our Artemis Crew," n.d., https://www.nasa.gov/feature/our-artemis-crew/; Kenneth Chang, "NASA Names Diverse Astronaut Crew for Artemis II Moon Mission," *New York Times*, April 3, 2023.

17 SpaceX sent a commercial crew of two women and two men on the Polaris Dawn mission around the Moon in 2024, making them the first humans near the Moon since 1972.

18 Roger Handberg, "Déjà vu as Space Policy," *Space Review*, April 1, 2019, https://www.thespacereview.com/article/3685/1.

19 Ellen Ochoa, interview with author, March 15, 2023; Anne Roemer, Johnson Space Center Human Relations Office, interview with author, May 8, 2023.

20 NASA, "Want to Be an Astronaut? You'll Need These 4 Expeditionary Skills," n.d., https://www.nasa.gov/stem-content/want-to-be-an-astronaut-youll-need-these-4-skills/; Albert W. Holland and James J. Picano, psychologists in NASA Johnson Space Center's Space Medicine Division, Behavioral Health and Performance Operations group, interviews with author, May 11, 2023, and April 3, 2024.

21 Katie Hiler, "NASA's New Class of Astronauts Gives Parity to Men and Women," *New York Times*, June 19, 2013.

22 Anne Roemer interview; Duane Ross, former head of Astronaut Selection in NASA JSC Human Relations office, interview with author, May 18, 2023.

23 NASA Media Advisory M11-234, "NASA Kicks Off Application Process For New Astronauts," November 14, 2011, https://www.nasa.gov/news-release/nasa-kicks-off-application-process-for-new-astronauts/; Adam Hadhazy, "Popular Science Q&A: How NASA Selected The 2013 Class Of Astronauts," *Popular Science*, January 31, 2013, https://www.popsci.com/technology/article/2013-01/popsci-qampa-choosing-2013-class-nasa-astronauts/.

24 Hadhazy, "Popular Science Q&A."

25 NASA News Release 21-167, "NASA Selects New Astronaut Recruits to Train for Future Missions," December 6, 2021, https://www.nasa.gov/news-release/nasa-selects-new-astronaut-recruits-to-train-for-future-missions/.

26 NASA News Release 15-234, "Be an Astronaut: NASA Accepting Applications for Future Explorers," December 14, 2015, https://www.nasa.gov/news-release/be-an-astronaut-nasa-accepting-applications-for-future-explorers/.

27 NASA News Release 03-183, "NASA Opens Applications for New Astronaut Class," May 27, 2003, https://www3.nasa.gov/home/hqnews/2003/may/HQ_03183_astronaut_applications.html.

28 NASA News Release 03-016, "NASA Administrator Announces Out-of-This-World Opportunity for Teachers," January 21, 2003, https://www3.nasa.gov/audience/formedia/archives/MP_Archive_2003_page2.html.

29 NASA News Release 04-152, "NASA Introduces The Next Generation Of Explorers," May 6, 2004.

30 Profiles of the astronauts are informed by NASA sources in the public domain and these well-researched books: Umberto Cavallaro, *To the Stars: Women Spacefarers' Legacy* (Chichester, UK: Springer-Praxis Books, 2023); David Shayler and Colin Burgess, *NASA'S Scientist-Astronauts* (Chichester, UK: Springer-Praxis, 2007); and David J. Shayler and Ian Moule, *Women in Space: Following*

Valentina (Chichester, UK: Springer-Praxis, 2005).

31 Dottie Metcalf-Lindenburger, "Dottie Metcalf-Lindenburger," n.d., https://www.nasa.gov/people/dottie-metcalf-lindenburger/; Cavallaro, *To the Stars*, 153–59.

32 Dottie Metcalf-Lindenburger, interview with author, March 26, 2023.

33 Ibid.

34 Andrea Leinfelder, "Houston-Born Astronaut Shannon Walker Ready for Her Flight to the Space Station," *Houston Chronicle*, November 12, 2020; "Meet Shannon Walker, Crew-1 Mission Specialist," YouTube, November 15, 2020, https://www.youtube.com/watch?v=9zj5-KzHJrY.

35 Cavallaro, *To the Stars*, 420–28.

36 NASA News Release J21-002, "Shannon Walker to Become First Native Houstonian to Command International Space Station," April 13, 2021, https://www.nasa.gov/news-release/shannon-walker-to-become-first-native-houstonian-to-command-international-space-station/; Ben Evans, "Walker Becomes Third Female ISS Skipper, Expedition 65 Set For Three Commanders," *AmericaSpace*, April 16, 2021, https://www.americaspace.com/2021/04/16/walker-becomes-third-female-iss-skipper-expedition-65-set-for-three-commanders/.

37 "Shannon Walker Neighborhood Library Ribbon Cutting 2024," YouTube, June 17, 2024, https://www.youtube.com/watch?v=gl5ST9gg5Gw.

38 NASA News Release H07-196, "NASA Opens Applications for New Astronaut Class," September 2007.

39 "Help Wanted: Astronauts," September 18, 2007, http://www.nasa.gov/astronauts/recruit.html.

40 NASA News Release 09-149, "NASA Selects Nine New Astronauts for Future Space Exploration," June 29, 2009, https://www.nasa.gov/news-release/nasa-selects-nine-new-astronauts-for-future-space-exploration/.

41 Cavallaro, *To the Stars*, 173–80.

42 Astronaut Biography Serena M. Aunon-Chancellor, https://www.nasa.gov/wp-content/uploads/2016/02/aunon.pdf.

43 Cavallaro, *To the Stars*, 173–80.

44 NASA, "In Their Own Words: Serena M. Aunon," www.nasa.gov/astronauts/2009_aunon.html, June 29, 2009, https://web.archive.org/web/20091028205915/http://www.nasa.gov/astronauts/2009_aunon.html.

45 The primary source of information for this profile is Jeanette Epps, NASA Johnson Space Center Oral History Project interview, February 16, 2012, https://historycollection.jsc.nasa.gov/JSCHistoryPortal/history/history_collection.htm.

46 NASA News Release 18-004, "NASA Announces Updated Crew Assignments for Space Station Missions," January 18, 2018, https://www.nasa.gov/news-release/nasa-announces-updated-crew-assignments-for-space-station-missions/; Eric Berger, "NASA Has Pulled Jeanette Epps Just Months before Her First Flight," *Ars Technica*, January 19, 2018, https://arstechnica.com/science/2018/01/nasa-has-pulled-jeanette-epps-just-months-before-her-first-flight/.

47 Berger, "NASA Has Pulled Jeanette Epps"; Eric Berger, "After Being Pulled from a Spaceflight in January, Jeanette Epps Speaks Up," *Ars Technica*, June 22, 2018, https://arstechnica.com/science/2018/06/after-being-pulled-from-a-spaceflight-in-january-jeanette-epps-speaks-up/.

48 Epps, JSC interview.

49 Cavallaro, *To the Stars*, 165–73.

50 ESA, "Kate and the Moon," *Caves & Pangaea* (blog), September 16, 2021, https://blogs.esa.int/caves/2021/09/16/kate-and-the-moon/.

51 "UCSD Alumna First to Sequence DNA in Space," *San Diego Union-Tribune*, August 29, 2016, https://www.sandiegouniontribune.com/news/science/sdut-kate-rubins-dna-2016aug29-story.html.

52 Javier Orona, "Astronaut Commissions into the U.S. Army Reserve," https://www.army.mil/article/252002/astronaut_commissions_into_the_u_s_army_reserve.

53 NASA News Release 11-336, "NASA To Seek Applicants For Next Astronaut Candidate Class," October 3, 2011,

http://www.nasa.gov/home/hqnews/2011/oct/HQ_11-336_Astronaut_Candidate_Class.html.

54 Hadhazy, "Popular Science Q&A."

55 NASA News Release 13-177, "NASA Selects Next Generation of Space Explorers," June 17, 2013, https://www.nasa.gov/news-release/nasa-selects-next-generation-of-space-explorers-google-hangout-today/; Hiler, "NASA's New Class of Astronauts Gives Parity."

56 "NASA Holds First All-Female Spacewalk in 54 Years of Spacewalking," amightygirl.com, October 18, 2019, http://www.amightygirl.com/blog?p=23616.

57 Cavallaro, To the Stars, 251–59.

58 Frank Rubio spent 371 consecutive days in space in 2022–2023; two other US men, Mark Vande Hei and Scott Kelly, came close with 355 and 340 days respectively, and several male cosmonauts have exceeded a full year in space.

59 "NASA Holds First All-Female Spacewalk."

60 Elizabeth Howell, "NASA Astronaut Celebrates Women's Equality Day in Space at Midpoint of Record-Breaking Flight," Space.com, August 26, 2019, https://www.space.com/astronaut-christina-koch-womens-equality-day-in-space-2019.html.

61 Naval Academy Athletics, "Nicole Aunapu Mann Selected as 2024 NCAA Silver Anniversary Award Recipient," December 26, 2023, https://navysports.com/news/2023/12/26/womens-soccer-nicole-aunapu-mann-selected-as-2024-ncaa-silver-anniversary-award-recipient.aspx.

62 Kavi Mookherjee Amodt, "Q&A: Nicole Aunapu Mann Talks Moon Mission, Challenging Barriers to Diversity in Space Travel," Stanford Daily, January 12, 2021, https://stanforddaily.com/2021/01/12/qa-nicole-aunapu-mann-talks-moon-mission-challenging-barriers-to-diversity-in-space-travel/; Jeffrey Kluger, "Nicole Mann, the First Native American Woman in Space, on Gardening in Zero Gravity," Time, December 1, 2022, https://time.com/6238064/nicole-mann-interview-zero-gravity-gardening/.

63 Cavallaro, To the Stars, 486–95.

64 Hanneke Weitering, "All-Female Spacewalk Was Cancelled on 'My Recommendation,' Astronaut Anne McClain Says," Space.com, April 2, 2019, https://www.space.com/anne-mcclain-spacesuit-response-video.html; Matthew S. Schwartz, "NASA Scraps First All-Female Spacewalk for Want of a Medium-Size Spacesuit," NPR, March 26, 2019, http://www.capradio.org/news/npr/story?storyid=706779637; Eric Berger, "It's Unfortunate NASA Canceled the All-Female EVA, but It's the Right Decision," Ars Technica, March 26, 2019, https://arstechnica.com/science/2019/03/yes-nasa-has-canceled-an-all-female-spacewalk-no-its-not-a-conspiracy/.

65 Gonzaga Prep, "Prep Alumna Returns to Spokane: Anne McClain," n.d., https://www.gprep.com/apps/pages/index.jsp?uREC_ID=258129&type=d&pREC_ID=1790033; Mathew Callaghan, "Spokane Astronaut Anne McClain Shares Her Thoughts on Upcoming Artemis 1 Mission: 'There's Really Nothing Not to Be Excited About,'" Spokesman-Review, August 28, 2022.

66 "Meet Artemis Team Member Jessica Meir," YouTube, December 9, 2020, https://www.youtube.com/watch?v=u1stlxjQV-Q; "7 Questions for Astronaut Jessica Meir," Britannica.com, February 17, 2022, https://www.britannica.com/story/7-questions-for-astronaut-jessica-meir.

67 Cavallaro, To the Stars, 259–69.

68 "NASA Holds First All-Female Spacewalk."

69 "7 Questions for Astronaut Jessica Meir."

70 "Be an Astronaut."

71 NASA News Release 16-018, "Record Number of Americans Apply to #BeAnAstronaut at NASA," February 19, 2016; NASA News Release 17-054, "NASA's Newest Astronaut Recruits to Conduct Research off the Earth, for the Earth and Deep Space Missions," June 7, 2017, https://www.nasa.gov/news-release/nasas-newest-astronaut-recruits-to-conduct-research-off-the-earth-for-the-earth-and-deep-space-missions/.

72 Robert Pearlman, "'Turtles' on a Really High Post: NASA's New Astronauts Get Their Nickname," CollectSpace.com,

September 18, 2017, http://www
.collectspace.com/news/news-091817b
-turtles-astronaut-class-nickname.html.

73 NASA News Release 20-002, "NASA's
Newest Astronauts Ready for Space
Station, Moon, and Mars Missions,"
January 10, 2020, https://www.nasa.gov
/news-release/nasas-newest-astronauts
-ready-for-space-station-moon-and-mars
-missions/.

74 NASA, "Meet Artemis Astronaut Kayla
Barron," YouTube, December 9, 2020,
https://www.youtube.com/watch?v=N7mz
_xz5RN0; "Expedition 66 Astronaut Kayla
Barron Answers U.S. Navy, Student
Questions," YouTube, March 16, 2022,
https://www.youtube.com/watch?app
=desktop&v=TXqUKNLTUmw.

75 Cavallaro, To the Stars, 496–503.

76 Matt Carroll, "Penn State Graduate
Student Selected for NASA Astronaut
Program," Penn State University,
https://www.psu.edu/news/impact/story
/penn-state-graduate-student-selected
-nasa-astronaut-program/.

77 NASA News Release 24-108, "NASA
Decides to Bring Starliner Spacecraft
Back to Earth Without Crew," August 24,
2024, https://www.nasa.gov/news
-release/nasa-decides-to-bring-starliner
-spacecraft-back-to-earth-without-crew/;
NASA News Release 24-111, "CORREC-
TION: NASA's SpaceX Crew-9 Changes
Ahead of September Launch,"
August 30, 2024, https://www.nasa.gov
/news-release/correction-nasas-spacex
-crew-9-changes-ahead-of-september
-launch/.

78 NASA News Release 25-010, "NASA
Shares SpaceX Crew-11 Assignments for
Space Station Mission," March 27, 2025.

79 "Meet NASA's New Mighty Women
Astronauts"; Robin Wright, "Jasmin
Moghbeli, Badass Astronaut," The New
Yorker, July 2, 2017, https://www
.newyorker.com/news/news-desk/jasmin
-moghbeli-americas-badass-immigrant
-astronaut.

80 MIT Athletics, "Jasmin Moghbeli '05
Graduates From NASA Astronaut
Candidate Training Program," Massachusetts Institute of Technology, January 16,
2020, https://mitathletics.com/news/2020
/1/16/1_16_2020_6821.aspx.

81 Loral O'Hara, "From Space to Seafloor:
The Joy Is in the Journey," TEDxNewBed-
ford, December 2, 2015, https://www
.youtube.com/watch?v=fgU0vhnVoYI.

82 Shafiq Najib, "NASA Astronaut Jessica
Watkins Makes History as First Black
Woman on International Space Station
Mission," People, April 27, 2022,
https://people.com/human-interest/nasa
-astronaut-jessica-watkins-makes-history
-as-first-black-woman-on-international
-space-station-mission/.

83 Dante A Ciampaglia, "10 Questions for
Jessica Watkins," Time for Kids, December 8, 2017, https://www.timeforkids.com
/g34/10-q-jessica-watkins/; "Meet Artemis
Team Member Jessica Meir"; NASA, "To
the Stars with Jessica Watkins," NASA
podcast, April 11, 2023, https://www.nasa
.gov/podcasts/curious-universe/to-the
-stars-with-astronaut-jessica-watkins/.

84 NASA Media Advisory M20-024,
"Explorers Wanted: NASA to Hire More
Artemis Generation Astronauts,"
February 11, 2020, https://www.nasa.gov
/news-release/explorers-wanted-nasa-to
-hire-more-artemis-generation
-astronauts/; NASA News Release
20-022, "#BeAnAstronaut: NASA Seeks
Applicants to Explore Moon, Mars,"
March 2, 2020, https://www.nasa.gov
/news-release/beanastronaut-nasa-seeks
-applicants-to-explore-moon-mars/.

85 In 2023, needing more time to complete
work on the rocket and landing vehicle,
NASA postponed that target date to
2025. The Artemis III date was again
postponed to 2027 at the earliest after
Artemis II was postponed to 2026.

86 "NASA Selects New Astronaut Recruits to
Train for Future Missions."

87 NASA Media Advisory M24-021, "New
Artemis Generation Astronauts to
Graduate, NASA Sets Coverage,"
February 7, 2024, https://www.nasa.gov
/news-release/new-artemis-generation
-astronauts-to-graduate-nasa-sets
-coverage/.

88 "NASA Opens Astronaut Applications as
Newest Class Graduates."

89 Stephanie Butzer and Deb Stanley,
"Colorado Woman Selected amid
Thousands as New NASA Astronaut
Candidate," Denver7, December 6, 2021,

https://www.denver7.com/news/local-news/colorado-woman-selected-amid-thousands-as-new-nasa-astronaut-candidate; CBS News Colorado, "Colorado's Nichole Ayers Aims To 'Execute That Mission Well' As New NASA Astronaut," December 7, 2021, https://www.cbsnews.com/colorado/news/colorado-nichole-ayers-nasa-astronaut/.

90   Mike Williams, "12 Questions for Nichole Ayers, Rice's Newest Astronaut," Rice University, December 7, 2021, https://news.rice.edu/news/2021/12-questions-nichole-ayers-rices-newest-astronaut.

91   NASA News Release 24-099, "NASA Shares Its SpaceX Crew-10 Assignments for Space Station Mission," August 1, 2024, https://www.nasa.gov/news-release/nasa-shares-its-spacex-crew-10-assignments-for-space-station-mission/.

92   "Christina Birch's Race History at Crossresults.Com," 2016, https://www.crossresults.com/racer/37532.

93   Betsy Welch, "Christina Birch's Journey from MIT Bioengineer to Olympic Hopeful," Velo (blog), November 21, 2019, https://velo.outsideonline.com/road/road-racing/christina-birchs-journey-from-mit-bioengineer-to-olympic-hopeful/.

94   Elizabeth Howell, "Splashdown Practice for Artemis 2 Moon Mission 'an Incredible Experience,' New NASA Astronaut Says," Space.com, March 13, 2024, https://www.space.com/artemis-2-moon-mission-splashdown-practice-christina-birch-astronaut-interview.splashdown.

95   Sandi Miller and Mit Department of Physics, "NASA's 10-Member Astronaut Candidate Class of 2021 Includes 3 MIT Alumni," SciTechDaily (blog), December 9, 2021, https://scitechdaily.com/nasa-10-member-astronaut-candidate-class-of-2021-includes-3-mit-alumni/.

96   "Deniz Burnham and Flying EVERY-THING," YouTube, December 5, 2023, https://www.youtube.com/watch?v=79s1YrxEnhE.

97   "Profiles in Professionalism: Lt. Deniz Burnham," January 20, 2022, https://www.navyreserve.navy.mil/News/Article-View-News/Article/2905528/profiles-in-professionalism-lt-deniz-burnham/https%3A%2F%2Fwww.navyreserve.navy.mil%2FNews%2FArticle-View-News%2FArticle%2F2905528%2Fprofiles-in-professionalism-lt-deniz-burnham%2F.

98   Katherine Connor, "Engineering Alumna Becomes Newest NASA Astronaut, Deniz Burnham," University of California San Diego, March 19, 2024, https://today.ucsd.edu/story/engineering-alumna-becomes-newest-nasa-astronaut.

99   "Faces of NPS: Lt. Cmdr. Jessica Wittner," n.d., https://www.npsfoundation.org//faces-of-nps/lt-cmdr-jessica-wittner; Fox Weather, "From Navy to NASA: Jessica Wittner 'Honored' by New Endeavor," https://www.foxweather.com/watch/play-55713b7f0000f48.

100   Nadjedja, "A Whole New World for Jessica," Caves & Pangaea (blog), September 29, 2023, https://blogs.esa.int/caves/2023/09/29/a-whole-new-world-for-jessica/; "AME Alum and Astronaut Awaits Her First Mission: Jessica Wittner," University of Arizona, March 8, 2024, https://news.engineering.arizona.edu/news/ame-alum-and-astronaut-awaits-her-first-mission.

101   Mike Wall, "Diversity Will Be Key to Artemis Moon-to-Mars Push, NASA Officials Say," Space.com, April 21, 2023, https://www.space.com/artemis-moon-mars-push-diversity-key.

102   NASA News Release 20-128, "NASA Names Artemis Team of Astronauts Eligible for Early Moon Missions," December 9, 2020, https://www.nasa.gov/news-release/nasa-names-artemis-team-of-astronauts-eligible-for-early-moon-missions/; "Meet the Artemis Team of NASA Astronauts Who Have a Chance to Walk on the Moon," CBS News, http://www.cbs.com/playlist?list=PL2aBZuCeDwlQqwFjezU4wJMqK11qpNpR-; White House, "Remarks by Vice President Pence at the 8th Meeting of the National Space Council," December 9, 2020, https://trumpwhitehouse.archives.gov/briefings-statements/remarks-vice-president-pence-8th-meeting-national-space-council-cape-canaveral-fl/.

103   Elizabeth Howell, "NASA Opens up Artemis Moon Missions to All Astronauts," Space.com, August 18, 2022, https://www.space.com/nasa-artemis-moon-missions-open-all-astronauts. Chief of the Astronaut Office Reid Wiseman made the announcement during a media briefing:

"NASA Briefing on Artemis Mission to the Moon," *C-SPAN*, August 5, 2022, https://www.c-span.org/video/?522191-1/nasa-briefing-artemis-mission-moon.

104 NASA News Release 24-038, "NASA Opens Astronaut Applications as Newest Class Graduates," March 5, 2024, https://www.nasa.gov/news-release/nasa-opens-astronaut-applications-as-newest-class-graduates/.

105 Tony Ho Tran, "Why Hasn't NASA Hired More Black and Brown Astronauts?," *Futurism*, March 24, 2022, https://futurism.com/nasa-black-brown-astronauts.

Chapter 5

1 David J. Shayler and Colin Burgess include a list of myriad technical assignments for the 1978 class in *NASA's First Space Shuttle Astronaut Selection: Redefining the Right Stuff* (Cham, Switzerland: Springer International, 2020), 198–200. These assignments immediately followed completion of the one-year-long basic training. A list for the most recent astronauts would look equally busy, although on different tasks.

2 Rhea Seddon, *Go for Orbit: One of America's First Women Astronauts Finds Her Space* (Murfreesboro, TN: Your Space Press, 2015), 338.

3 Margaret Rhea Seddon, Johnson Space Center (JSC) Oral History Project interview, May 20, 2010.

4 Kathryn D. Sullivan, Johnson Space Center (JSC) Oral History Project interview, May 10, 2007.

5 Seddon, *Go for Orbit*, 341–45.

6 Seddon, JSC interview.

7 Anna Fisher interview with author, August 8, 2023.

8 Janet Kavandi interview with author, April 20, 2023.

9 Peggy Whitson, Johnson Space Center (JSC) Oral History Project interview, October 7, 2015.

10 NASA News Release, "NASA Assigns Crew for Final Scheduled Space Shuttle Mission," September 18, 2009.

11 Jennifer M. Ross-Nazzal, ed., *Making Space for Women: Stories from Trailblaz-ing Women of NASA's Johnson Space Center* (College Station: Texas A&M University Press, 2022), 349–60.

12 Ibid.; National Research Council, *Preparing for the High Frontier: The Role and Training of NASA Astronauts in the Post-Space Shuttle Era* (Washington, DC: National Academies Press, 2011).

13 Seddon, *Go for Orbit*, 395–404.

14 David J. Shayler and Ian Moule, *Women in Space: Following Valentina* (Chichester, UK: Springer, 2005), 222–23.

15 Mary L. Cleave, NASA Johnson Space Center (JSC) Oral History Project interview, March 5, 2002.

16 Kathy Sullivan, "Waking up with Robin Williams," Kathy Sullivan Explores, March 2, 2022, https://www.kathysullivanexplores.com/podcast/tag/Memorable%20Moments. See video clip at https://www.youtube.com/watch?v=S7WJtQYU8i4.

17 Kathy Thornton interview with author, March 23, 2023.

18 Anna Fisher, NASA Johnson Space Center (JSC) Oral History Project interview, May 3, 2011; "No Room for Failure," *The Arrow of Pi Beta Phi*, Winter 2023, 10–14.

19 Sullivan, JSC interview.

20 Marsha Ivins correspondence with author, July 5, 2023.

21 Pamela A. Melroy, NASA Johnson Space Center (JSC) Oral History Project interview, November 16, 2011.

22 Fisher, JSC interview.

23 Shannon Lucid, *Tumbleweed: Six Months Living on Mir* (Privately published, 2020). Jerry Linenger also has written about his experiences in Russia and on *Mir*.

24 Seddon, *Go for Orbit*, 153.

25 Cleave, JSC interview.

26 Sullivan, JSC interview.

27 Kathy Sullivan, *Handprints on Hubble: An Astronaut's Story of Invention* (Cambridge, MA: MIT Press, 2019), 32–34.

28 NASA News Release, "NASA Administrator Names Whitson First NASA ISS Science Officer," September 16, 2002, https://www3.nasa.gov/home/hqnews/2002/02-175.txt.

29   Comments in interviews with author in February–September 2023. For a thoughtful perspective on leadership styles, see Melroy, JSC interview.

30   "NEEMO—NASA," https://www.nasa.gov/mission/neemo/; "About NEEMO (NASA ExtremeEnvironmentMissionOperations)—NASA," June 24, 2015, https://www.nasa.gov/missions/analog-field-testing/neemo/about-neemo-nasa-extreme-environment-mission-operations/.

31   NEEMO participants include Sunita Williams, Jessica Meir, Cady Coleman, Nicole Stott, Karen Nyberg, Serena Auñon, Kate Rubins, Jeanette Epps, and Jessica Watkins. Serving as NEEMO mission commanders were Peggy Whitson, Sandra Magnus, Heide Stefanyshyn-Piper, Shannon Walker, Dorothy Metcalf-Lindenburger, and Megan McArthur.

32   "NEEMO 12 Mission Journal, Day3," nasa.gov/mission_pages/NEEMO/NEEMO12/mission_journal_3.html; "Day 4," nasa.gov/mission_pages/NEEMO/NEEMO12/mission_journal_4.html.

33   Cleave, JSC interview.

34   Whitson, JSC interview.

35   Seddon, Go for Orbit, 400–404; Sullivan, Handprints on Hubble, 195–98; Cady Coleman, Sharing Space: An Astronaut's Guide to Mission, Wonder, and Making Change (New York: Penguin Random House, 2024), 43–51, 127–33.

36   Whitson, JSC interview.

37   Ibid.; Sullivan, JSC interview; Sullivan, Handprints on Hubble, 210–18.

38   Whitson, JSC interview.

39   Bonnie Dunbar, NASA Johnson Space Center (JSC) Oral History Project interview, June 16, 1998.

40   Wendy Lawrence, NASA Johnson Space Center (JSC) Oral History Project interview, July 28, 1998.

41   Lucid, Tumbleweed.

42   Seddon, JSC interview.

43   Whitson, JSC interview.

44   Seddon, JSC interview, May 21, 2010.

45   Margaret Rhea Seddon, interview with author, February 28, 2023.

46   "NASA Administrator Names Whitson."

47   Shayler and Moule, Women in Space, 485.

48   Marsha Ivins, interview with author, March 18, 2023.

49   Seddon, JSC interview.

50   Sullivan, JSC interview; "Astronaut Judy Resnik Interview 9th April 1981," YouTube, July 25, 2016, https://www.youtube.com/watch?v=EaafRyuwA8w.

51   Cleave, JSC interview.

Chapter 6

1   Author's interviews with thirty former US women astronauts conducted February–September 2023.

2   David Shayler and Colin Burgess, NASA's Scientist-Astronauts (Chichester, UK: Springer-Praxis, 2007). The men who resigned before flying in space were Duane Graveline (1965), Brian O'Leary (1968), John Llewellyn (1968), Curtis Michel (1969), Philip Chapman (1972), Donald Holmquest (1973), and Robb Kulin (2018). Women who left the astronaut corps did so only after one or more flights to take other jobs, focus on family life, or retire.

3   Smith L. Johnston et al., "Astronaut Medical Selection during the Shuttle Era: 1981–2011," Aviation, Space, and Environmental Medicine 85, no. 8 (August 2014): 824–25; NASA Office of the Chief Health and Medical Officer, "NASA Astronaut Medical Standards, Selection and Annual Recertification, OCHMO-STD-100.1A," May 10, 2021, https://www.nasa.gov/wp-content/uploads/2023/04/ochmo-std-100.1a.pdf?emrc=497c02.

4   Ibid.

5   David E. Longnecker, Institute of Medicine, et al., Review of NASA's Longitudinal Study of Astronaut Health (Washington, DC: National Academies Press, 2004).

6   Richard S. Johnston and Lawrence F. Dietlein, "Biomedical Results from Skylab," January 1, 1977, https://ntrs.nasa.gov/citations/19770026836.

7   Arnauld E. Nicogossian and James F. Parker, Space Physiology and Medicine,

NASA SP 447 (Washington, D.C: National Aeronautics and Space Administration Scientific and Technical Information Branch, 1982); Arnauld E. Nicogossian, Carolyn Leach Huntoon, et al., eds., *Space Physiology and Medicine*, 2nd ed. (Philadelphia: Lea & Febiger, 1989); Arnauld E. Nicogossian, Carolyn Leach Huntoon, and Sam L. Pool, eds., *Space Physiology and Medicine*, 3rd ed. (Philadelphia: Lea & Fibiger, 1994); Arnauld E. Nicogossian, Richard S. Williams, et al., eds., *Space Physiology and Medicine: From Evidence to Practice*, 4th ed. (New York: Springer, 2016).

8   Michael R. Barratt and Sam L. Pool, eds., *Principles of Clinical Medicine for Space Flight* (New York: Springer, 2008); Michael R. Barratt, Ellen S. Baker, and Sam L. Pool, *Principles of Clinical Medicine for Space Flight*, 2nd ed. (New York: Springer, 2019), xi and 395.

9   Longnecker et al., *Review of NASA's Longitudinal Study of Astronaut Health*.

10 A review of cardiovascular data through 2016 concluded that astronauts exposed to spaceflight had no increased risk of mortality by cardiovascular disease compared to a well-matched cohort, although there was evidence of increased incidences. Jacqueline Charvat et al., "Long-Term Cardiovascular Risk in Astronauts," *Mayo Clinic Proceedings* 97, no. 7 (July 2022): 1237–46.

11 Nicogossian et al., *Space Physiology and Medicine*, 4th ed., 108, 214–16.

12 National Academies of Science, *Space Radiation and Astronaut Health: Managing and Communicating Cancer Risks* (Washington, DC: National Academies Press, 2021). No author is listed; it was a consensus report by several committees. See https://nap.nationalacademies.org/catalog/26155/space-radiation-and-astronaut-health-managing-and-communicating-cancer-risks.

13 Francine E. Garrett-Bakelman et al., "The NASA Twins Study: A Multidimensional Analysis of a Year-Long Human Spaceflight," *Science* 364, no. 6436 (April 12, 2019): 144.

14 Patricia A. Santy, "Women in Space: A Medical Perspective," *Journal of the American Medical Women's Association* 39, no. 1 (January 1984): 13–17.

15 Amy E. Foster, *Integrating Women into the Astronaut Corps: Politics and Logistics at NASA, 1972–2004* (Baltimore: Johns Hopkins University Press, 2011), ch. 7. Astronauts Rhea Seddon and Kathryn Sullivan also commented on this in their oral histories: Don't assume something will be a problem unless it becomes an actual problem.

16 Ibid., 118–27.

17 Terminology evolved from "sex-related differences" to "sex and gender differences" or "gender differences," although physiological research is about biological male and female sexes rather than social identity genders. The term "gender" may be more apt in behavioral health than physiological health research, but "gender" has become the common term in contemporary biomedical studies.

18 Rhea Seddon et al., "Gender-Related Issues in Space Flight Research and Health Care Workshop Report," National Space Biomedical Research Institute, September 30, 1999.

19 Foster, *Integrating Women into the Astronaut Corps*, ch. 7; David Shayler and Ian A. Moule, *Women in Space: Following Valentina* (Chichester, UK: Springer-Praxis, 2005), 285–89.

20 Kathy Sawyer, "Thelma and Louise in Space?: Idea of All-Female Crew Could Redefine 'Unmanned' in NASA-Speak," *Washington Post*, April 13, 1999. The UPI and various newspapers ran stories and commentary about NASA considering an all-female crew.

21 Deborah L. Harm, Richard T. Jennings, and Peggy A. Whitson, "Gender Issues Related to Spaceflight: A NASA Perspective," *Journal of Applied Physiology* (2001): 2374–83.

22 Ibid., 2381.

23 Saralyn Mark et al., "The Impact of Sex and Gender on Adaptation to Space: Executive Summary," *Journal of Women's Health* 23, no. 11 (November 2014): 941–47.

24 Ibid.

25 Barratt, Baker, and Pool, *Principles of Clinical Medicine for Space Flight*.

26 Caroline Criado-Perez, *Invisible Women: Data Bias in a World Designed for Men* (New York: Abrams Press, 2019); Caroline Criado-Perez, "From ADHD to Endometriosis, Women Are Often Misdiagnosed. Why? The World Was Made for Men," *Washington Post*, March 21, 2019.

27 Serena M. Auñón-Chancellor et al., "Venous Thrombosis during Spaceflight," *New England Journal of Medicine* 382, no. 1 (January 2, 2020): 89–90.

28 Author interview with affected astronaut, March 12, 2023.

29 NASA News Release J12-003, "Astronaut Janice Voss Dies," February 6, 2012; Purdue College of Engineering, "Dream Realized—Scholarship Established in Memory of Astronaut Alum Janice Voss," College of Engineering - Purdue University, https://engineering.purdue.edu/Engr/Giving/Imprints/2012/dream-realized.

30 The seven deceased women astronauts are Judith Resnik (*Challenger*), Kalpana Chawla and Laurel Clark (*Columbia*), Patricia Robertson (aircraft accident), Sally Ride (pancreatic cancer), Janice Voss (breast cancer), and Mary Cleave (stroke).

31 Jon G. Steller et al., "Menstrual Management Considerations in the Space Environment," *REACH* 23–24 (September 1, 2021): 100044, https://doi.org/10.1016/j.reach.2021.100044.

32 Seddon et al., "Gender-Related Issues in Space Flight Research"; Richard T. Jennings and Ellen S. Baker, "Gynecological and Reproductive Issues for Women in Space: A Review," *Obstetrical & Gynecological Survey* 55, no. 2 (February 2000).

33 Richard T. Jennings, MD, interview with author, May 24, 2024. From 1987 to 1995 Jennings was chief of the Flight Medicine Clinic at JSC, chief of medical operations for space shuttle missions, and a crew surgeon for many space shuttle missions. He then moved to the University of Texas Medical Branch in Galveston to direct the aerospace medicine residency program and maintained a close connection with the NASA clinic for many years. He published often on gynecological and reproductive concerns in spaceflight.

34 Jennings and Baker, "Gynecological and Reproductive Issues for Women in Space:"

35 Harm, Jennings, and Whitson, "Gender Issues Related to Spaceflight: A NASA Perspective."

36 April E. Ronca et al., "Effects of Sex and Gender on Adaptations to Space: Reproductive Health," *Journal of Women's Health* 23, no. 11 (November 2014): 967–74.

37 Barratt, Baker, and Pool, *Principles of Clinical Medicine for Space Flight*, ch. 24.

38 Ronca et al., "Effects of Sex and Gender on Adaptations to Space"; Barratt, Baker, and Pool, *Principles of Clinical Medicine for Space Flight*, ch. 24.

39 A thorough discussion of these topics is Jennings and Baker's "Gynecologic and Reproductive Considerations," ch. 24 in Barratt, Baker, and Pool, *Principles of Clinical Medicine in Space Flight*, 2019.

40 Office of the Chief Health and Medical Officer, "NASA Astronaut Medical Standards, Selection and Annual Recertification, OCHMO-STD-100.1A."

41 Barratt, Baker, and Pool, *Principles of Clinical Medicine for Space Flight*, ch. 24, 749.

42 Author interviews with former women astronauts, February–September 2023.

43 Johnnie R. Betson Jr. and Robert R. Secrest, "Prospective Women Astronauts Selection Program: Rationale and Comments," *American Journal of Obstetrics and Gynecology* 88, no. 3 (February 1964): 421–23.

44 Richard T. Jennings, "Capt. Nels O. Monserud and Medical Consideration of the Women Air Force Service Pilots: An Enduring Legacy," *Aerospace Medicine and Human Performance* 88, no. 5 (May 2017): 516–17; N. O. Monserud, "Medical Considerations of the WASP," *Air Surgeon's Bulletin* 11, no. July (1945): 1–53, 214–16. Monserud's article provided by Richard T. Jennings.

45 Anna L. Fisher, NASA Johnson Space Center Oral History Project interview, March 3, 2011.

46 Jennings and Baker, "Gynecological and Reproductive Issues for Women in

Space"; Steller et al., "Menstrual Management Considerations in the Space Environment."

47   Steller et al., "Menstrual Management Considerations in the Space Environment."

48   Lynn Sherr, *Sally Ride: America's First Woman in Space* (New York: Simon & Schuster, 2014), 144–46; Meredith E. Bagby, *The New Guys: The Historic Class of Astronauts That Broke Barriers and Changed the Face of Space Travel* (New York: Morrow, 2023), 142; Kathryn D. Sullivan, NASA Johnson Space Center Oral History Project Interview, May 28, 2009.

49   Jennings and Baker, "Gynecological and Reproductive Issues for Women in Space."

50   Jennings and Baker; Barratt, Baker, and Pool, *Principles of Clinical Medicine for Space Flight*, ch. 24.

51   Barratt, Baker, and Pool, *Principles of Clinical Medicine for Space Flight*, ch. 24.

52   Rhea Seddon, *Go For Orbit: One of America's First Women Astronauts Finds Her Space* (Murfreesboro, TN: Your Space Press, 2015); Margaret Rhea Seddon, NASA Johnson Space Center (JSC) Oral History Project interview, May 20, 2010; Anna L. Fisher, JSC Oral History Project interview, February 17, 2009; Jessica Contrera, "She Was Pregnant When NASA Offered to Send Her to Space. Anna Fisher Didn't Hesitate," *Washington Post*, May 13, 2019.

53   Eileen Collins and Jonathan H. Ward, *Through the Glass Ceiling to the Stars: The Story of the First American Woman to Command a Space Mission* (New York: Arcade, 2021), 150–51.

54   As of 2024, these twenty-three women astronauts had borne or adopted children: Baker, Clark, Collins, Coleman, Currie, Fisher, Godwin, Jernigan, Lucid, Mann, McArthur, Metcalf-Lindenburger, Meir, Moghbeli, Morgan, Nowak, Nyberg, Ochoa, Stefanyshyn-Piper, Seddon, Still (Kilrain), Stott, and Thornton.

55   Interviews with author, February–September 2023.

56   Anna L. Fisher, JSC interview; interview with author, August 8, 2023.

57   Jennings and Baker, "Gynecological and Reproductive Issues for Women in Space", 113; Jon G. Steller et al., "Gynecologic Risk Mitigation Considerations for Long-Duration Spaceflight," *Aerospace Medicine and Human Performance* 91, no. 7 (July 1, 2020): 543–64; Barratt, Baker, and Pool, *Principles of Clinical Medicine for Space Flight*, 755.

58   Jennings and Baker, "Gynecological and Reproductive Issues for Women in Space."

59   "Assisted Reproductive Technology for Astronauts," 1999, unpublished, courtesy of Richard T. Jennings.

60   Seddon, *Go for Orbit*, 450.

61   Richard T. Jennings, MD, interview with author, May 20, 2024.

62   Office of the Chief Health and Medical Officer, "NASA Astronaut Medical Standards, Selection and Annual Recertification, OCHMO-STD-100.1A."

63   Walter E. Sipes, James D. Polk, Gary Beven, and Marc Shepanek, "Behavioral Health and Performance," ch. 14 in Nicogossian et al., *Space Physiology and Medicine*.

64   Andrew Chaikin, "The Loneliness of the Long-Distance Astronaut.," *Discover*, February 1985; Patricia A. Santy, *Choosing the Right Stuff: The Psychological Selection of Astronauts and Cosmonauts*, Human Evolution, Behavior, and Intelligence (Westport, Conn: Praeger, 1994).

65   An incident interpreted by journalists as a crew "strike" or "mutiny" occurred on the Skylab 4 mission in 1973–1974 but was not substantiated by records from the mission. John Uri, "The Real Story of the Skylab 4 'Strike' in Space—NASA," November 16, 2020, https://www.nasa.gov/history/the-real-story-of-the-skylab-4-strike-in-space/.

66   Patricia A. Santy, "The Journey out and in: Psychiatry and Space Exploration," *American Journal of Psychiatry* 140, no. 5 (May 1983): 519–27.

67   Mary M. Connors, Harrison, Albert A., and Akins, Faren R., *Living Aloft: Human Requirements for Extended Spaceflight*, SP-483 (Washington, DC: National Aeronautics and Space Administration, 1985); Nicogossian, Huntoon, and Pool,

eds., *Space Physiology and Medicine*, 3rd ed., included a chapter on "Psychologic and Psychiatric Considerations"; Douglas A. Vakoch, ed., *Psychology of Space Exploration: Contemporary Research in Historical Perspective*, NASA SP 4411 (Washington, DC: National Aeronautics and Space Administration, Office of Communications, History Program Office, 2011). See also Douglas A. Vakoch, ed., *On Orbit and Beyond: Psychological Perspectives on Human Spaceflight*, 2nd ed, Space Technology Library 29 (Heidelberg: Springer-Verlag, 2013); H. S. Cooper, "The Loneliness of the Long-Duration Astronaut," *Air & Space Smithsonian* 11, no. 2 (1996): 37–45.

68 Namni Goel et al., "Effects of Sex and Gender on Adaptation to Space: Behavioral Health," *Journal of Women's Health* 23, no. 11 (November 2014): 975–86.

69 Jack Stuster, "Behavioral Issues Associated with Long Duration Space Expeditions: Review and Analysis of Astronaut Journals," July 8, 2010, https://ntrs.nasa.gov/citations /20100026549.

70 Note the difference in chapter titles between the third and fourth editions of *Space Physiology and Medicine*: "Psychologic and Psychiatric Considerations" (1994) and "Behavioral Health and Performance" (2016). Chapters on "Behavioral Health and Performance Support" appeared in both the 2008 and 2019 editions of *Principles of Clinical Medicine for Space Flight*.

71 Albert W. Holland, Behavioral Health and Performance Center, NASA Johnson Space Center, interview with author, May 18, 2023.

72 Walter E. Sipes et al., "Behavioral Health and Performance," in Nicogossian et al., *Space Physiology and Medicine*.

73 Albert W. Holland and James J. Picano, psychologists in the Behavioral Health and Performance Center at NASA Johnson Space Center, interview and correspondence with author, May–June 2023.

74 Shannon Connellan, "How Do Astronauts Practice Self-Care in Space?," Mashable, September 29, 2019, https://mashable .com/article/astronauts-self-care.

75 Shannon Lucid, *Tumbleweed: Six Months Living on Mir* (United States: MkEk Publishing, 2020).

76 Seddon et al., "Gender-Related Issues in Space Flight Research"; Ronca et al., "Effects of Sex and Gender on Adaptations to Space."

77 Patrick Paolantonio, "Shuttle Columbia Tragedy 20 Years Later, Husband Talks Loss and Discovery," WISN, February 3, 2023, https://www.wisn.com/article /shuttle-columbia-tragedy-20-years-later -husband-talks-loss-and-discovery /42750278.

78 Marriages within the astronaut corps include Anna and William Fisher, Rhea Seddon and Robert L. "Hoot" Gibson, Sally Ride and Steven Hawley, Bonnie Dunbar and Ronald Sega, Jan Davis and Mark Lee, Linda Godwin and Steven Nagel, Tammy Jernigan and Jeffrey Wisoff, Karen Nyberg and Douglas Hurley, Shannon Walker and Andrew Thomas, and Megan McArthur and Robert Behnken.

79 Seddon, *Go for Orbit*, 356-59; "First Mom in Space! Anna Fisher Tells Her Amazing NASA Story," Space Showcase, http:// videos.space.com/m/iU3S58C6/first -mom-in-space-anna-fisher-tells-her -amazing-nasa-story-video.

80 Kathryn Thornton, interview and correspondence with author, March 23, 2023, and February 27, 2024; Cady Coleman, *Sharing Space: An Astronaut's Guide to Mission, Wonder, and Making Change* (New York: Penguin Random House, 2024), 91–99.

81 Sara Shookman, "Sara's Circle: Launching Success with NASA's Janet Kavandi," May 12, 2016, https://www.wkyc.com /article/features/saras-circle-launching -success-with-nasas-janet-kavandi/95 -187720907.

82 Collins and Ward, *Through the Glass Ceiling to the Stars*, 180, 186–87.

83 Ronca et al., "Effects of Sex and Gender on Adaptations to Space."

84 Seddon, *Go for Orbit*, 339–40.

85 Author interview with affected astronaut, February 19, 2023.

86 Collins and Ward, *Through the Glass Ceiling to the Stars*, 188.

87 Seddon, *Go for Orbit*, 347.

88 Interview with author, February 22, 2023.

89 Kate McCann, "I'm an Astronaut, I Need Some Space: Cady Coleman," *The Guardian*, August 6, 2010, sec. Life and style, https://www.theguardian.com/lifeandstyle/2010/aug/07/nasa-astronaut-cady-coleman; Anna Moselein, "What Is It Like to Actually Parent From Space?," Glamour, September 4, 2020, https://www.glamour.com/story/away-netflix.; Mark Wilkins, "The Astro-Couple," *Smithsonian*, February 14, 2018, https://www.smithsonianmag.com/air-space-magazine/astro-couple-180968166/.

90 Ido Mizrahy, dir., *Space: The Longest Goodbye*, Independent Lens, 2024, https://www.pbs.org/independentlens/documentaries/the-longest-goodbye/; Hannah Berryman, dir., *Spacewoman*, https://spacewoman.film/.

91 Seddon, *Go for Orbit*, 348.

92 Sara W. Bock, "Pushing the Boundaries: Marine Astronauts Nicole Mann and Jasmin Moghbeli Are at the Forefront of Space Exploration," *Leatherneck*, October 2021, 36–43.

93 Credit for the phrase "Being Small in a Big Suit World" belongs to Christina Koch and Jessica Meir, "Our All-Female Spacewalk Is a Sign of the Times," *Washington Post*, November 11, 2019.

94 Kathy Sullivan, *Handprints on Hubble: An Astronaut's Story of Invention* (Cambridge, MA: MIT Press, 2019), 34–38.

95 Santy, "Women in Space: A Medical Perspective."

96 Seddon et al., "Gender-Related Issues in Space Flight Research."

97 Astronaut Michael Lopez-Alegria, interview with author, April 28, 2023.

98 Recollections in this discussion are from interviews with former women astronauts, February–September 2023, unless otherwise noted from published sources.

99 Kenneth S. Thomas and Harold J. McMann, *U.S. Spacesuits*, 2nd ed (Chichester, UK: Springer-Praxis, 2012), ch. 10.

100 Coleman, *Sharing Space by Cady Coleman*, 136; Nell Greenfieldboyce,

"When It Comes to the Spacewalk, Size Matters," WUNC, December 15, 2006, https://www.wunc.org/2006-12-15/when-it-comes-to-the-spacewalk-size-matters.

101 Coleman, *Sharing Space*. See ch. 7, "If the Suit Doesn't Fit, Wear it Anyway . . . and Wear it Well."

102 Interview with author, May 23, 2024.

103 David R. Williams, MD, and Brian J. Johnson, "EMU Shoulder Injury Tiger Team Report" (NASA Johnson Space Center, September 2003).

104 NASA News Release 97-126, "Spacewalkers Named for Space Station Assembly Flights," June 9, 1997.

105 Greenfieldboyce, "When It Comes to the Spacewalk, Size Matters."

106 Hanneke Weitering, "All-Female Spacewalk Was Cancelled on 'My Recommendation,' Astronaut Anne McClain Says," April 2, 2019, https://www.space.com/anne-mcclain-spacesuit-response-video.html; Matthew S. Schwartz, "NASA Scraps First All-Female Spacewalk For Want Of A Medium-Size Spacesuit," March 26, 2019, http://www.capradio.org/news/npr/story?storyid=706779637.

107 Meghan Bartels, "Spacesuit Sizing Stymied a Historic NASA Moment, and It May Always Be Tricky," Space.com, April 3, 2019, https://www.space.com/astronaut-spacesuits-future-sizes-challenges.html.

108 Jennings and Baker, "Gynecological and Reproductive Issues for Women in Space."

109 Coleman, *Sharing Space*, ch. 7.

110 The fifteen spacewalking women who followed Kathy Sullivan, Kathy Thornton, and Linda Godwin conducted EVAs at the International Space Station: Tammy Jernigan, Susan Helms, Peggy Whitson, Heide Stefanyshyn-Piper, Suni Wiliams, Nicole Stott, Tracy Caldwell Dyson, Kate Rubins, Anne McClain, Christina Koch, Jessica Meir, Kayla Barron, Nicole Mann, Jazmin Moghbeli, and Loral O'Hara. Linda Godwin also did her second spacewalk at the ISS.

111 Koch and Meir, "Our All-Female Spacewalk Is a Sign of the Times."

112 For example, as collegiates, about one third of the women astronauts were selected into prominent academic honor societies, including Sigma Xi for scientific achievement and Phi Beta Kappa for the humanities. Others received recognition and awards for their academic, athletic, or military performance.

113 Mary Cleave, Mae Jemison, Sandra Magnus, Ellen Ochoa, and Kathryn Sullivan have been elected to one or more of the academies of arts and letters, engineering, medicine, and public administration; Ochoa and Sullivan are fellows of both the American Association for the Advancement of Science and the American Institute of Aeronautics and Astronautics.

114 See Sally Ride cover and story, *People*, June 2, 1983; see also Jennifer M. Ross-Nazzal, "You've Come a Long Way, Maybe: The First Six Women Astronauts and the Media," in *Spacefarers: Images of Astronauts and Cosmonauts in the Heroic Era of Spaceflight*, ed. Michael J. Neufeld (Washington, DC: Smithsonian Institution Press, 2013), 175–201.

115 Interview with author, April 11, 2023.

116 Collins and Ward, *Through the Glass Ceiling to the Stars*, 177–78.

117 This commentary is based on interviews with the author in February–September 2023.

118 Jim Bridenstine, introducing Group 22 in June 2017.

INDEX